Merchant Princes
of the East

Merchant Princes of the East

CULTURAL DELUSIONS, ECONOMIC SUCCESS AND THE OVERSEAS CHINESE IN SOUTHEAST ASIA

Rupert Hodder

Chinese University of Hong Kong

JOHN WILEY & SONS

Chichester · New York · Brisbane · Toronto · Singapore

Other Wiley Editorial Offices

John Wiley & Sons, Inc., 605 Third Avenue,
New York, NY 10158-0012, USA

Jacaranda Wiley Ltd, 33 Park Road, Milton,
Queensland 4064, Australia

John Wiley & Sons (Canada) Ltd, 22 Worcester Road,
Rexdale, Ontario M9W 1L1, Canada

John Wiley & Sons (SEA) Pte Ltd, 37 Jalan Pemimpin #05-04,
Block B, Union Industrial Building, Singapore 2057

Library of Congress Cataloging-in-Publication Data

Hodder, Rupert.
     Merchant princes of the east: cultural delusions, economic success and the overseas
Chinese in Southeast Asia / Rupert Hodder.
          p.   cm.
     Includes bibliographical references and index.
     ISBN 0-471-96230-9
     1. Asia, Southeastern—Commerce.   2. Chinese—Commerce—Asia, Southeastern.
3. Merchants—Asia, Southeastern.   4. Success in business—Asia, Southeastern.
5. Chinese—Asia, Southeastern—Economic conditions.   6. Chinese—Asia,
Southeastern—Social conditions.   7. Chinese—Asia, Southeastern—Societies, etc.
I. Title.
HF3790.8.Z5H63   1995
382'. 0951059—dc20   95–33477
                                                                    CIP

British Library Cataloguing in Publication Data

A catalogue record for this book is available from the British Library

ISBN-0-471-96230-9

Typeset in 10/12pt Palatino by Saxon Graphics Ltd, Derby
Printed and bound in Great Britain by Biddles Ltd, Guildford and King's Lynn
This book is printed on acid-free paper responsibly manufactured from sustainable foresta-
tion, for which at least two trees are planted for each one used for paper production.

To Jonathan and Rosabelle

Contents

Preface

This book presents an interpretation of Overseas Chinese economic success in Southeast Asia. No attempt is made to provide a formal explanation of their success: indeed, this book argues strongly against scientific and theoretical analyses of the Overseas Chinese and their economic activities, and is critical of the importance frequently attached to 'culture'. The view taken in these pages is that Chinese societies or groups overseas are not unique, nor profoundly different from those of any other people: their material success is an outcome, not of 'Chineseness', but of multidimensional values, institutions and actions which have been consciously manipulated and 'turned' towards the extension and institutionalisation of trade.

This is not to say that obvious differences between groups do not exist. But such differences, it is suggested here, do not reflect or constitute a deterministic culture. Certainly there may appear to be profound differences in institutions, values and behaviour if the perception chosen is unidimensional and if the observer chooses to concentrate upon unique cosmetic details which are an inevitable consequence of the human mind's spontaneity and its variety of responses to circumstance. The apparent profundity of these differences is deepened further by individuals who create, adapt, alter or adopt institutions, values and behaviour which they view or present as symbols of their uniqueness. But if these phenomena are 'turned' so that their other aspects may be viewed, then the significance of their cosmetic details begins to recede, and similarities among groups begins to emerge. As the perspective rotates and as other aspects begin to slide into view, it becomes clear that differences between 'cultural' or 'ethnic' groups reflect the manner in which institutions, values and exchange are used, and the purpose to which they are directed. The redirection of reciprocity from its conduct as an end in itself towards its use to construct relationships which facilitate trade requires only a subtle alteration in values and attitudes. The family whose status and reputation depend upon the conduct of reciprocity for its own sake may appear to be very different from the family company, yet each requires only a shift in aspect to become the other. Even the very concepts of 'culture' and 'ethnicity' constitute part of the institutional patterns and moral imperatives associated with a form of exchange. 'Chinese' and 'Filipino', for instance, are often deliberately presented as unidimensional phenomena in order that they may be used as positive kernels to help establish reciprocal relationships for the conduct of business, or as an antitype against which the individual may create the 'cosmopolitan' and thereby expand opportunities for trade.

To help focus, and add greater force, to these arguments, and to demonstrate that the multidimensional values, institutions and actions of individuals are not mere abstractions, I have chosen in Chapter 6 to concentrate upon a Filipino community and its merchants*—both Filipino and Chinese. This may perhaps seem a little unusual but it does help illustrate in some detail that whatever designated group they may be assigned to, individuals, and the values, institutions and actions (such as exchange) which they create or adopt, possess many aspects, and, if viewed from different perspectives or if directed towards a common end, may therefore reveal striking similarities. For the same reason, while the term 'Chinese' in this book generally refers to the Overseas Chinese in Southeast Asia, I do not always make a distinction between the Chinese in China and the Chinese overseas. This is particularly evident in Chapter 3 and does, I believe, help emphasise the argument that the multidimensional institutions, values, actions and behaviour created or adopted by individuals called 'Chinese' may not always be directed towards the prosecution of trade and the realisation of sustained material progress. If a definition is needed, culture, or, more accurately, the individual's own perception of culture, is merely an imperfect expression of the direction of multidimensional values, institutions and forms of exchange towards particular, or many diverse, ends.

The material in Chapter 6 is based upon a number of lengthy visits to Davao City between 1985 and 1994. The people I have written about may seem rather colourless: there is much which I have chosen to omit. Even so, in view of the very personal nature of this material, names have been replaced with letters. I hope that I have done no injustice to those who have shown me extraordinary kindness, patience and generosity. There is often a very fine distinction between emotions, the values

---

*The term is used in this book to refer to 'traders' or 'businessmen', and should not be understood to refer only to retailers or wholesalers, nor should it be understood to indicate a distinction between 'modern' and 'traditional' practices or forms of organisation.

which individuals believe to be absolute, and intellectualised debates. It may be partly because of the absence of a clear distinction between the intellect and the individual's many other aspects that meanings and implications which a writer did not intend are found by others among his[†] words. It is also true, perhaps, that words in print often take on a permanence and absoluteness which is mostly undeserved and which usually distort reality. But in this book I have argued against the pretence at objective, precise, rigorous analyses of individuals and the societies they create; for individuals, the values they hold, and the actions they take, all possess many dimensions. A writer must be aware of these aspects and of the different possible interpretations. These pages are merely an attempt at one such interpretation.

The arguments enunciated here strongly favour a movement away from a method of thought which I have termed, perhaps unfairly, the scientific approach. Undoubtedly, part of the reason for the popularity of the cultural explanation is its romance, for as *The Economist* (1994a) suggests:

'Civilizations on the rise like to explain their economic success not through some dreary shift in comparative advantage or the pattern of technological change but by pointing to a set of virtues unique to their culture' (28 May, p. 9).

But science and the romance of culture are symbiotic. The choice between explanations which this quote implies is too stark, but it indicates well enough a way of thinking which underlies much of the debate on material success and which adds to the popularity of culture. There is in some respects little to choose between the two for both are essentially deterministic explanations. One resorts to an ultimate force that is economic, the other to the equally abstract force of culture: both are searching for the idol of origins. If one rejects economic, political and historical forces then what is left? And since the Chinese are generally far stronger economically in proportion to their numbers in many countries in Southeast Asia, what else other than culture explains their success? The logic is soothing for there exists strong mutual support between the presumption that culture is a deterministic force and, on the other hand, the application of the scientific approach with its emphasis on the search for origins, on chains of cause and effect, and on the fragmentation of phenomena into discrete variables within an ordered, predictable and explicable whole. Culture is a useful variable or, for some writers, represents the idol of origins around which may be constructed a scientific explanation. And for the cultural determinists, science provides a fashionable respectability.

[†]When no particular individual is specified the use of 'he', 'his' or 'him' may also be understood to read as 'she', 'hers' or 'her'. In these contexts too, 'man' or 'men' are used as generic terms.

The desire to produce neat explanations and conceptual frameworks with the power of prediction and which conform to what is expected of thorough 'scientific' analysis, requires the careful selection, and the selective interpretation, of institutions, forms of exchange, values, beliefs, actions, decisions, choices and desires, in order to create uni-dimensional phenomena linked together by chains of cause and effect. A complex, fluid reality becomes falsified and distorted for the sake of scholarly convenience and fashion. Values, forms of exchange, institutions (such as the family, the company or the association), beliefs, philosophies, and patterns of behaviour, although created by individuals, now simply 'exist': they comprise a structured mass, and so firmly rooted are they in purported cultural elements (such as Confucianism) that together the structured expression of the underlying cultural elements and the cultural elements themselves are thought to be capable of determining and explaining individuals, their business activities and their material success. For many writers, cultural determinism, the romance of culture, and science work together well. The explanations constructed possess a certain 'weight' and a satisfying roundness. Better still, they are easily drawn into epic stories and detailed accounts of the Overseas Chinese and their economic success, and may thereby achieve canonical status (Seagrave, 1995; Cragg, 1995; Kotkin, 1993; and East Asia Analytical Unit, 1995). Yet such explanations also possess disturbing undertones. There are those who believe their own 'culture' to be an absolute measure of worth and excellence; and those who perceive values, beliefs and institutions as relativistic. The first of these beliefs breeds arrogance and, worse still, contempt for others; the second allows that arrogance and contempt to flourish. The writer who withdraws into a flat unidimensional world runs the danger of playing to the darker side of human nature.

Rupert Hodder
Hong Kong, May 1995

Acknowledgements

In preparing this book I have been helped by a large number of people in the UK, Hong Kong and the Philippines. Many of the ideas presented in this book, its initial outlines, and plans for work in the field were begun while I was Lecturer in Human Geography in the Department of Geography at the London School of Economics. I remain very grateful to the Department, and most especially to R. J. Bennett, D. K. C. Jones and D. Diamond for their support and encouragement during the early stages of this book's preparation. I would also like to acknowledge a very great debt of gratitude I owe to Frank Leeming, my supervisor at the University of Leeds, under whose guidance a number of the ideas presented here were initially developed. While I was writing this book, Frank died suddenly. He was a kind man, highly intuitive and wonderfully untroubled by all the artificial techniques and pretensions of modern research—something I found refreshing and liberating.

In Hong Kong I have been fortunate to receive support from many different quarters. In particular, I would like to acknowledge the generous and uncomplicated funding I have received from the Faculty of Social Sciences and the Hong Kong Institute of Asia-Pacific Studies at the Chinese University of Hong Kong, and for the freedom with which I have been allowed to conduct this research. Special thanks, too, are due to Professor Yeung Yue-man, the Institute's Director, for his help in so many ways. Thanks are also due to the Overseas Chinese Archives (HKIAPS), the University Services Centre (CUHK), to Mr Chang Kim-ho (President of Taclon Industries), to Stewart Chan and Henry Chow, and to the officers and members of numerous Filipino associations for all their help and advice. I would also like to thank Mr Tou for his exceptional hand-drawn maps.

In the Philippines I was supported by a large number of people: the Borrillos, who put me up and put up with me; the Mendozas and

Martinez; the Siazons; the Castañedas and the Castillos; Juliet and Tommy Pamintuan; Rigoberto Tacan; Antonio Mendoza; members of the Aboitiz, Gaisano and Aportadera families; James Lee; Karen Vertido of the Davao City Chamber of Commerce; Benjamin Duterte (private secretary to the Mayor of Davao); Dominador Zuño; Joaquin Ong and his family; Lualhati Malata; Ernesto Corciño; the directors and members of the Business Bureau and the Bureau of Internal Revenue; officers of the Armed Forces of the Philippines; officers and members of the Filipino–Chinese Association, the Long Hua Temple, the Chinese Family Associations, and the Filipino–Chinese Charitable Association; 'Ondoy' and his family; Vice-Governor Turan of Cotabato; and many others. I am extremely grateful to all of them. Finally, thanks to San, my wife, without whose patience and charm this book could not have been written—*salamat sa imong gibuhat ug pagantos ning tanan.*

List of abbreviations

FDJYZ   Fudan Daxue Jingji Yanjiu Zhongxin (Fudan University
          Economic Research Centre)
FJDWJM   *Fujian Duiwai Jingmao (Fujian Foreign Economic*
          Relations and Trade)
FJJJB   *Fujian Jingji Bao (Fujian Economic Daily)*
GDQB   *Guangdong Qiaobao (Guangdong Overseas Chinese Daily)*
GDDWJM   *Guangdong Duiwai Jingmao (Guangdong Foreign*
          Economic Relations and Trade)
GGJ   Gongshang Guanliju, Tianjin (Industrial and Commercial
          Administration Bureau, Tianjin)
GGJGTS   Guojia Gongshang Xingzheng Guanliju Geti Jingjisi
          (State Industrial and Commercial Administration Bureau,
          Department for Private (Individual) Economy)
JJRB   *Jingji Ribao (Economic Daily)*
SHJJXX   *Shanghai Jingji Xinxi (Shanghai Economic News)*
ZG   Zhonggong, Tianjinshi (Communist Party, Tianjin)
ZHGSSB   *Zhonghua Gongshang Shibao (China Business Times)*
ZJRB   *Zhejiang Ribao (Zhejiang Daily)*

1

Dimensions of culture

The Chinese in Southeast Asia

Chinese commercial influence overseas goes back at least to the third century AD, when official missions were despatched to report on countries bordering *Nanyang* (the South Seas), to be followed by Buddhist pilgrims and later, during the Sung dynasty, by traders. After conquering China the Mongols also traded in the South Seas and with Arab traders in the area. This sequence of contacts, believes Winchester (1991), 'explains why the attitudes and sympathies of these mercantile classes underlie all Overseas Chinese life today' (p. 238). Most of the migrants came from three provinces in southeastern China—Guangdong, Fujian and Guangxi. Poverty, recurrent famines, internal strife, physical insecurity, political intolerance, and, for many, the seemingly impermeable social barriers to advancement gave the Chinese strong encouragement to look overseas for a better life.

The effects of Chinese economic activities in Southeast Asia were felt well before colonisation by the Europeans, for as Dixon (1991) points out, 'Chinese trade, exploitation of resources and introduction of technology made a major contribution to the region's economies long before the nineteenth century' (p. 45). But it was during the European colonial period that a marked and sustained movement of Chinese into Southeast Asia took place. The Chinese originated from a society that was in many ways more sophisticated technically, commercially and administratively than those in which they settled. When the Europeans arrived they found that the indigenes had little interest or experience in such matters as administration, supervisory work, mining, merchandising, or supplying goods to more remote or newly developing areas. The colonial administrators believed the indigenes to be unsuited to modern economic activity and unwilling to make the changes in their societies which a modern economy would demand of them. The Chinese, on the other hand, appeared, to the European at least, to be far more enterprising and hard-working and, being outsiders, seemed to care little for changing political frontiers or for the indigenes' localised social boundaries and religious differences.

Table 1.1   The Chinese in Southeast Asia

	Total population (millions)	Chinese (millions)	% Chinese*
Indonesia	147.00 –182.65	4.16 – 4.93	2.7 – 2.8
Philippines	46.00 – 61.48	0.69 – 0.74	1.2 – 1.5
Malaysia	12.77 – 18.00	4.88 – 5.96	30.9 – 33.1
Singapore	2.41 – 2.75	2.04 – 2.09	76.00
Brunei	0.24 – 0.3	0.05 – 0.085	18.3 – 28.0
Vietnam	65.1 – 66.473	0.99 – 1.8	1.5 – 2.76
Thailand	55.80 – 56.4	4.46 – 6.37	8.0 – 11.29
Cambodia	7.87 – 8.4	0.2 – 0.46	2.5 – 5.47
Burma	40.77 – 41.61	0.004 – 0.65	1.0 – 1.59
Laos	4.186	0.008	0.19
Total	382.146–442.249	17.482–23.093[†]	3.95–6.04[‡]

* Highest and lowest figures given in sources.
[†] If the populations of Hong Kong and Taiwan are included, the total number of overseas Chinese in East and Southeast Asia (excluding North and South Korea and Japan) amounts to somewhere between 43 and 50 million.
[‡] Figures represent smallest estimated total population of Chinese expressed as a percentage of largest total population, and largest population of Chinese expressed as a percentage of smallest total population.
Sources: adapted from Rao Meijiao (1993), Somers Heidhues (1992), Amer (1992), M. Smith (1992), Rigg (1991), Suryadinata (1989).

At first the movement of Chinese was largely voluntary, but as the need for labour grew—especially in the mines and plantations of the European colonies, for the colonialists were frequently operating in areas with low population densities and immigration was thought essential—Chinese were recruited more formally and in larger numbers or, in some instances, press-ganged by Chinese marauders into the service of European merchants. In pre-colonial Malaya, large numbers of Chinese were already engaged in a wide range of activities, including tin mining on the peninsula and gold in Sarawak. When the British arrived and established more technically advanced mining operations and rubber plantations, the Chinese were encouraged to arrive in much larger numbers, although the occupations now open to them were restricted by the interests and competitiveness of the colonial power. Manual and skilled labour, food production and merchandising were occupations in which the Chinese soon made themselves indispensable throughout the European colonies of Southeast Asia and, numerically, the Chinese soon came to dominate the major urban centres.

It seems beyond question, and certainly it is generally argued, that the economic significance of Overseas Chinese in Southeast Asia is considerable and that in many respects they provide a catalyst for economic growth. Depending on definitions and on the countries included, there are thought to be somewhere between 43 and 50 million Overseas Chinese in the region (see Table 1.1). Collectively they control a larger share of regional trade than do other ethnic groups and they generate the equivalent of a GNP two-thirds the size of China's (Redding, 1990). Of the four newly industrialised countries (NICs) in Southeast Asia, three (Hong Kong, Taiwan and Singapore) are populated largely by Chinese, and together they possess larger foreign reserves than either Japan or the United States. The 'funds' of Overseas Chinese in the region (excluding Hong Kong and Taiwan) are conservatively estimated at US$400 billion; and in each country in which the Overseas Chinese constitute a minority group they control a disproportionate share of economic activities. In Indonesia, Overseas Chinese number around 5 million (or about 3% of the population), yet more than half the country's trade and about three-quarters of private domestic capital is in their hands. In Thailand the majority of corporate assets, nine-tenths of investment in commerce and manufacturing, and half the financial resources of Thai banks are owned by Chinese. In the Philippines, 40% of the assets of private domestic banks are owned by Chinese, and of the largest 300 enterprises in the mid 1980s, two-fifths were Chinese companies which together generated 35% of the sales of all domestic firms. In Malaysia, 44.9% of companies are owned by Chinese, despite the implementation of the New Economic Policy. In Sarawak, too, where most of the non-indigenous people are Chinese, they occupy the more densely populated coastal and lower valley strips in the relatively well-developed west. The Chinese, too, are the main urban people in Sarawak, as they are throughout most of Malaysia: 60% of the population of

Kuching, the capital of Sarawak, is Chinese (Liang Yingming, 1993; Rao Meijiao, 1993; Kraar, 1993; L. Lim, 1992; Kotkin, 1993).

Explanations of Chinese economic success are many and varied. Few writers would deny that an explanation of Chinese economic success must also include the broader macro-economic policies, institutional arrangements and international trading strategies pursued by governments in the Far East. Any explanation must also consider the suggestion that the Chinese happened to be in the right place at the right time; that they had the skills, the temperament, drive, experience of commerce and urban life that coincided with the needs of the European colonialists; that today they have developed complex business networks; and that they are patient, hard-working, frugal, and possess sharp business acumen. But surely this cannot be the whole story? Policies have to be made; drive, business acumen and business networks must derive from something; opportunities have to be taken advantage of; and, after all, in many instances the Chinese have achieved economic success despite the inappropriate or often hostile actions of host states, many of which have not performed as successfully as most of those dominated by Chinese. Does all this not indicate that there is something about Chinese culture which explains material success? Before proceeding any further with this question, however, it is important to address four issues.

(i) Although much of the discussion is concerned with the Overseas Chinese in Southeast Asia, arguments and observations are not restricted to that region. Moreover, the wider political and economic context of the Far East as a whole is also pertinent to an understanding of Chinese economic success (Fig. 1.1). It is for these reasons that the terms Far East or West Pacific Rim are also used occasionally, and interchangeably, in these pages. Analysis and interpretation are rarely served well by the imposition of geographical limits (which are, to some extent, inevitably rather loose and arbitrary), and this is particularly true for the arguments ranged here against the link between 'culture' and material progress. No attempt, then, has been made to give Chinese communities in each of the countries of Southeast Asia balanced coverage, and while it is common in some of the literature (most especially in Chinese-language materials) to make a distinction between Chinese communities in Southeast Asia and those in Hong Kong and Taiwan, no such distinction is made in this book.

(ii) Wider analyses of economic growth in the Far East deserve to be kept in mind, not just because they help 'set the scene' but also, perhaps more importantly, because the reality of growth serves as a useful lifeline between the unidimensional world which exists only as an abstraction

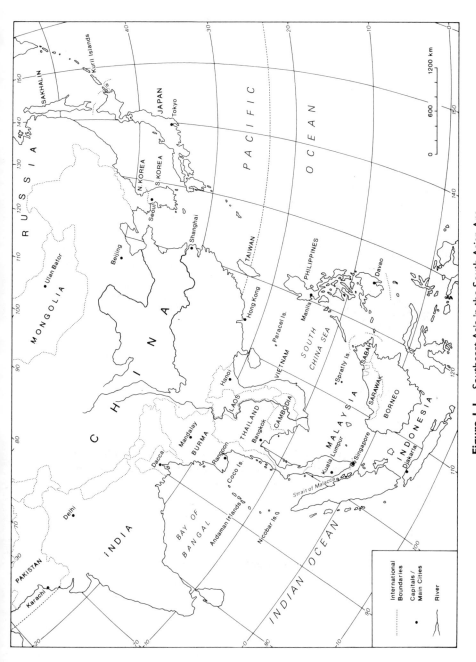

Figure 1.1 Southeast Asia in the South Asian Arc

within the intellect of the social scientist, and the complex multidimensional world created by the individuals whom the social scientist studies. Cultural determinists may consider wider political and economic policies and international trading strategies, but only in so far as they help isolate those seminal 'cultural' traits which are presumed to exist.

It is interesting to observe that although many of the peoples and nations of the region worked to discredit much of the conventional wisdom about economic development current in the 1960s and 1970s, their relative economic success is now attracting commentators determined to hold these countries up as a vindication of particular and often widely contrasting ideals, beliefs and theoretical prescriptions. The Far East has something to offer to everyone, or so it would seem. It is significant, for instance, that, with few exceptions, there has never been in the countries of the region the same intense ideological critique of colonialism as that experienced in Africa and Latin America. Why this should be so is an interesting question. Some writers have noted the 'unfortunate failure' of scholarship in the region to take up the 'vigorous and theoretically innovative debates that have been part and parcel of the analysis of change in Latin America . . . and Africa'. This failure 'is largely the consequence . . . of the extraordinary influence of positivist and empiricist traditions upon Southeast Asian studies' (Higgott and Robinson, 1985, p. 3). The theoretical 'successes' (and the emotions which are their foundation) have now been recycled under different guises such as environmentalism and 'sustainable' development. Then there are those who present the Far East as an illustration of the importance of strong, authoritarian government; those who argue that the experience of countries in the region indicates that democracy is best for economic growth; those who believe that economic success in the region confirms the vital importance of government intervention and the establishment of focused institutions which can direct or guide public and private entrepreneurs; and those who believe the peoples of the Far East have, in essence, always been free marketeers—a fact which is gradually becoming more and more apparent as they opt for minimal government and let the markets do what they are supposed to do. And then there are those who suggest the Far East owes much of its success to the family, self-reliance and other social characteristics which allow welfare costs to be kept to a minimum; those who point to the state's active role in the provision of housing, education, training and employment in many of the most successful countries; those who hold up many countries of the region as an example of the evils of individualism, materialism and capitalism; and those who see these nations' economic success as an expression of the virtue of collectivism (see, for example: Burnett, 1993; Dixon and Drakakis-Smith, 1993; *The Economist*, 1994b; Wade, 1990; Williamson, 1994; Olson, 1993; Revel, 1994; Rigg, 1991; White and Wade, 1988; World Bank, 1991; Gourevitch, 1989; Gereffi, 1989; Elster and Moene, 1989).

Many writers find in the countries of the Far East useful material for the support of their own theories, ideologies and prejudices. And even when attempts are made to present a genuinely balanced view of events, the general level of analysis inevitably requires a vocabulary which gives the impression that the nations of the region are mere scenery, the governments and their peoples symbolic characters. Nevertheless, the growing and varied literature, the different institutions and values developing and changing in each country, and the different experiences of each country, seem to indicate not that all countries or any particular country serve as a special case or a model for the rest of the world, but rather that there is no formula, no set of determining variables or factors which leads inevitably to economic success (R. Hodder, 1992; World Bank, 1993). Economic growth derives from a determination to achieve such growth, and whatever is expedient in encouraging and facilitating the extension and institutionalisation of trade will permit rapid and sustained material progress.

(iii) The third issue, which concerns definitions of 'Chinese' and 'Chineseness', is mentioned here partly to caution against the unthinking collection and interpretation of statistics and observations, but mainly to introduce doubts and questions about the very meaning of 'Chinese'—a theme which is developed throughout this book. It has already been noted that 'the Chinese' exert a pervasive influence over much of Southeast Asia: indeed, one writer argues that the main link between the various countries, cities and islands of the region is the presence of the Overseas Chinese, 'the greatest diaspora in the world' (Winchester, 1991, p. 220). However, their numbers and their proportion of the total population in the countries of the region vary widely (Table 1.1). In those countries where the Overseas Chinese form a minority, doubts surround these figures and it is still more difficult to come by precise information about the role played by the Overseas Chinese in economic development. All that can be said with some confidence is that in these particular states such information is probably more reliable for Malaysia—though even here figures on ownership are disputed—and possibly Brunei than for elsewhere. The political sensitivity of the information partly explains its dubious reliability. But there is another more important reason, and that is the uncertainty surrounding the definition and identification of 'Chinese'. Chinese do not form a homogeneous group. Overseas Chinese originate from many parts of China, though mostly from southeastern China. They speak different dialects, most of which are mutually unintelligible (though the script can be understood by the literate regardless of their dialect), and, as in any society, they comprise individuals who possess different beliefs, values and predilections, as indeed the fusion of some Chinese with the indigenes, who are themselves often extremely fragmented, indicates. The distinction between 'the Chinese' and the indigene Malays is often very blurred as in the case of the *babas* in Malaysia and *peranakan* in Indonesia. In the Philippines, too, many individuals identified as 'Chinese'—both ethnic Chinese and mestizos—have adopted Filipino names and Catholicism or even, in some cases, Protestantism; their children are educated in Filipino schools; they speak one or more of the many languages of the Philippines; and they are Philippine nationals.

There are also difficulties with the notions of 'ethnic', 'pure' and 'true'. How are 'true' Chinese to be identified in the first instance? The identification of 'Chinese' implicitly assumes inherent distinctions. The notion of profound genetic differences between peoples is now little regarded, and yet its echoes are still to be heard within the fable of cultural determinism—a fable which has proved both popular and influential. The identification of Chinese, then, may generally be taken to assume cultural distinctiveness, although notions of racial and cultural differences may be, and occasionally are, used in support of each other. But what if cultural differences, distinctions and uniqueness, are merely unidimensional and static perceptions of the shifting aspects of multidimensional institutions, values, acts and individuals? The identification of a group (by an 'outsider' or an 'insider'), the derivation of statistics, the interpretation and the conclusions would be no more than the creation of an intellectualised image. It would be an image which has no substance, not because the techniques used to create it are faulty, but because in reality there is no group, no 'Chineseness': there is only the *idea* of a group, culture or society, and the limited, unidimensional and static *presentation* of changing and multidimensional contexts.

(iv) The fourth issue concerns this book's dissatisfaction with the 'scientific approach'—a way of thinking which is used to create 'cultural' explanations of relative economic success. It is not intended that the whole of social science should be tarred with the same brush. But the 'scientific' approach is not restricted to the study of Overseas Chinese nor to cultural explanations of their economic success. Even those who do not in any case find the cultural explanation convincing may still feel it reasonable to ask that if there is no attempt in these pages to identify, construct or manipulate generalisations (the term is used here in its formal sense) or theoretical frameworks, then what purpose can be served by a one-off description of the Chinese in East and Southeast Asia? Such a description does not involve any attempt to identify, isolate and systematically analyse the nature and interaction of those variables that together comprise the phenomena with which the social sciences are concerned. It cannot begin to delineate a clear, if simplified, outline of phenomena; it cannot hope to explain these phenomena; nor can it add to an accumulated store of knowledge arrived at by the logical formulation of theory and disciplined empirical investigation. At worst, interpretation and understanding will provide little more than a confused account of a mass of detail, much of which may be irrelevant and misleading. At best, it will represent a clumsy, mechanistic reflection of reality (see Cohen, 1974, pp. 45–47).

Specific criticisms of the 'scientific approach' and its application to the analysis of the Overseas Chinese are developed in the following chapters. However, given that advocates of the cultural explanation frequently draw on the wider acceptability and applicability of the scientific method to legitimise conclusions and interpretations which help mystify Chinese society, it is worth briefly summarising in more general terms objections to the application of 'science'.

Two fundamental criticisms of the 'scientific approach' are made in these pages. The first is that the search for origins and for chains of cause and effect inevitably creates a distorted and 'flattened' (or unidimensional) interpretation which is often portrayed, misleadingly, as an 'explanation'. Moreover, observations interpreted in accordance with formal intellectual constructs are often presented as factual evidence in support of those same constructs and the resulting 'explanation'. This criticism is not restricted to a specific philosophical approach: the scientific method is not the exclusive preserve of the positivists. Many of those writers who adopt different philosophical approaches—such as structuralism and humanism and their variations—share a belief in the crucial importance of attempts to identify generalisations and to engineer theoretical frameworks. Even if not expressed as a rigid, overarching body of deterministic laws and structures, these generalisations and theories may take the form of a dialectic between man and the structures, values, behavioural environments, components of pure consciousness, the world of the routine, and the frameworks for action which he creates. This dialectic, which itself implies the existence of some guiding force external to man (or programmed within him), can then be used to account for particular phenomena, for without such a force it becomes impossible to incorporate explanations or understandings of the actions and behaviour of many individuals within a wider set of generalisations and theoretical constructs. The discussion then begins to reach further and further back into a profound philosophical or psychological enunciation of being, perception and ideas, and thereby transforms itself into a search for the idol of origins.

Some may argue that the presentation of the social scientists' methodology and underlying philosophy in this book is crude and overstated. Indeed, it is easy to point to social scientists who acknowledge a fundamental difference in the nature of the natural and social sciences, and who explicitly reject the notion of external, disembodied forces or programmed minds. Yet while they may believe that there are fundamental differences between the natural and the social sciences, and that the nature of their generalisations is logically discrepant, they still maintain that the notion of generalisations is relevant to the analysis of phenomena by social scientists. Put another way, many social scientists explicitly reject the notion of natural, disembodied laws governing human behaviour and organisation, and yet simultaneously believe that there are certain 'forces' or 'structures' which arise from man, which are altered by him, and which in turn affect him. Carrithers (1992), for instance, who shares Hacking's (1983) and M. Polanyi's (1958) scepticism of 'science', and who sees no deterministic culture, portrays a constantly changing world comprising the flow of human relationships—relationships which may be interpreted in many different ways. And yet he argues that this 'sociality' itself comprises the play of certain traits universal to all people, and that it is this universality which, in conjunction with the conscious decisions, judgements and actions of individuals, produces extraordinarily diverse cultures. So, too, he argues that it is individuals who comprise a legal system and that the 'system' and legal reasoning lead the conduct of our lives only if we let them; and yet he believes in human causation—vast networks of relationships constructed by man, creating an environment which the individual experiences as given, ineluctable and inescapable.

Many social scientists, then, have proved themselves adept at deriving a form of words which allows the explicit rejection of natural laws (because such a notion, as Giddens (1987) puts it, runs against common sense), and yet allows the scientist within the context of detailed analysis to permit the covert existence of such laws for the sake of theoretical constructs and their constituent generalisations. There is, then, an inherent contradiction in the scientific approach. Even when laws are explicitly rejected, the scientific method leads inexorably to the acceptance and manipulation of imagined forces. The truth at present, however, is that scholars of the humanities—the self-styled 'social scientists'—have yet to demonstrate the existence of any such generalisation (cf. Bottomore, 1972). And if such laws cannot be inferred, and if there are no idols of origin, no recursive structures, no human 'causation', and no chains of cause and effect, then all that is left is the spontaneity of the human mind, the play of circumstances and chance events, the individual's judgements, desires, choices and actions. One might begin by asking why should there not be 'cultural' diversity? Values, relationships, institutions, regular actions (such as exchange), and patterns of behaviour are multidimensional and, therefore, if viewed from changing perspectives or if directed towards different ends, or towards the same ends, create both similarities and differences. It is precisely because individuals are spontaneous, unpredictable and multidimensional, and it is because no assumptions can be made about the existence and operation of guiding laws, of chains of cause and effect, of pre-programmed traits and of recursive structures, that there are no set behavioural, institutional and moral patterns. It may perhaps be suggested, then, that an obsession with the identification and classification of universals, of defining characteristics, and of unique and profound differences can be abandoned, for all this obsession implies is that such traits pinpoint the beginning or end of some pathway between the essence of human nature (the ultimate origin) and the societies which man creates. What is of far greater interest and significance is the rotation of multidimensional actions, institutions and values, and the manner in which, and the ends to which, they are directed by individuals, and the motivation and purposes for which individuals create unidimensional presentations.

The second objection to the scientific approach is that emotion, judgemental moralising and ideology are frequently used to make concrete the perceived objectivity of observations interpreted in accordance with formal intellectual constructs. The perceived objectivity of these imaginings is further legitimised by a belief that the scientific approach must be relevant to an explanation of society, and by the knowledge that greater status and credibility are attached by some to the scientist than to the intuitive pragmatist. Emotions strengthened by intellectual credentials become a powerful magic (Polanyi's 'dynamo-objective coupling'). So strong is this magic that the notions of forces and traditions, social and cultural (but always fundamentally political) have been invoked to criticise the purportedly scientific and objective 'Western' perceptions and analyses of the Orient (Said, 1991, pp. 202–206).

The scientific charade, then, is harnessed to legitimise certain beliefs and to provide an objective foundation for the arbitrary selection of an arbitrarily determined variable used to conduct analysis, to define problems artificially, and to delimit a field of study. Social scientists move inexorably from description of their impressions of the general to the description of generalisations. In their world, which they believe has presented itself to them in a coherent totality, an individual is permitted to exhibit a degree of choice, but only within the strict limitations fixed by those 'variables' which constantly buffet him. These 'variables'—once just words—become discrete phenomena in their own right. A discipline then becomes wrapped and sealed in the emotions which are its foundations. Armed with the right conceptual frameworks and the other trappings of science, the promotion of prejudice and the exclusion of contrary opinions and views are made respectable and acceptable. Emotion, moral righteousnes and science reinforce each other; and the moral authority of the new discipline becomes unquestionable.

Overseas Chinese and cultural determinism

Clearly these four issues have far-reaching implications for the cultural explanation of Chinese economic success. The presumed deterministic nature of Chinese culture is sometimes made explicit, but more usually it is preserved implicitly within the 'scientific approach'. In the same way that many social scientists deny the existence of laws while simultaneously allowing mechanisms and forces to operate within conceptual frameworks, so the complex differences among individuals labelled as Chinese are explicitly recognised, and yet those individuals are presented and treated in analysis as a unidimensional phenomenon—'the Chinese'—which is allowed to operate within the intellect and to predetermine interpretation.

Terms such as the 'Overseas Chinese' or, as Wang Gungwu (1991) would have it, the 'Chinese overseas', are frequently taken to represent a people who share a distinct culture, and who can be defined by reference to that culture. And if a people and their subsets can be identified loosely as a group because they share a common 'culture', then the discussion begins to move quickly towards the notion that somehow this 'culture' defines a people and thus can provide an explanation of why those individuals are as they are. This 'culture' may be intimately related to a people—for it was created or adopted and adapted by them, and now, as if part of a complex dialectic, shapes them—but it is nevertheless 'apart' from them. Once again, thought becomes wrapped in the social scientists' all-purpose argument: individuals are subordinate to institutions, beliefs and values—phenomena which may have arisen from individuals, but which have since become detached from, and now revisit, those same individuals. This is a particularly useful line of thought because it creates and legitimises the notion of 'the Chinese' as a distinct entity which can be explained by the implicit application of laws and forces which are presumed to exist. 'Chinese' is not treated or used as an imperfect adjective or noun, but as a cloak which, if placed upon individuals, transforms them into something which can be more easily explained and understood: it is a neat, defined variable or system comprising subvariables and subsystems which can be slipped into a greater whole.

One inevitable outcome of this scientific approach is that the discussion begins to drift further and further back along chains of cause and effect in the search for origins. If the cultural mantle explains a people, then what is the origin of this mantle? The discussion now becomes even more tortuous. Either its origins are to be found within the psyche of a people, in which case there is a strong and logical suspicion that the cultural mantle must be rooted, at least in part, in a genetic code. Or its origins must be understood to comprise social, political, economic and historical forces which emanate from individuals, and then remain about them, reshaping their lives. The term 'the Chinese', then, has become symptomatic of the pervasive influence of the scientific approach. It depicts a mechanistic and unidimensional world of variables or systems which are reasonably well defined and well understood and which may therefore serve as a template upon which other explanations of other phenomena (such as economic success) can be hammered out.

Even for those who are concerned about the ease with which some writers claim that Chinese culture is unique, the scientific approach and the assumption that Chinese culture is quite distinctive still forms the terms of reference for debate. As Nathan (1993) points out, all too often interpretation is presented as fact; 'difference' is commonly and misleadingly, read as 'unique', and apparent differences, which comparison with other culture may uncover, are immediately declared the prime cause of an outcome (such as economic progress or democracy). Yet for Nathan too, there is no question that Chinese culture is peculiarly 'Chinese'. The issue for Nathan is methodological. Where are the tools that will enable a writer to state clearly, and then to prove empirically, the exact ways in which Chinese culture is distinctive? One technique, which Nathan appears to find attractive is 'the ladder of abstraction'—a convenient device that allows any particular cultural attribute of the Chinese simultaneously to be unique and to differ only in degree from that of another culture.

The notion of 'the Chinese', with its roots in a spurious science, cultural determinism, disguised emotions, elitism and defensiveness, has obvious implications for analysis and interpretation. It may also be turned to darker purposes. The deterministic culture by which this conceptual mass—'the Chinese'—is defined, may be drawn on to reinforce and legitimise feelings of cultural or even racial superiority. It is interesting to note that the deliberate manipulation of institutions, values and behaviour in order to create or express feelings of superiority and to present an image of a unique and clearly demarcated group, is not a particularly remarkable observation with respect to other parts of the world (see, for example, Maquet, 1961). Yet the Chinese scholar who claims that individuals belonging to 'the Chinese race' fortunately possess an inherent aptitude for economics (see Chapter 7) will find it easy to reinforce his attempt at mystification with the beliefs of more 'scientific' and 'rational' European and American cultural determinists who use Confucianism, cultural maps programmed within the mind, psyches, values, institutions, and modes of organisation (all of which are said to be specifically 'Chinese') to 'explain' Chinese economic success. It is as if scholars and many other commentators have adopted 'Chinese' symbols of uniqueness in order to mystify their own field of academic inquiry. In those generations which have not known war and the destructiveness of poverty from which their forefathers raised themselves, or which have been isolated from the world for so long only to be made bitter and resentful by their neighbours' affluence, the realisation of material progress may soften attitudes. But it might also help breed intolerance and contempt for people of different hues. In a nation which is searching for cohesion as economic power is decentralised and which is far from ready for democracy, cultural determinism may feed a dangerously militant nationalism. Alternatively, the notion of 'the Chinese' may be used to portray the Chinese as a single body working in unison towards a common goal and thereby to create an imaginary economic and military threat; or it may be used to legitimise and justify duplicity, dishonesty and corruption in business and public life.

These objections to the notion of 'the Chinese' do not imply that the term should be used pedantically. Concern here lies with the concept and its implications, not with the semantics of the term. There is nothing to be gained by taking exception to cautious descriptive generalisations about the Chinese: rather, exception is taken to the use of the term to indicate the application of generalisations to explain the actions, beliefs, values and institutions of individuals who happen to be Chinese. Coercion, choice, chance, desire and self-interest, whether defined by survival or by needs which stem from the isolation of intelligent beings, may give the appearance of a human tide of activities, but there is no force. The need to explain change is an artificial problem. It implies that the norm is timeless stagnation. The question is not how and why a society changes, but in what ways and for what ends individuals alter their opinions, values and beliefs, and in what ways individuals agree or disagree with each other, and whether they say yes or no. Institutions and values do not exist as distinct, separate entities or structures. They have no power, no means of influencing human action and decisions. They are an expression of human will, desire, beliefs and manipulation, and are shaped and directed by the individuals of which they are comprised, and by the actions, beliefs and desires of others. The observer is faced with a fluid mix of institutions, values and actions which possess many different aspects. Formalised procedures, institutions, codes of behaviour and regularised actions are thrown up, but they are only temporary expressions which alter or collapse as individuals come and go or change their minds.

Multidimensional Chinese

It has been argued that no assumptions should be made explicitly or implicitly about the existence or operation of laws, structures and given contexts. Instead, analyses of the Overseas Chinese and their relative economic success should be informed by the notion of constantly changing and multidimensional institutions, values, actions and behaviour, conducted by and comprising individuals who also possess many different aspects. In the analysis of problems the observer must be aware of these different aspects, of the different interpretations possible, and of constant change.

But how, then, may cohesion be brought to an interpretation of the desires, actions and decisions of multidimensional individuals and to an understanding of their economic success? It is suggested in these pages that the individuals and the fluid and multidimensional phenomena which make up societies and groups may be thought of as comprising wealth values and their associated institutions, moral values and forms of exchange. The significance and implications of these ideas for an understanding of material progress in China have been discussed elsewhere (R.N.W. Hodder, 1993a, b). They are reiterated here (in a slightly modified form) since they give shape to—and are, in turn, given much greater specificity by—the analysis of Overseas Chinese in the following chapters.

The notion of wealth values begins with the premise that wealth is a subjective abstraction. Analyses of societies with very different forms of economic, political, legal, kinship and religious organisation show that the items (goods and services) which comprise wealth vary tremendously; but these analyses also reveal other interesting dimensions to the notion of wealth. Items which comprise wealth may, for instance, be regarded as symbols of an individual's ability to manage social relations; they may be symbols of an individual's or group's political domination over others; or they may be symbolic of an individual's ability to create, to innovate, to provide and to contribute.

It is important to be wary of an obsession with tracing cause and effect when considering the possibility of a connection between wealth values and those items which, if possessed or held, are regarded as a mark of wealth. Items may be regarded as a mark of wealth for many reasons; and ability and industry may be admired for themselves; or it may be that the need or desire for an item and the admiration for an ability reinforce each other. While theoretically items which are regarded as a mark of wealth may comprise anything, in practice human needs, desires and perceptions may largely determine such items. Comfort, beauty, an appreciation of skill and craftsmanship, rarity, necessity, strategic importance, fashion, sentimental associations of a religious, historical, mystical or mythical nature, and a wide range of other emotions and practical considerations will all play their part. These considerations, which impart an emotional and, to the producer and the user, an objective aspect to these items, add to their importance as a mark and conveyor of wealth values.

There is, then, no certain connection between the amassing of items acknowledged as a mark of wealth and the regard with which the individual is held. The creation of wealth is the act of exchange, for it is the act of exchange which is the realisation, affirmation and acceptance of the prevailing values by which an individual's worth is judged; and it is the act of exchange which endows with the prevailing wealth values those items entered into exchange. Thus items which may be regarded as a mark of wealth but which were obtained through a form of exchange which does not predominate or which is regarded as inferior, immoral or illegal, will not endow the individual with status and prestige as understood according to the prevailing wealth values.

This discussion raises the question of the origin of wealth values. It has already been argued that the notion of a principle (behavioural or otherwise) and the implicit or explicit assumption of deterministic chain reactions between wealth values and their associated logistical and moral imperatives are misplaced. Values arise from within the human psyche—they are an example of the spontaneity of the human mind. Wealth values may represent a response to a mixture of chance events, existing forms of exchange, institutional patterns and moral imperatives, and other practical considerations; they may arise from an instinctive drive for survival intensified by feelings of insecurity and accompanied by a desire for acceptance and to demonstrate worth and superiority. Whatever the truth, the circumstances which lead to their formation are perhaps unknowable. Furthermore, even though the inculcation of wealth (and other) values and the need to observe and comply with these values should not be underestimated, motivations and perceptions will vary: there is no reason why individuals should feel the need to match their achievements with a commensurate level of wealth, or feel that there is a clear connection between the wealth they have amassed and their achievements; values will alter over time within an individual and within a notional group or society. It is also unlikely that different wealth values will exist in isolation from each other within a society or within the individual. It has already been implied that individuals or groups may engage in other forms of exchange or in illegal activities in order to obtain the wealth items which ordinarily would symbolise different but predominate wealth values. And different wealth values may have a place within the same society. The ability to manage social relations may well be (or made to be) an important part of the ability to improve, to innovate and to create, thereby forming an institutional support and moral imperative for trade. Or merchants may be made subordinate to the ruling elite which has the power to procure and distribute at will, which despises trade and profit, and yet which is to some extent dependent upon the goods and services generated by trade. Wealth values and forms of exchange, then, may not simply coexist; they may complement each other or create tension.

The second question concerns the regular association which, it is suggested here, exists between specific wealth values and specific forms of exchange. Phrased in such a way the problem is, perhaps, a little misleading. In any tangible sense, a form of exchange, it is argued here, does not exist: it is simply a description for an act which may be imbued with intentions which express practical considerations and the desire to demonstrate worth, and which therefore possesses many different aspects. Exchange is, or may become, an expression of wealth values, but it also has or conveys many other meanings and purposes. There is little point in attempting to uncover the essence or cause or effect of exchange, let alone a particular 'form' of exchange. For that act is at once, or may be perceived as or may become, many different things. However, if intentions and meanings are to be regularised, then certain imperatives (or practical requirements) will have to be met, though the nature, details and pattern of these institutions and values are not predetermined and all phenomena will retain many different aspects.

The 'loose association' between specific wealth values and forms of exchange, then, is not a product of chains of cause and effect: it is intentions and desires expressed through an act. Without the regularisation of intention, the significance of exchange will remain highly ambiguous, yet even when exchange is regularised, intentions may still retain ambiguities, for exchange, institutions and values are created by multidimensional individuals and may therefore be redirected towards other ends. For instance, the production or collection of goods and their movement is essential for survival, and without some regularised and institutionalised means of exchange there would be strife; exchange is also a means to establish and cement kinship and friendship, to make peace, or to effect and institutionalise marriage; and, as already noted, different forms of exchange may be redirected in support of or in opposition to each other.

Why it should be that wealth values find expression in acts imbued with different intentions and meanings which are described here by the phrase 'forms of exchange' is, again, perhaps unknowable. The association between wealth values and forms of exchange may be a consequence of chance, habit and the transmission of practices and ideas. Or it may be that a particular form of exchange strikes a particular emotional and instinctive chord within the psyche. To give, knowing that this act will not be reciprocated, is easily understood as acknowledging a willingness to subjugate oneself to another who is implicitly regarded as worthy and capable of ruling. To engage in reciprocity—that is to give in the expectation that one will, sooner or later, receive something in return—is easily understood as an acknowledgement and confirmation of mutual respect and trust, and both respect and trust support the act of exchange. To give is to be generous, and to be generous is to give, and both this act and its supporting emotions or values demonstrate that one believes the other to be worthy of help. To build up a network of relationships based on or strengthened by reciprocity is easily understood to demonstrate intrinsic worth, character, patience, status and power. To receive is to acknowledge the generosity and worth of the other, and it implies that one is worthy to be part of a wider network. To engage in procurement and distribution—to produce for the whole group, which is represented by the central authority or state, and to receive from the group—is easily understood as a willingness to subject oneself to the will of the group, to accept dependence on the group, and to confirm the leader's ability and right to achieve and retain control. Individuals or independent organisations which engage in exchange for profit—that is, to engage in trade—and whose survival is dependent upon such exchange are easily understood as capable of creating, providing and contributing goods and services distinguished from and better than those produced by others.

It is true that much depends upon context and perception: to give, knowing that this act will not be reciprocated, could also be regarded as charitable and patronising; exchange for profit could be interpreted as greed and an expression of disdain for others; reciprocity could be thought of as emotional blackmail which hinders the advancement of the individual. These different, conflicting interpretations serve to confirm the tension between different wealth values, but they also help to emphasise that the significance of intention and the aspects of a form of exchange may alter as the wider context changes. In other words, the same act in different circumstances takes on a different meaning. Thus, for example, gifts made to an association in a society dominated by trade, even if motivated by altruism, serve to legitimise economic success.

It may also be suggested that exchange and the logistics which regularise, facilitate, safeguard and legitimise that exchange, enable individuals to achieve certain goals and to exercise power in a manner which happens to coincide with the respective wealth values. Thus, for example, tribute may provide the psychological and physical means to exercise power; procurement and distribution enable control to be effected over the whole group, for each unit or individual is only part of the whole by virtue of their dependence upon those who procure and distribute goods and services; reciprocity enables the establishment of complex networks which make it possible for individuals to bypass and override formal institutions and bureaucracy; and trade means profit, and profit is the means to improve and innovate. Wealth values, then, are a justification of actions motivated by desire, need or practical considerations (such as order) other than the demonstration of worth.

A 'form' of exchange, then, merely describes the intentions and significance of the act of exchange, and these intentions and meanings may exist irrespective of the logistical imperatives. But if the intentions and meanings are to be regularised and institutionalised, then certain logistical imperatives must be met, though the patterns and details of the institutions and moral debates which evolve are not predetermined. Whether that intention and meaning arise from wealth values, or those wealth values justify actions which are essentially pragmatic (such as realising material progress or maintaining centralised control), the significance of giving and taking reinforce, and are reinforced by, institutions and moral arguments.

The logistics accompanying the operation of a particular form of exchange bring the discussion to the question of institutional patterns and moral imperatives. It has been suggested that wealth values are communicated by the act of exchange and that a form of exchange is but part of the logistical and moral imperatives required for the confirmation, affirmation, communication and institutionalisation of wealth values. These imperatives are, by definition, indispensable circumstances. However, this should not be taken to imply that these circumstances, nor the way in which they are met, are universal. The decentralisation of economic power, private property and independent economic units and the acceptance of profit are imperative for the institutionalisation, safeguard, legitimisation and prosecution of trade. But moral debates permitting and justifying profit and private ownership, and the nature of the political system are not predetermined. In short, although there are broad circumstances (or imperatives) essential for the institutionalisation of a specific form of exchange, and although the act of exchange is essential for the communication, affirmation and institutionalisation of wealth values, there is no single institutional formula or pattern or set of moral values which alone are capable of facilitating, supporting, safeguarding and institutionalising a specific form of exchange and its associated wealth value.

It must also be reiterated that institutions may include another form of exchange, and that institutions and values each possess many different aspects. Institutions and values associated with a specific form of exchange are unlikely to be associated only with that form of exchange. Whether by chance or by design, institutions and moral debates may also safeguard and legitimise different forms of exchange simultaneously, and they may, intentionally or unintentionally, be redirected to ends quite unrelated to the institutionalisation of any particular form of exchange.

Conclusions

The social scientist's perceptions, assumptions and methods belong to the structured, tutored mind; but human societies belong to the structured spirit, to the spontaneity of the human mind and to its multiple intelligence.[1] The Nuer, 'are undoubtedly a primitive people by the usual standard of reckoning, but their religious thought is remarkably sensitive, refined and intelligent. It is also highly complex' (Evans-Pritchard, 1956, p. 311). Nigerian students learning geometry some 30 years ago found it difficult to picture in their minds or to draw free-hand a reasonably accurate right angle or straight line. And yet their indigenous musical rhythms are extraordinarily intricate and micro-tones are precise: songs, though partly influenced by the tonal language, still possess a distinct melodic life of their own (Richards, 1972), for pitch warping is used to minimise conflict between text-tones and melody.

In their studies of the Chinese, at least, many social scientists often allow their thinking to be chiselled, chipped, straightened, bound and channelled by their adherence to a shimmering science. Conceptual frameworks and mysticism; the tenets and techniques which help define and legitimise an academic discipline; the desire to create and to be shaped by a self-conscious store of rigorous scientific knowledge; and the desire to place oneself within a specific approach: all this is now taught increasingly as a formal prerequisite or accompaniment for successful research. In many instances, analyses have become dominated by a vocabulary infused with complex, semi-technical jargon and politically correct terms, all designed to sanitise, to make banal, and to convey an air of objectivity, while the stock phrases and terms of the social scientist have been allowed to determine thought. Markets, marketing systems, urban systems, counterurbanisation, market forces, gender, social structures are, for some, no longer descriptive devices—they take on a life of their own. And because they now exist as discrete phenomena, they must have a definite function, a definite purpose, a definite effect, and a definite place within a greater whole.

The obsession with the scientific approach has made it difficult to comprehend or accept or embrace in analysis the flexibility and adroitness of the human mind; indeed, the social sciences have often attempted to limit and confine it. If there is no structure, no social force, no order as the trained social scientist understands it, then there must be chaos; and if there is no chaos, then there must be order. Structures, forces and variables interacting in a way that is ultimately predictable are assumed to form part of a greater whole which forms the context of thorough analysis and about which generalisations can be made.

This book is essentially an attempt to understand the relative economic success of the Chinese in Southeast Asia. Although the word 'description' is more often than not used disparagingly by social scientists, it has been argued in this chapter that self-styled 'scientific' analyses are little more than description warped by preconceptions, rigid opinion, and emotion. All that is suggested in this book is quite simply a way of thinking about phenomena which are of the humanities and not of the sciences. It is argued that these phenomena make up a fluid and transitory world: they possess many different aspects; they may, occasionally and temporarily, complement each other or create tension and conflict; they are expressions of the spontaneity of the human mind and are directed only by human will. But while there is order, there is no overarching structured whole—only a mosaic of possibilities which, by chance or by design as individuals attempt to fulfil their desires and achieve their goals, presents a sense of pattern which then alters as individuals and their ideas, desires, values and beliefs change. This mosaic, then, possesses many dimensions. If the aspect presented is unidimensional, the mosaic takes on the appearance of a specific, structured pattern; viewed from many different angles it presents a complex, ever-changing picture of numerous, and seemingly contradictory, patterns, interactions, tensions or conflicts. The observer is enveloped in a world that constantly changes its aspects as it moves about him, and as he himself alters his perceptions and beliefs.

The Chinese, like any other people who create complex societies, possess a wide range of values, ideas, beliefs and desires, and exercise many different choices, decisions and actions. This is as true of the individual as it is of the group. There is among the Chinese a blend of wealth values, forms of exchange, institutional patterns and moral imperatives. It follows that the Chinese cannot be considered, nor even in analysis treated implicitly, as a monolithic bloc. This latter condition would be regarded as a 'doctrine of perfection' by some (Mackie, 1992b). But then many of the arguments pursued in these pages conflict quite sharply with the views of a number of writers on the Overseas Chinese, largely because the ideas and interpretations presented here are not based upon the conventional assumptions of social science. More specifically, the views expressed in this book confront a solid and influential body of opinion which holds that the institutions, values and practices of the Chinese are essentially 'Chinese' in nature.

The economic success of the Overseas Chinese, it is suggested here, may be understood partly by circumstances and events, among the most significant of which were European colonialism, the communist revolution of 1949, and Japan's economic growth, itself built upon American strategic concerns in the Far East. But of crucial importance to an understanding of Overseas Chinese economic success is the desire of many Chinese to achieve material progress, or, more accurately, the desire of many Chinese to 'turn' institutions towards the extension, safeguard, legitimisation and institutionalisation of trade and its associated wealth values. There were no formulae which they had to work to, nor were there particular institutions, sets of moral arguments or values which they had to adopt or develop to enable or facilitate the prosecution of trade and thereby to permit sustained material progress. The complex multidimensional institutions considered in the next five chapters are the market, reciprocity, the company (a nexus of reciprocity), and an even more ambiguous institution—the association.

Note

1. The term is taken from Gardner (1983) but is not used here to imply that the neuro-scientists' views of multiple or modular intelligence explain the multidimensional nature of individuals and their actions, institutions and values. The term is used here simply to describe the individual's many different and constantly changing aspects.

2

The Chinese and markets

Introduction: trade, markets and 'Chineseness'

It is argued by some authorities that markets in China are endogenous phenomena which originate and develop according to the prescriptions dictated by a particular variant of central-place theory (Skinner, 1964, 1965). Skinner's arguments are mentioned here partly because they have gained wide acceptance, and Skinner's terminology has been applied to markets established by Chinese elsewhere in the West Pacific Rim (see, for example, Baker, 1968; Ng and Ng, 1992; Crissman, 1973; Sangren, 1984, 1987); and partly because elements of these arguments have exerted some influence on analyses of other 'Chinese' institutions, notably the Chinese associations examined in Chapter 5. Criticisms of Skinner's analysis have been made elsewhere (R. N. W. Hodder, 1993a, b) and need not be elaborated upon here. Suffice it to say there is no convincing evidence nor any strong or reliable argument which leads to the conclusion that markets established by Chinese in China or in other parts of the West Pacific Rim are either particularly remarkable or different from those established by other people in other parts of the world. Any insistence that markets are either more true to theoretical models, or are in some other way distinct, because they are 'Chinese', would amount to little more than a dogmatic rejection of alternative interpretations and contrary evidence. Moreover, as the similarity between the case of Hong Kong examined below and the case of Davao City in the Philippines presented in Chapter 6 demonstrate, particularly when considered in the context of the broader literature on Africa and Europe, the link between trade, markets and the physical growth of settlements is clearly not conditioned by 'Chineseness'.

As argued elsewhere (Hodder, R. 1993b), the extension and institutionalisation of trade itself is a highly flexible and exogenous phenomenon which proceeds from simple periodic markets to sophisticated commercial activities that may exhibit no geographical form; and from itinerant traders to complex business organisations. Progression along this abstract continuum is not predetermined or inevitable; it does not necessarily occur in logical steps; and both simple and complex expressions of this continuum may exist together. They may also possess different aspects: a market, for instance, may be 'turned' towards different ends and may thus represent a focus of social and political, as well as economic, activities.

Moreover, the extension and institutionalisation of trade itself may be considered to exist within a much wider assortment of supporting, protective and legitimising institutions and moral values—such as private property, a belief in the virtue of profit, welfare services, charity, the freedom of choice, and complex arrays of checks and balances among government, civil organisations, private enterprises and individuals. These institutions and values may also possess different aspects simultaneously and may therefore only subsequently facilitate trade and may do so merely by chance. They may also simultaneously be 'turned' against trade or in support of other forms of exchange (reciprocity, tribute, and state procurement and distribution) and their associated institutions and moral values, and thereby create tension.

There are, then, four arguments which, it is suggested here, are of particular significance for an understanding of markets. As argued elsewhere (Hodder, 1993b), the first is that markets are, by definition, exogenous phenomena. It is external contact—between different tribal and sub-tribal groups, between different ecological zones, or over long distance—which provides the stimulus and opportunity for trade and markets. The second is that markets are little more than constantly changing streams of people and goods focused on a specific place and at a certain time. Markets are not 'things', nor part of any 'system' or 'mechanism'. Markets do not serve any 'function', nor do they 'do' anything. They are simply one expression of the institutionalisation of trade.

The third is that no assumptions, explicit or implicit, should be made either about the existence of forces external to man or present within him, or about the existence of 'structures' capable of determining human action, regardless of whether or not it is claimed that these 'structures' emanate from man and are intimately related to him as part of some complex dialectic. The conduct of trade and its extension and institutionalisation are a consequence of conscious judgements, choices and decisions; the transmission of ideas; and common-sense responses to practical and dynamic circumstances, chance events, and the technical and material progress which trade itself brings. Save for those impera-

tives which must be met if trade is to be regularised, there are no formulae governing the way in which trade is prosecuted, nor are there any
evolutionary stages through which the extension and institutionalisation
of trade must pass. Once a market-place or even the *idea* of a market is
established and transmitted, the development of this institution is not
predetermined; nor is it inevitable. In one instance change may be
blocked by local social conflicts; in another instance, change is prevented
because tensions generated by conflicting values find expression in the
political economy of a nation.

The fourth argument is that markets may exhibit different aspects simultaneously. While the stimulus and opportunity for trade and markets are
external, the decision to establish a market (and to trade in the first
instance) may be as much an expression of a desire to demonstrate the
value and worth of an individual and community, or to achieve certain
political or social objectives (such as securing peace and stability, providing a revenue for the maintenance of a temple or for furthering political
ambition, or providing a place and an excuse for young men and women
to meet) as it is a response to the practical need, or desire, for trade.

Market Hong Kong

Markets: their origins and other aspects

The markets of Hong Kong are exogenous phenomena which possess
many different aspects. This holds true not just for the last 150 years
since the beginning of British administration: a number of settlements in
what is now Hong Kong were engaged in trade with Southeast Asia,
Japan, India, the Middle East, Europe and possibly Africa, as well as with
many other parts of China, for at least eight centuries before 1841 (Lo,
1963; Wheatley, 1964). The creation of new markets and market settlements and the growth which subsequently took place around them
were stimulated, and occasionally revitalised, by contact between different linguistic, clan and lineage groups, between different ecological
zones, and across political boundaries, often over long distances.

The earliest trade settlement was Hong Kong *cun* (village) in Shih-pai
Wan (Aberdeen Harbour) on the southern shore of Hong Kong Island.
Around the ninth or tenth century Hong Kong *cun* was the sole export-
ing agent for incense grown in Dongguan (in Canton (Guangdong)
province), at Shatin and on Lantau Island. The incense was processed in

Tsimshatsui, bulked at Hong Kong *cun*, and then shipped on to Canton (Guangzhou) (Lo, 1963; K. K. Siu, 1984).

It was Tuen Mun, however, that was regarded as Canton's outer port-of-call for foreign and Chinese ships. From about the tenth century onwards, trading vessels arriving from Arabia, Persia, India, Indochina and the East Indies with the summer's southwest monsoon would call twice at Tuen Mun—once before proceeding up the Pearl River to Canton, and then again before their return journey with the northeast monsoon at their backs. The port also served as a military outpost which came to form part of an elaborate system designed to monitor and control foreign shipping (Lo, 1963).

On the east side of the Kowloon Peninsula, Kowloon Bay served as a shelter and thoroughfare for Chinese and foreign ships navigating the coast from Jiangsu, Zhejiang and Fujian. In the thirteenth century remnants of the Sung court, pursued by the Mongols, established a residence at, or very near the site of, Kowloon City. During the seventeenth century, the site was occupied by a garrison and the local magistrate for Kowloon and what is now the New Territories. Although primarily an administrative and military outpost, villages existed in and around Kowloon City. It would seem unlikely that no markets developed here, for the settlement was at that time also the final destination of land routes along the east side of the New Territories to southern China (Lo, 1963).

The most important markets of the New Territories, however, were not established until the seventeenth century or later. All these markets were on land routes from Tuen Mun and Kowloon City to Canton and other markets to the northeast, north and northwest of Shamchun (Shenzhen), itself an important market in the seventeenth century. Moreover, all the markets shown in Figures 2.1 and 2.2, with the exception of Shekwuhui, were on the coast or, as in the case of Yuen Long and Shamchun, could be reached from the sea by river.

Shamchun lies in the middle of a fertile plain surounded by a range of mountains and steep rolling hills which separate the market from the agricultural and population centres to the east, north and west (Baker, 1968). It also lies at the entrance to the shortest and least difficult route through the northern highlands to Canton and to other parts of China to the northeast, and on a river connecting the settlement with the Pearl River. The plain on which Shamchun is situated is the same as that on which Shekwuhui (established sometime during the seventeenth century[2]) stands. A land route passes through Shekwuhui and runs along the

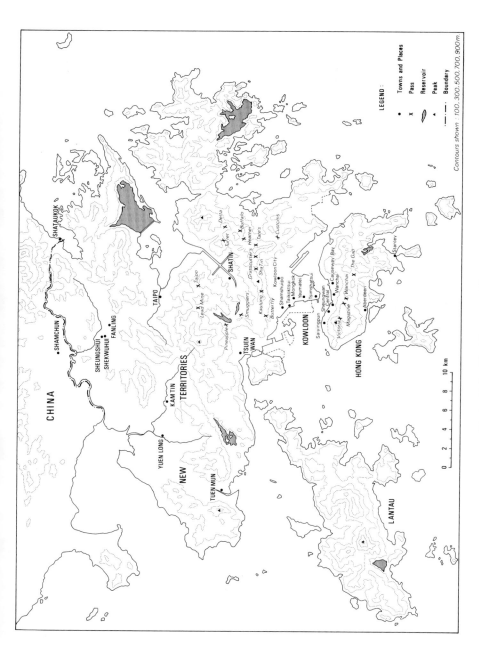

Figure 2.1　The Territory of Hong Kong

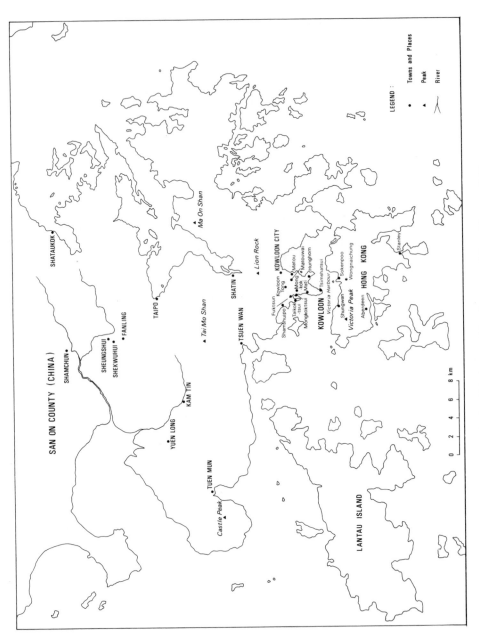

Figure 2.2 The Markets and Settlements of Hong Kong, c. 1840

eastern side of the New Territories through Taipo and Shatin; it then divides and runs on, via passes in the Kowloon Hills, either to Kowloon City or southwest to Tsuen Wan, Shamshuipo and the eastern side of the Kowloon Peninsula. A second route strikes southwest from Shekwuhui through Yuen Long to Tuen Mun. Shamchun, then, lay on long-distance trade routes into China, and, in addition, might have acted as a sentinel protecting Canton's southern flank and its approaches along the Pearl River.

Another important market on the southeastern route from Shamchun was Taipo. The origins of Taipo market, or at least the initial stimulus for the market, may have been the trade routes by land and by sea. But a deciding factor in the establishment of the market was the tension between two clans. It was said that the first market (the old market) was set up at some time during the sixteenth century, possibly earlier, to pay for a temple dedicated to the memory of a man who had committed suicide after failing to retrieve his father who had been kidnapped by pirates. Whatever the truth of this tale (for it must be remembered that legitimising myths and stories were tools used by lineages to build up status and influence), filial piety helped justify both the establishment of the market and the payment of dues to the temple and the market owners by traders using the market. Later, during the 19th century, the market became the focus of tension between the Tang clan, who controlled the market, and the Man. In 1893, the Man, after considerable trouble, finally persuaded the local magistrate to allow them to establish a second market just to the north of the old market. This was an expensive undertaking and politically dangerous without wider support. The Man succeeded in stitching together a pact among seven groups of villages (the ts'at yeuk) which formed a semicircle around the markets. The Man donated the land for the new market in which each group of villages was allowed to set up shops (Groves, 1964).

Another example is Yuen Long which lies on the southwestern route from Shamchun to Tsuen Wan and Kowloon. This market was established at some point during the seventeenth century (possibly earlier) by the Tang of Kam Tin. Together with their land, their concoctions of myths and legends, and the presence of a degree-holder among their numbers, the addition of a market was a symbol of—as well as a valuable source of revenue to finance—their rise to power and influence in medieval Chinese society (Faure, 1984a).

The importance of tension and competition among clans and lineages as a stimulus to the formation of markets should not be underestimated. With the exception of Shataukok, all the markets of the New Territories were organised and controlled by one of the main clans (Baker, 1966, 1968; Ng and Ng, 1992; Faure, 1984a). Indeed, it may be suggested that in south and southeastern China more generally, conflict among innumerable communities divided by clan, lineage and dialect created ideal opportunities and stimuli for trade which were intensified by long-distance contact over sea and land with Vietnam, Burma, and many other parts of Southeast and East Asia, India, the Middle East, Europe and possibly Africa. Markets were often generated among different communities, and within settlements any resident merchants or artisans were unlikely to be members of the settlements' lineage or lineages.

Trade was at best a lowly activity and at worst a despicable profession, but while successful military activities (either soldiering, which might attract the state's patronage, or banditry) could provide the means to amass land and to prepare oneself or one's sons for entry into the bureaucracy, trade was, for many, the only realistic path of transition from the peasantry to the gentry. Trade produced an income, and that income could be used to elicit the power and prestige derived from land-holdings and from the establishment of connections with, or, better still, from entry into, the ranks of the bureaucracy. These were accomplishments which possessed obvious practical advantages. However, the accumulation of land also required considerable social and political skill; entry into the bureaucracy could be achieved by merit alone, but fruitful connections with the bureaucracy and successful operation within it required more than high intellectual calibre; and while material possessions could be amassed through trade (and yet even then, in many instances, merchants were forbidden to display these symbols such as wearing silk and riding in horse-drawn carriages), they meant little unless merchants were permitted (and this was not always certain, Loewe, 1985) to use their profits to win prestige and status as judged according to the prevailing wealth values.

There was, then, an uneasy alliance between trade and the bureaucracy. The market may have been an expression of the institutionalisation of trade, but few of the values commonly associated with trade and entrepreneurship were attached to the market. Markets were a symbol of, and could be used to support and amplify, the power and prestige derived from the maintenance of the status quo and state power: the dues which the owner of the market-place levied on each trader generated a revenue; and to have permission to establish a market would have required favourable contacts with officialdom, and suggested that officialdom had a measure of confidence in the owner of the market-place. Bringing together lineages and villages, ensuring the survival of the market, and perhaps suppressing the establishment of competing ones, all required and demonstrated political acumen. Running a market also necessitated careful administration. Transport (such as a ferry service across a river) might have to be arranged, goods had to be weighed, dues had to be estimated, and disputes would have to be mediated.

Partly because of their symbolic and practical advantages, and partly because of the nature of individuals who had set up the market and of the circumstances in which those individuals found themselves, markets were also a potential source or focus of conflict as well as a means of bringing about some degree of stability. This was particularly worrying for the state bureaucrats who relied heavily on the landowners and their tenants to regulate themselves. In a world that was already fractious and violent (the tablets in the ancestral halls listed the names of individuals elevated in the ancestral hierarchy by virtue of their death in battle against rival clans), it was important that markets were organised and managed properly.

There are, then, three observations which may be of significance for an understanding of Hong Kong's markets. First, the initial and most important of the markets in the New Territories were not established until the sixteenth or seventeenth centuries or later, well after the rise of the ports of Tuen Mun and Hong Kong and the use of Kowloon Bay as a shelter for trading vessels. Second, all the markets of the New Territories lay on land routes between Shamchun (with Canton to the northwest, and other large markets to the north and northeast) and the initial ports. Third, all the markets, with the exception of Shataukok, were organised or controlled by competing clans and lineages.

It is true that not all long-distance traffic (whether by land or sea) was a consequence of trading activity: some of the most important goods produced and transported were at different times state monopolies or subject to varying degrees of state control. It is also true that the importance of Tuen Mun and Hong Kong was in decline from the seventeenth century onwards, possibly because of the rise of markets in the New Territories, or perhaps because of changes in shipping technology, or siltation in the waters around Tuen Mun, or improved access to Canton which the European powers forced China's government to accede to. Moreover, by the mid-nineteenth century the population of what is now the Colony of Hong Kong reached no more than a few thousand, and even the largest settlements each contained no more than a few hundred inhabitants. The trade which took place, then, did so on a small scale: it was almost an incidental offshoot of trading activities focused on Canton. Yet trade there undoubtedly was, and, despite their decline, both Hong Kong and Tuen Mun still retained markets, shops and a busy fishing port when the British took over Hong Kong Island in 1841.

It may be surmised that the markets which had been established by the mid-nineteenth century were forestalling points for foreign trade and long-distance domestic trade with Canton. But long-distance trade was only part of the reason for the establishment of markets. Another important stimulus was contact between rival clans and lineages as each attempted to build up its own power and status. Moreover, the values by which individual worth was judged, the institutions through which they were expressed, and the power and *éclat* which were thereby derived, could also be turned against trade and markets, rendering them very secondary considerations, or suppressing them altogether. By the opening decade of the twentieth century it would seem that little had changed in the New Territories. In the summer of 1912, District Officer G. N. Orme wrote:

'The Chinese villager . . . finds himself entirely unable to understand our aims and ideas, and our dismal condition of unrest: he frankly dislikes our iconoclastic spirit . . . but he does recognise some solid advantages from British rule—chiefly in the security of life and property. . . . The domestic life of the villager does not differ much from that of Chinese in other parts of China, nor has it altered much during the few years of British occupation: if anything it falls rather behind the general standard of freedom and enlightenment in Canton province. . . . In the New Territories as elsewhere continuous descent in the male line is the paramount object in the life of the Chinese, and the necessity for this is the foundation for many of their habits and customs' (p. 57).

Even so, important changes had already occurred in Hong Kong Island and Kowloon. When the Chinese immigrants arrived in Hong Kong under British rule, they found themselves released from the exigencies of reciprocity. Trade had been a means to an end in China—an imperfect way of achieving political and social status for oneself, for the immediate family and for the lineage. It was certainly the case that in many parts of Southeast Asia the indigenes measured status and power largely by the construction of social and political networks of reciprocity. But east of India and south of China there was no extensive, pervasive and sophisticated political state which could even begin to compare with that of China. Moreover, there was the strong influence of the Europeans involved in a complex mixture of global and regional strategic machinations and in long-distance trade with each other and, later, with Russia, Japan and the United States. The Overseas Chinese found themselves not only less constrained by bureaucratic and social politicking, but also under colonial authorities who, initially at least, managed their territories in a rather haphazard, distracted fashion, and yet, more often than not, barred the Chinese from entering the bureaucracies and many of the professions. The Overseas Chinese had not only to survive: they had also to maintain their self-esteem. Left to their own devices they chose to trade and to advance themselves materially.

They were well suited to the circumstances in which they found themselves. For they were experienced in the use of trade for social and political ends and in the handling of money, and they were well used to their inferior social position, to walls of intellectual and social snobbery, to whimsical government, and to inter-clan hostility and violence. In Southeast Asia they adapted the life which they had known to the realities into which they had been forced by labour gangs or persecution, or which they had chosen to accept in their search for a better life. Their experience was directed towards the extension and institutionalisation of trade and, thereby, the realisation of sustained material progress—a single, clear purpose which enabled their physical survival and according to which they could define their own standards of achievement. Material advancement was a symbol of the ability to create, to improve, to provide and to contribute—the basic values by which an individual's worth was judged.

These values have been perpetuated, but their acceptance from one generation to the next was not inevitable. The Chinese overseas had no one to rely on but themselves. The materialism they became engaged in, often obsessed by, symbolised and affirmed their own worth and abilities, and served very practical ends. Material success was often displayed in a way that was gaudy and distasteful to European sensibilities. But by what other means and values were they to judge themselves and express their own worth and abilities? The hard realities of survival did not change; and the desire for self-improvement according to the values which they had evolved, and which they had need of, were legitimised and strengthened by the self-confidence which material success engendered.[3]

Markets: anchor settlements, immigration and industry

After 1841 the creation of new markets and market settlements in Hong Kong and the invigoration of existing ones (all of which subsequently acted together as mutual external stimuli) were brought about by entrepot trade; by direct trade, primarily with China, Southeast Asia and, after 1949, with America, Japan and other parts of the West Pacific Rim; and by immigration. The sea of immigrants which flooded into the

Territory, most especially during and after the Taiping movement in the latter half of the nineteenth century, and then again during the upheavals following Liberation in 1949, simultaneously enlarged the pool of labour, skills, capital and range of contacts with China and other parts of Southeast Asia; increased demand for goods and services produced in Hong Kong; and, together with the immigration of Chinese to Southeast Asia and America, swelled demand for products from China for which Hong Kong was the main outlet. Trade enabled sustained material progress and thereby affected the level of sophistication of trade's extension and institutionalisation which, in its turn, profoundly affected Hong Kong's physical growth.

There was no predetermined order of development which markets, shops, services and industry, in simple or more complex forms, would have to follow. Both the British and the Chinese were practised in the institutionalisation of trade, for both had experience in retailing, handicrafts, industry and the establishment and operation of extensive and complex trade networks. In some instances (such as at Taikoktsui and Mongkoktsui), dockyards, wharfs and go-downs which coated the shores of the Island and the Kowloon Peninsula after 1841 served as points of external contact around which industry, services, shops and markets gathered. But given the impecunious circumstances of the Chinese immigrants, their experience in trade and markets, and the existence of long-established markets and market settlements, it was hardly surprising that the market would be a striking and consistent feature of Hong Kong's economy.

Markets, whether new or long-established (and around which clusters of shops, handicrafts, industry and residences had often already gathered), became the foci of more shops, services, handicrafts, industry and residences. The markets themselves also gathered around the more sophisticated expressions of trade's institutionalisation which had formed the initial or secondary foci; in some instances, then, markets only subsequently became additional foci of growth. These clumped 'anchor' settlements, which later agglomerated, stimulated the creation of new anchor settlements by acting as centres of demand, sources of goods, and reservoirs of experienced traders with capital who were willing and able to move to less developed, less expensive, less competitive parts of the colony.

In 1841 the British military first established themselves at Saiyingpun and West Point, and at Shungwan (Central) on the east side of a *nullah* which later became Garden Road. The majority of the European merchants (most of them British) settled in Shungwan, the Chinese at

Saiyingpun. In between the two groups a bazaar was set up on a beach head which later became a part of Jervois Street (Leeming, 1977). That same year work began on a road which followed a well-marked track along the north shore of the island. The road (named Queen's Road) was completed in mid-1842, by which time there were some 12 000 Chinese on the island—mostly workers and traders attracted by the possibility of employment, by the stability, freedom and opportunities for trade which existed under the British flag, and by the knowledge that stability and trade would soon push up land prices. Shops quickly clustered around the bazaar and then spilt out along the seaward side of Queen's Road. The area between Saiyingpun and Shungwan—which centred on what were to become Bonham Strand West, Wing Lok and adjacent streets—formed the heart of Chinese business activity and the centre of trade links with China and Southeast Asia until after the Second World War (Leeming, 1977).

This first bazaar was not the only one. Markets appeared and grew so quickly that the British colonial authorities frequently stepped in to regulate and relocate them, anxious to ensure that the health risks they associated with the markets were minimised, and to ensure that the Chinese were kept out of European areas while serving those same areas with a market. Indeed, growth was so rapid that one writer, describing the city of Victoria as early as 1845, observed:

'The ruins of a market with an old military hospital and a magazine come first in the central division of the town' (Chinese Repository, 1845, p. 294, cited in Sayer, 1937).

Throughout the nineteenth century growth remained largely concentrated at Saiyingpun and Shungwan, though beads of minor settlements were strung out along the north coast to Shaukiwan. The most important of these settlements were at East Point: to the immediate east of the point, on the shore of Causeway Bay, was the small village of Sokenpoo; to the west lay Wongnaichung—a larger village set about half a mile (just under a kilometre) back from the coast at the head of a valley of paddy fields (which later became the Happy Valley racecourse); and on a path through the Gap to Stanley and Aberdeen on the south side of the island. Wongnaichung had been established in the eighteenth century. It was one of a number of the larger coastal villages which included Aberdeen (Hong Kong cun), Stanley and Shaukiwan. It is difficult to say in what way Wongnaichung had been involved in trade before the British arrived. Although the incense industry had long since declined, Aberdeen still had a population of around 500 and a variety of shops; the path through the Gap from Wongnaichung was but one of many which criss-crossed the island and which had been used by merchants; the north coast was pitted with granite quarries where slabs for export were cut; and at Shaukiwan fish were dried, stored and shipped on to other parts of China (Lo, 1963; Hayes, 1984; *Hong Kong Guide*, 1893; Hurley, 1920).

It may have been the presence of a settlement (perhaps with a market and shops) at East Point and at the head of the Wongnaichung Valley, and the existence of flat land for possible further expansion which initially attracted Jardine, Matheson and Co. They established the company village at East Point in 1841 with the expectation that this would be the main area of growth on the island. The majority of Chinese, however, refused to set up businesses there, preferring Saiyingpun (Wood, 1940): malarial fever, which infected the whole Territory, was particularly severe at East Point, possibly because of the paddy fields' still waters. Even so, Chinese labourers, artisans and shopkeepers followed the company, and a bazaar was set up in, or just to the west of, Sokenpoo village, a little way behind the go-downs. The bazaar may have been set in, or near, an existing market at Sokenpoo. For although the market on the south side of Jardine's Bazaar at its eastern end is marked clearly only on maps from the 1880s, an empty but enclosed square appears on maps from the early 1870s. Whatever the case, it was from Jardine's Bazaar and the Sokenpoo market that the settlement spread over the next 50 years outwards along Irving Street and Keswick Street to the east, and westwards along Yeewo Street which later became the business heart of Causeway Bay.

Over the same period, shops, industries and services also spread outwards from two additional foci, both a little further to the west of East Point and Jardine's Bazaar. One was the Bowring Canal—a channel left running through land reclaimed from the sea at the mouth of the Wongnaichung Valley. Alongside Canal Road East, which ran next to the channel, junks took on and unloaded their cargoes. The other focus of growth was positioned still further west. At the foot of Morrison Hill where Queen's Road East meets Wanchai Road lay Wanchai Market. The market is named on maps dating from 1863, but at the site of the market an area bordered by streets existed in 1850, by which time a string of small terraced buildings—very much to the plan of Chinese shophouses—ran along Wanchai Road a little to the north of the market-place.

Initial development at East Point was slow. Indeed, to the east of Wellington Barracks and the Eastern Market there was little growth on the landward side of Queen's Road East during the nineteenth century. Even in the 1930s Causeway Bay was little more than a company village and bazaar with a few go-downs, small factories, shops and tenements attached (Leeming, 1977). The most developed area east of Central at this time was centred on Canal Road East and Wanchai. Nevertheless, it was around these anchor settlements that the further institutionalisation of trade occurred.

On the Kowloon Peninsula the growth of settlements was limited until the end of the nineteenth century. It was only in October 1860, during the uncertainty and instability brought about by the Taiping movement, that Kowloon was ceded to Britain, and even then development was, for several years after cession, impeded by a dispute between the military and civil colonial authorities over the use of land (Bristow, 1987).

The main settlements, such as they were, lay on the eastern side of the peninsula. They comprised a few small villages and industries, particularly those associated with incense and salt, and an assortment of forts and smaller military outposts (built during the seventeenth century as a measure against piracy and uncontrolled foreign trading vessels) coordinated by the administrative centre at Kowloon City. To the south of Kowloon City lay Matou and Matouwai, To Kwa Wan, Hunghom, Tsimshatsui and Chopae (where incense had been processed). On the west side of the peninsula were a few even smaller settlements including Shamshuipo and, to its south, Fuktsun and Taikoktsui. Kowloon Tong lay further inland to the northeast of the latter two settlements. To the south of Kowloon Tong and immediately east of Taikoktsui across an inlet (later reclaimed from the sea) was the village of Mongkok. A little to the south of Mongkok were the settlements of Mongkoktsui and, still further south, Mati.

The main foci of initial growth during the second half of the nineteenth century were here on the west side of the peninsula (where the water was shallow enough for reclamation and deep enough for ships) at Taikoktsui, Mongkoktsui and, most importantly, at Yaumatei, almost directly opposite Shungwan and Saiyingpun on Hong Kong Island. The Cosmopolitan Docks were established at Taikoktsui in 1875, by which time go-downs, wharfs and boatyards had begun to scatter along the coastline. Sandwiched between Shanghai Street (then known as Station Road) and Temple Street—two of the main streets running north–south in a grid-plan—was a market. To the north, immediately opposite the market and across Market Street, was a public square. On the east side of the square was a temple, and on the west side lay the police station. A little to the north of the police station was a pier. To the north of the square was a small, haphazard collection of buildings—a group of hovels associated with the shipping trades. Shops, services and industries spread from the market southwards along Shanghai Street, Temple Street and adjacent roads, particularly Canton Street, gradually filling up the street grids (Leeming, 1977).

The main additional points of growth to the north of Yaumatei and its original collection of hovels were at Mongkoktsui—where about 30 maritime lots had been laid out by the late 1880s—and to the north of the Cosmoplitan Docks. Here again, expansion followed a grid-street plan dicing the villages of Taikoktsui and Fuktsun which, by the late 1890s, had either set up or added to a number of lime kilns[4], probably in order to furnish the dockyard and local construction industries. Squatter settlements had also appeared in and around these two villages which began to merge with Shamshuipo just across the Kowloon–China boundary immediately to the north.

Shamshuipo lay on the junction of three important land routes: one followed the coast northwest via Tsuen Wan to Tuen Mun and from there ran northeast through a valley to the Yuen Long Plain and the market at Yuen Long; another ran a little to the northwest of Shamshuipo and then struck northeast through the Kowloon Hills (via Butterfly Valley) to Shatin, Taipo and beyond; and the third ran straight across the peninsula to Kowloon City on the eastern fringe where development remained straggled and beaded until the 1930s. If any market existed at Shamshuipo before the 1860s it was either coincident with or had later been shifted to straddle the boundary between China and British jurisdiction over the peninsula. This market which, like that at Yaumatei, also lay at the entrance to a pier, was initially held periodically. By the 1870s shops had gathered around the market-place and the pier and stretched along Market Street and other adjacent streets. The market and its adjoining roads formed the trade centre of Shamshuipo[5] (C. T. Smith, 1984).

The squatters who had begun to collect at Fuktsun in the 1890s perhaps marked the beginning of more rapid development along the western edge of the peninsula following the leasing of the New Territories to Britain in 1898. After this date, squatting gathered pace and land speculation became more intense, most especially around Shamshuipo. Yaumatei and Mongkok were still very modest, but established, trade centres which had already begun to absorb the spillover from Hong Kong Island. More importantly, Yaumatei, Mongkok, Shamshuipo and, a little further along the coast to the northwest, Tsuen Wan, all lay at the junction between the New Territories and the business heart of Hong Kong Island. Over the next 20 years existing routes through this junction to the New Territories and China were upgraded. The road via Butterfly

Valley was improved, and the road to Taipo was completed by 1900. The next few years saw the completion of the railway which began in Tsimshatsui and passed just to the east of Yaumatei, Mongkok and Shamshuipo before it cut through the Kowloon Hills and ran on to Taipo and China. The road to Tuen Mun was finished in 1919, and extensions from the western side of the peninsula to Kowloon City followed. There were good military and administrative justifications for these improvements; but they also greatly facilitated the movement of goods and people and thereby the speed and intensity of trade (Orme, 1912). This had obvious implications for the growth of existing markets.

When the New Territories were ceded to Britain, Tsuen Wan was already an established mart. Indeed, by the early nineteenth century a variety of shops—some run by successful traders who subsequently set up businesses on Hong Kong Island supplying Hong Kong *cun* with firewood, fresh vegetables and meat—had collected in and around the old marketplace. The shops later spread along adjacent streets and merged with what later became another 'new' market (with stalls) which still operated during the first two decades of the twentieth century. With the cession of the New Territories and the improvement of transport links between Hong Kong Island, Kowloon, the New Territories and China, the bulking up and processing of goods produced in the New Territories was given further stimulus. By the time the Castle Peak Road (to Tuen Mun) was finished in 1919, Tsuen Wan's long-established handicraft and processing workshops were being replaced by small factories manufacturing a range of goods including soya sauce, ginger and joss-sticks (Faure, 1984b; Lai, 1963).

Housing and the 'new' towns

The extension and institutionalisation of trade itself existed within a wider association of supporting institutions among which the most important—paradoxically for this *laissez-faire* economy—has been government. Despite the unpredictable and often momentous events which buffeted the colony, the administration ensured a flexible and yet stable domestic institutional framework which consistently supported and legitimised trade. More specifically, the Hong Kong government provided a stable legal framework; it planned for urban growth; it influenced or directly controlled utility companies and their profits; and it built and operated modest systems of public health and education (Donnithorne, 1983).

Another activity of great importance which the government has influenced or in which it has directly intervened is housing. After 1945, government and businessmen had to deal with a series of events outside their control but which profoundly affected the colony's development: the Communist victory in China in 1949; the trade embargo imposed by the UN and the USA on goods originating from China; China's growing self-imposed isolation and hostility towards domestic and international trade; the economic failure which followed collectivisation and the Great Leap Forward in China; the resulting tides of immigrants which swept into Hong Kong; and a decline in entrepot trade and direct trade with the new Communist state. Places had to be found for the immigrants to live and for the manufacturing industries which now had to be developed.

The immigrants were naturally attracted to the existing settlements with their services, outlets, workshops and limited manufacturing, and perhaps their sanctuaries of kin and wider social networks. It was in and around the settlements of Shamshuipo, Yaumatei, Kowloon City (Kai Tak), Hunghom and Western that small industries were initially set up, often in residential buildings. Even by the end of the 1970s small manufacturing enterprises accounted for about 40% of total manufacturing employment, and for 10% of GDP (Lai and Sit, 1984; Sit et al., 1979).

One of the largest squatter settlements was at Shepkipmei just to the north of Mongkok which acted as the settlement's shopping and service centre. Here at Shepkipmei in 1953, a fire left more than 50 000 people homeless. The government was now forced to intervene directly to house the immigrants. To a large extent the public housing scheme (or Resettlement Programme as it was known originally) was at first probably a reflex action. But the authorities could not have been blind to the fact that government housing would do much for the provision of a relatively passive and cheap pool of labour which, with the support of an educational system limited in outlook yet clear in its goals, would also be of good quality. Whether by chance or by design, government housing became an important institutional arrangement supporting and legitimising the extension of trade in Hong Kong.

During the 1950s and 1960s, when immigration reached its peak, most government housing was sited on the periphery of existing settlements, most especially at Shepkipmei, Cheungshawan and Tsuen Wan in the west, and at Lei Yuemun, Kai Tak and Ngau Tau Kok in the east. A few schemes were also run in the New Territories, but it was not until 1973 that a start was made to decentralise population to the old settlements of Tuen Mun, Shatin, Taipo, Fanling, Sheungshui and Yuen Long. It was also during the 1970s that the government quickened the pace of sales of industrial land and made lease controls more detailed, increasing the variety of industries which could be accommodated (Bristow, 1987). Two special-purpose industrial estates were set up: one at Taipo, and the other near Yuen Long.

It is important to emphasise that growth in the old market settlements—which had now come to be known as the 'new' towns—was not entirely a consequence of the decentralisation of population and industry. Growth in the New Territories had continued throughout the twentieth century, although at a comparatively slower pace, stimulated by the expanding markets, settlements and industries of Kowloon and Hong Kong Island, and by trade between the colony and China which continued even after 1949. Shekwuhui, moribund at the end of the nineteenth century, and not revitalised until 1925 (when civil war had made it unsafe to market in Shamchun), was stimulated by the lorry-transport business during the 1950s and 1960s. The market was the last on the route to China (Baker, 1968). By the mid-1960s Shekwuhui had merged with the village whose people had originally established the market; both merged with Fanling; and all three merged with the new market of Luenwohui established shortly after the Second World War.

Old and new

Although material progress engendered by trade has altered the ways in which trade's institutionalisation are expressed, market-places and markets have not died out. Long-established markets continue to operate in Hong Kong, and there are demands for new ones; traders gather round the fringes of existing markets and at any other strategic points (such as in the corners of government housing estates); new markets, legal and illegal, permanent and temporary, have been set up; and itinerant traders and, indeed, entire markets, have been redirected and relocated into modern accommodation. Three striking contemporary examples are Taipo, Shekwuhui and Luenwohui.

At Taipo the small government market at the north end of Fushin Street—the site of the market established by the Man in 1892—has been closed since 1990. But just outside the government market building is the Chepin Hawker Bazaar—a ramshackle collection of wooden stalls, workshops and mah-jong parlours—which stands on land now owned by the real estate company Heng Kei. Half-way down Fushin Street is the Manmo Temple, on either side of which the very first shops were built, and in which were kept the scales used to weigh the goods brought to market. Both sides of the street are now lined with shops.

At the south end of Fushin Street is a junction where trucks load and unload baskets of vegetables, and where a few traders sell vegetables from makeshift stalls. Two streets which run parallel on either side of Fushin Street run back to the north from this junction.The junction also leads south and east into a maze of roads and alleys, all lined with shops and workshops. To the southwest of the junction is the old railway station; to the southeast, on the other side of Sun Tak Street and On Fu Road, is an alley which serves as the back entrance to the Temporary Market. Established in 1983, this market was the destination of the Po Yick Street Hawker Bazaar which formerly lay a little to the east of the present Chepin Bazaar. The traders in the Temporary Market will be rehoused in modern accommodation similar to the small concrete shops at the Plover Cove Road market and Taiyuen market in the new housing estates to the east and north of Fushin Street.

Near the government offices on the north bank of the river, and immediately opposite the Chepin Bazaar, is another, smaller market. It is held from 4.00 a.m. to 7.00 a.m. each day. The only traders allowed are those from the settlements of the *ts'at yeuk*. The goods to be sold are still weighed, and dues are still paid.

A similar agglomeration of markets, shops and workshops is found at Shekwuhui. In and around the market-place (which was moved to its present site in the 1950s after two fires) there are now three markets: the original (but relocated) market; and two temporary markets. When the original market was relocated to its present site nearly 40 years ago, it was marked by a wooden shed which stood in the middle of vegetable fields. Near by were a few residential houses. The market grew, and the houses were replaced by permanent concrete shops which now line a market square full of wooden stalls arranged into neat rows. Traders selling goods from mats spread on the ground or from small makeshift stalls tumble out into the surrounding streets and alleys which are also lined with shops and workshops. It is not possible to distinguish one market from another: clearly the original market-place has had to absorb a growing numbers of traders which could not be accommodated elsewhere. All three markets are to be relocated to the Sheungshui New Market Centre.

Luenwohui exhibits the same tell-tale signs of market-settlement growth. The market comprises some 400 stalls: some are wooden; some have corrugated second storeys; and some are now fairly respectable concrete shops. Shops also border the market square, and surrounding the square are networks of streets lined with shops and small work-shops. About 10–15 minutes' walk from Luenwohui is Tianguanghui—a market-place for wholesalers. This is a large, fenced, concrete square (which doubles as a basketball pitch) with administrative offices and a toilet just outside the gate. The market opens from 2.00 a.m. to 8.00 a.m. every day during which time traders from surrounding areas and from more distant parts of Hong Kong bulk up goods which then find their way to Luenwohui, Shekwuhui and Taipo.

There is nothing particularly remarkable about the coexistence of 'old' and 'new' expressions of trade's institutionalisation. Not only has the appearance of shops around markets been noted in many parts of the world, but so, too, has the continuation of markets and even market periodicity within modern urban settlements.[6] Periodicity was still exhib-ited by markets such as Shekwuhui and Luenwohui in the 1960s; indeed, even today there are still vestiges of periodicity in Hong Kong.

In Kwai Chung the stained, muddy-green blocks of the estate, complet-ed in the early 1960s, now look much older than their 30 or so years. The blocks are due to be demolished, but a few shops which line their ground floors remain open and some of the tiny flats are still occupied: birdcages, pot plants and clothes hang from grimy and peeling window grills. On the ground floor of one of the blocks the childrens' nursery, its windows still covered with brightly coloured paper animals, has now closed. Running between some of the monuments are wooden stalls (many as large as the ground-floor shops) and street restaurants—the largest with 25 or more tables. At dawn these thoroughfares are filled with the smells of freshly brewed tea, deep-frying dough, coarse ciga-rettes, fish gutted by blood-soaked hands, spices and tamarinds from shops selling dried groceries, and mothballs from old winter coats brought out of storage for the cold November morning. Behind the blocks are the factories with blackened windows and dirty, streaked walls draped with thick webs of piping. And behind the factories are rough, steep hills, their ridges and boulders precisely defined and rust-reddened by the sharp, early morning sun. The estate has an extraordi-nary atmosphere: it is as if each building is impregnated with the now-calmed memory of the ruthless energy and singleness of purpose of some uncompromising struggle.

A market was first established by the people of the Kwai Chung estate a little over 30 years ago. It was first set up near the residential blocks on open ground. When the land was needed for a factory, the market was moved to a playground and then to Block 27 in the estate proper. It was moved yet again to a basketball pitch near Block 42 while a proper market-place was constructed in between Blocks 29 and 30. By the time the market-place was completed, however, the basketball pitch had attracted many regular traders who set up a market each morning. This market ran at right angles to the legal market. The legal market also grew as more stalls (some with a floorspace as large as if not greater than the nearby ground-floor shops) set up by illegal traders became attached as permanent features. At weekends the small affair set up each morning, continuing to midday or later, grew into a market of some 300 or more traders, fanning out on either side of the basketball pitch along the open area at right angles to Blocks 30, 31, 37, 39 and 42.

The illegal market operates every morning, beginning shortly before dawn; by 7.30 a.m. it has completely disappeared. Arguably there are three markets operating here: the first is the legal market comprising stalls to which a number of illegal traders have become permanently attached; the second is a small, illegal, weekday market; and the third is an illegal weekend periodic market which still comprises around 300 traders but which nowadays breaks up by 8.45 a.m. Some of the individuals who attend the weekend market are hardened traders from Yuen Long and Tsuen Wan, and the goods sold originate from Mong Kok, the New Territories and Hong Kong Island. A few attend the weekend market regularly, but most do not. Their movements depend upon whether or not business is thought to be good, the weather, the vigilance of officials from the department of Urban Services, and the operation of other markets.

This last point might suggest that schedules exist. However, it must be emphasised that there is no evidence for periodic rings, and that, in any case, work conducted in other parts of the world suggests periodicity is determined by the people who set up the market rather than by groups of urban-based traders moving en masse in a circuit. Even so, it would seem that there are other markets which appear regularly at certain times, usually outside office hours, and at certain places. One such market is held every Sunday afternoon (beginning shortly after 12.00 p.m.) near the two-storeyed, concrete Shatin market which lies in the shadow of the modern shopping malls. Traders who attend the Kwai Chung market on Sunday may also attend the Shatin Sunday market.

Similar rhythms exist at other markets. At Taipo for example, the Chepin Bazaar, a rather sad affair during the weekdays when many of the stalls are shut up, becomes extremely lively on Saturday and Sunday. All the stalls open, the mah-jong tables clatter, and Fushin Street becomes choked with traders selling anything from snakes to pyjamas from their stalls or from mats spread on the ground outside the shops. And on

Sunday afternoons traders fill Tai Wing Lane (which lies opposite the Taipo Temporary Market)—a thoroughfare lined with modern shops but which is for the rest of the week empty save for a handful of traders who may gather if there are no officials around.

These vestiges of periodicity seem to derive from fluctuations in demand and from the working habits of the traders. But social considerations may also be relevant. At Kwai Chung many of the traders at one time lived on the estate and have friends living near by. At Luenwohui, which used to follow a 2–5–8-schedule, some of the older traders still follow the schedule, not for any economic reason, but because it has become a social calendar.

The existence of markets and itinerant traders alongside modern retailing, services and industry has been the subject of much confusion. Theories positing divisions in 'Third World' economies between 'firms' and 'bazaars', 'peasant' and 'capitalist' production, 'upper' and 'lower' 'circuits' have been developed; notions of technological and social dualism have been suggested; a seemingly interminable stream of overlapping categories of hawkers, pedlars and types of stalls, businesses and markets has been generated; and constant refinements of the divisions, terms, categories and definitions have been made to help take into account conflicting evidence and new intellectual fashions.

This confusion has arisen partly from a host of artificial dichotomies spawned by analyses which pit colonial powers against the rest of humanity; and partly from a tendency to think of national, regional and urban economies as complete, structured and discrete 'systems' whose evolution is linear and predictable. If there coexist institutions and behaviour which should, according to theoretical prescriptions, belong to different stages of this linear progression or to different 'types' of economies, then it is immediately concluded that these institutions, and the people who comprise them, must be part of separate, discrete and structured systems. The questions which then become the focus of research are: why do these separate systems coexist, how do they work, and how do they interact, if at all?

The interpretation advanced in this chapter makes such dichotomies and categories redundant. The so-called 'traditional' and 'modern' institutions are all part of the extension and institutionalisation of trade. They coexist as a loose assortment of phenomena with different aspects. Provided there is sufficient demand and there are individuals willing to participate in markets—either because they are seen as a refuge for the poor, or as a training ground for commerce, or as an excuse and opportunity for establishing social contacts—there is no reason why old and new expressions of trade's institutionalisation should not coexist. It is only the preconceived theoretical prescriptions which make the fact of a market unlikely, unexpected or a departure from the 'norm'. Indeed, it may be argued that once markets, permanent shops, small industries and housing have been established, the appearance and evolution of markets could accelerate because demand increases, trade over long distances becomes faster and easier, and greater varieties, and wider sources, of goods and services become available. After all, markets are not a product of local surplus and local trade; rather, they are, by definition, exogenous phenomena.

The continuation or reappearance of periodicity merely helps to reinforce the argument that periodicity is not a phenomenon in its own right which defines a distinct type of market: it is no more than an adjective describing the responses of individuals who set up the market to prevailing circumstances, such as fluctuations in demand and the existence of other markets. Nor are the relocation of markets and the redirection of traders a new phenomenon. As already noted, the establishment, location and continuation of markets has always been subject to the wishes of people who set up the market and to the political and practical considerations of government.

It should be of no surprise, then, that classical examples of markets generated by external contact still exist in Hong Kong. Zhongying Street in Shataukok is perhaps the clearest example of a market formed between two different groups. It is particularly ironic that here, at Shataukok, China has swapped its comparative listlessness and numbing timelessness for Hong Kong's energy. The extraordinary difference is accentuated by a distance between Hong Kong and Mainland China which measures no more than a few paces. In Shataukok, on the Hong Kong side of the border, a few shops and stalls cluster around the entrance to the border post. To the east—lining Cheping Street, the chicane, and the path which leads straight to the pier—are a few tattered shops and stalls, hoping to pick up a little extra business from the Mainland Chinese who stray across the border with affected distraction, searching for more exotic goods. A few yards away across the border lies Zhongying Street and, attached to this street, a network of roads and alleys lined with many hundreds of shops and stalls selling goods—including clothes from Hong Kong and Japan, fruits from Australia, New Zealand and Israel, and dried foods from Japan—brought in by traders (many of whom are residents of Shataukok) from Hong Kong. Jewellery and gold—of a purity difficult to obtain in China—are especially popular while inflation remains high in the mainland. Here, in Shataukok on the mainland side, thousands of people from as far away as Lanzhou, Beijing, Sichuan, Henan, Hunan and Shanghai parade themselves, dressed in their finest clothes and adorned with gold accessories, expensive watches and glasses. A few minutes' walk from these streets and beyond the buildings under construction to the north, are splashes of forestalling markets.

Conclusions

The views presented in this chapter have two important implications. First, the emphasis given to the analysis of trade and the way in which it is extended and institutionalised, places trade networks, markets and the more complex expressions of trade's institutionalisation, right at the heart of an understanding of material progress. Although such a view does have its forceful advocates (see, for example, Bauer, 1991; Price, 1989; Grassby, 1970; Pirenne, 1925), it has been ignored by those obsessed with central-place theory, the professed explanatory powers of 'culture', and the search for any empirical evidence which appears to support prejudiced ideological beliefs, most especially those associated with the outdated concept of a dual economy.

Secondly, the common and popular notion that sustained material progress in Hong Kong has been so recent and so fast that the colony may be considered to have performed an economic miracle may be questioned. As Leeming (1975) points out: 'the modern economy of Hong Kong did not arise transformed within a year or two in the late 1940s, but was already geared to transformation before the Second World War, on the basis of an industrial inheritance of expertise in labour and capital going back for a century' (p. 342). Hong Kong's success was dependent upon three facts: the international trading activities of the British and, later, the Americans; a government which kept its aims simple and, rather than allow individuals to descend once again into parochial and narrow politicking, ensured (partly by design and partly by default) that individuals 'turned' their institutions, values and actions towards the extension and institutionalisation of trade; and the honing of values and expertise necessary for the prosecution of trade, and, thereby, the conversion of labourers, farmers, artisans and a few mandarin-industrialists into a proficient and formidably competitive workforce with crystal-clear ambitions.

A central theme of this book is that institutions, values and actions (such as exchange) are multidimensional. If viewed from different perspectives, or if 'turned' towards a common end, such as trade, then similarities between apparently different 'cultural' or 'ethnic' groups begin to emerge. These similarities are likely to be reinforced or conditioned by the logistical imperatives for the institutionalisation, regularisation and legitimisation of a specific form of exchange, and, most especially in the case of trade, by the transmission of ideas. It is important to emphasise that the precise manner in which these imperatives are met is not predetermined. Thus, the redirection of particular institutions, values or actions towards a specific form of exchange is not inevitable. Private ownership, individual economic entities, and the legitimisation of profit are imperative for trade; but the family, the association or reciprocal exchange, for instance, are not essential for satisfying those imperatives. But if so 'turned' or directed, institutions, values and actions then reveal similarities between apparently different groups of individuals. This chapter has attempted to illustrate these arguments by an interpretation of the rise and development of markets—one important expression of trade's extension and institutionalisation—in Hong Kong. The following chapters go on to consider the complex interaction among other multidimensional institutions, values and forms of exchange, and, in particular, the redirection or 'turning' of reciprocity by the Overseas Chinese towards trade.

Notes

1. It is important to emphasise that markets and settlements should not be considered as one and the same. This discussion is bound up with the debate over the legitimacy of central-place theory and is pursued elsewhere (R.N.W. Hodder, 1993b).
2. The market was originally called Tin Kong Hui, the location of which is unknown.
3. For a wider discussion on these issues see, for example: Freedman (1958, 1960a, b); Grimm (1985); Metzger (1970); Morse (1966); Yu Ying-shih (1967); Purcell (1965).
4. Lime kilns and handicraft industries had been a common feature of the larger settlements in south China at least since the seventeenth century. See P. L.Y. Ng (1961).
5. In 1918 the market was moved a few hundred metres to the northwest, thereby following the expanding trade centre.
6. See, for example: Broadbridge (1966); Hodder and Ukwu (1969); McGee and Yeung (1977); Tse (1974); Yoshino (1971); Ishihara et al. (1980).

3

Reciprocity and 'Chineseness'

Introduction

In Chapter 2 it was suggested that the extension and institutionalisation of trade is an expression of human will, and that, while there is no governing formula, it occurs in a way broadly similar in many different societies where there is the desire, need or circumstances conducive for trade. These broad similarities derive partly from decisions formed around the logistics of conducting and furthering trade; and partly from the transmission of ideas which trade (by its very nature as an exogenous phenomenon) will facilitate.

The present chapter, and the three chapters which follow it, go on to consider other institutions, values and actions (such as exchange) manipulated by the Overseas Chinese in the conduct of trade. Again it is suggested that these actions, institutions, moral values and arguments, and their manipulation, are an expression of practical responses to logistical and moral imperatives which have to be met if sustained material progress is to be permitted; that the precise manner in which these imperatives are met is not predetermined; and that these institutions and moral values each possess many different aspects. These different aspects, which derive from the spontaneity of the human mind, also help towards an understanding of the importance of chance events and of 'circumstances' or 'context'. While institutions and values may be manipulated to extend, institutionalise, safeguard and legitimise trade by design, they may also do so by chance: chance in the sense of unknowingly or by accident; and in the sense that as changes occur in the wider context there may be a shift in the significance of the individual's actions, decisions, choices and desires which may therefore require, or seem to require, adjustment, modification or rejection. Individuals, then, may act to regularise, safeguard and legitimise trade not unconsciously, but accidentally or unknowingly, following changes in the desires, perceptions, choices and decisions of other individuals, or they may, following such changes in the wider context or circumstances, act consciously in support of trade even though the individual's primary objectives and motives are quite unrelated to the prosecution of trade. In other words, the same action, decision, choice, desire or belief in different circumstances takes on several different meanings. While the institutionalisation and legitimisation of trade (and the creation of a context in which trade is the dominant form of exchange) may be driven by conscious will and desire, it may also be that a symbiosis is allowed to develop between the support of trade and the realisation of other objectives and the preservation of opposing values. In this way institutions, actions and values may be allowed to 'turn' gradually towards and, eventually, focus upon the prosecution of trade as an end in itself.

Changes from one form of exchange to another, in institutions and values, in the nature of their different aspects, and in the significance and relative importance and emphasis given to these aspects, are constant. This shifting context is an expression of the replacement of individuals, of spontaneous changes in beliefs, choices and desires, and of varied responses to logistical and moral imperatives and to chance events or, more specifically, to prior changes within the wider context. Thus, while constraints upon the individual's actions may derive from a will imposed, they may also derive, not from recursive social structures, but, on the contrary, from fluid and often unpredictable change.

Reciprocity

Guanxi (or reciprocity) is not a 'thing', or 'variable' or 'channel'. It does not characterise 'the Chinese', nor is part of a cultural mantle by which individuals can be identified as Chinese. It does not serve any function. It is simply a word for an act: it is the give and take of objects, favours, obligations, debts, responsibilities and loyalties. Sometimes actions and intentions are clearly defined, but often they are left unspoken and 'felt' rather than formed precisely within the intellect. Reciprocity may be closely entwined with friendship, compassion and affection and their associated emotions which together form part of its soft-framed institutions, but none of this denies the simultaneous existence of simple friendship and compassion without ulterior motives: feelings, emotions, values, desires, beliefs and the act of exchange, and their manipulation, exist in a complex, often contradictory, mix.

Nevertheless, to engage in reciprocity—that is, to give in the expectation that one will, sooner or later, receive something in return—is easily understood as an acknowledgement and confirmation of mutual respect and trust. To give is to be generous and demonstrates that one believes the other to be worthy of help. To receive is to acknowledge the generosity and worth of the other, and it implies that one is worthy to be part of a wider network. To build up a network of relationships based on, or strengthened by, reciprocity is easily understood to demonstrate intrinsic worth, character, patience, status and power.

But reciprocity and its associated wealth values—the ability to manage social relations—may also be directed towards another purpose: the extension and institutionalisation of trade and of trade's associated wealth values. Reciprocity and the networks built up through or strengthened by reciprocity are a realisation, affirmation and symbol of the ability to manage social relations, but they do not in themselves necessarily endow prestige, social status or power. They are not, then, taken to be the creation of wealth or as a mark of wealth: they are part of trade's extension and institutionalisation and therefore enable material progress. It is material progress which is a mark of the predominant wealth values which convey prestige, and it is only by realising material progress that reciprocity and its associated networks may become a reflection of this status as well as a means to material progress.

Whether practised as an end in itself or in the support of trade, reciprocity is used to achieve a specific purpose: the construction of social networks and personal status. Reciprocity, then, is not simply a dyadic or group phenomenon, nor is it restricted by kinship. Relationships extend far beyond the individual and the immediate family, and both complex and profound ramifications may stem from exchange, from failures to meet obligations, to pay debts, to discharge responsibilities and to maintain loyalties, and from good or negative assessments of character according to attitudes and values which are often judgmental. An individual who is not merely aware of the repercussions which radiate from his dealings with others, or from dealings among other individuals, but who can also work these effects to his advantage or, indeed, to the disadvantage of others, either now or in the long term, is regarded as an individual to be reckoned with.

It has already been noted that a range of soft-framed institutions (and moral imperatives) are associated with reciprocity. Together with the exchange of favours, obligations, debts, loyalties and responsibilities, they form an amorphous flow of relationships. These soft-framed institutions describe emotions (some genuine, some only demonstrated because they are expected by others) and their manipulation. They comprise the play of egos, jealousies, envy and confidence; friendships, some genuine, some 'informal formalities', some temporary expedients, some built up over many years; the ability to make others feel at ease, to exhibit the common touch, and to transmit commitment, feelings and enthusiasm; judgements of character and worth; and value placed on personal characteristics such as thrift, quickness of mind, the ability to perceive immediately opportunities whether in the present or in the long term, the capacity to learn quickly, sagacity and, above all, expedient trust and reliability.

There are no clear types or categories of relationships, no black or white—all is shade. The formation and conduct of relationships are a traffic in glances, expressions, mannerisms, conversations with layers of meaning, and emotions: all is choice and judgement about feelings, about conflicts and about the possible or eventual ramifications of actions taken. These hazy, flickering, transient webs of relationships throw up groupings which may take on formalised expressions, such as the association or *barkada* (see Chapters 5 and 6). Yet even these formal expressions are in essence shifting phenomena which take on a variety of different aspects depending upon the actions and desires of individuals.

Taken as a whole, this dull swell of emotions, choices, actions and professed values puts to sleep the senses. Yet this is the very stuff of business. The apparently rigid and hard-framed institutions—such as the bureaucracy, the company and the association—and their formal procedures are often no more than elaborate but imperfectly fitting and gimcrack façades. These may be used as points of reference or reminders of status, position and authority which may be brought into play as inflexible determinants when it is convenient so to do or when individuals get out of hand and personal authority must be reinforced. Status and prestige, then, often lie with the individual, not with the office, and many will attempt to create and operate their own, perhaps competing, networks. Consequently, power-brokers may not always be those in formal positions of authority.

Although it is tempting to categorise relationships and to precisely define their rules of conduct and the pathways of their formation, this only serves to create wooden performances and therefore to misrepresent the nature of these relationships. In fact, relationships which comprise the social networks associated with reciprocity are as much 'felt' as reasoned intellectually, and they possess many different aspects. For the practised, dealing in obligations, friendships, emotions and reputations becomes an easy, natural and intuitive response to conscious will. There may be no ulterior motive underlying these relationships—just affection or altruism. But equally, relationships may, when desired or as circumstances require, take on a different aspect as they are effortlessly manipulated (perhaps ruthlessly so) or 'turned' in order to fulfil a specific purpose, whether that be the conduct of trade or the establishment of social networks as an end in itself.

More will be said about the discriminating use of reciprocity in Chapter 6, but it is worth noting here that, from the point of view of some Filipino businessmen in Davao City, the Philippines, Chinese business contacts are blatantly manipulative, not only in their dealings with Filipinos, but also with other Chinese: and, from the Filipino merchants' point of view, herein lies a Chinese merchant's strength and weakness. Another way of expressing these observations is that some individuals are more careful to limit reciprocity, to direct its use and to separate its practice from the wealth values with which it is commonly associated. The manipulation of relationships therefore appears less sophisticated and more awkward, especially from the point of view of those individuals for whom reciprocity and trade, and the wealth values and institutions associated with these forms of exchange, are intimately entwined. A similar point has been made of Chinese and indigenes in Papua New Guinea where, according to Wu (1982, p. 105), Chinese economic success has been realised mainly through the creation of new strategies in manipulating kinship to suit the local situation:

'The New Guinea Chinese see their kinship behaviour, which deviates so markedly from tradition, as an advantage in commercial development. When I discussed this with them they often pointed out to me that a lot of the bankruptcy among native storekeepers was caused by the automatic claims of kinsmen on goods and the refusal of the kinsmen to pay. The Chinese say, "The native trade store can never survive long simply because it usually is 'eaten' by the relatives of the storekeeper before he can make any profit." The native entrepreneurs' problems, of course, are not so simple, but what the Chinese point out has considerable truth' (p. 106).

Indeed, the limited and discriminating use of reciprocity, though not described in such terms, has been observed in several studies.[1] In his study of Lukang, Taiwan, DeGlopper (1972) notes that to do business:

' . . . one must know people and establish relations of mutual confidence, but one need not know people terribly well, and it is better to have limited relationships with a lot of people than very close ties with only a few. Close, particularistic relations are important only in special circumstances. The small businessmen of Lukang desire to maximise their autonomy and freedom of choice, and prefer limited, functionally specific relations to diffuse ties fused with personal relations' (p. 323).

'Orientals' and reciprocity

In many ways the interpretations expressed in these pages are at odds with much of the literature on reciprocity (or *guanxi*, social connections, interpersonal relations or particularistic ties) which is nearly always mentioned in, and is very often of central concern to, studies of Chinese business. Although some writers (Freedman, 1957; Fried, 1953, Lim, 1983; Omohundro, 1981, 1983) have taken a very pragmatic view of reciprocity, few have done so all of the time, and a great many more reject or marginalise the notion that *guanxi* is consciously manipulated: in so far as reciprocity and its associated social networks and soft-framed institutions are 'used', this 'use' is taken to be inevitable or unconscious because the practice of reciprocity is rooted in Chinese culture. Reciprocity, then, is taken, not as an expression of the extension and institutionalisation of trade, but rather as an expression of cultural characteristics which define the Chinese—and this is a view which is strong in those few analyses specifically concerned with reciprocity. It is perhaps surprising that, rather like the markets[2] set up by Chinese, the establishment and use of complex webs of reciprocity should be regarded as a peculiarly 'Oriental', let alone Chinese, characteristic. Indeed, the similarities between Chinese 'cultural traits' and practices and those of other societies will be referred to on numerous occasions in this book. It might still be suggested, nevertheless, that reciprocity can be regarded as peculiarly Chinese since it derives from cultural traits which are believed to be peculiarly 'Chinese' in nature: the outcome—reciprocity—may be similar to the practices of other 'cultural' groups, but its origins and the characteristics of its practice are essentially and distinctively 'Chinese'. Such an argument, however, is difficult to accept unless a valid link is first established between reciprocity and those values, institutions, behaviour and practices which are unquestionably and definitively 'Chinese'.

Yet despite the obvious weakness of the initial assumptions, terms such as *guanxi, xinyong, ganqing* and *mianzi* (the latter three terms may be described as soft-framed institutions) seem to have about them a curious *gravitas* and resonance which lead many scholars to treat them as indicators of 'Chineseness'. As part of the scholarly exercise to demonstrate causal links assumed to exist between reciprocity and the purported underlying cultural traits, *guanxi* and its soft-framed institutions have been codified into manuals of behaviour, etiquette, performance and structure (Yengoyan, 1983; Omohundro, 1981, 1983; Young, 1971; DeGlopper, 1972) or arranged into rings of social distance (Omohundro, 1981; Landa, 1983). This makes it easier to pin down and analyse a phenomenon which is amorphous and which may only exist as a casual act and unspoken understandings. The outcome of such analyses, however, is to create a behavioural master plan for unidimensional and mechanistic beings.

Jacobs's (1979, 1980, 1982) analysis of *guanxi* in Taiwan is just such an example. Jacobs (1979) provides a model of *guanxi* which attempts to account for variations in the hierarchical level of the political system, types of persons tied, types of political arena, and change over time. Such an approach, he argues, is 'important to the study of cultural influences on politics', and 'may help in the development of a theory of Chinese politics' (p. 240). Jacobs further maintains that 'even assuming generalisations to Chinese culture, cross-cultural studies must be undertaken before we can determine which aspects of the analysis are specifically Chinese and which are of a more general nature' (pp. 240–241). At the very outset, then, Jacobs portrays a world of distinct variables: a culture and a political system which are peculiarly Chinese; and unspecified cultural and political elements which are probably universal. The aim of comparison is merely to isolate something which already exists— 'Chineseness'. On top of these assumptions a number of others are built.

Jacobs develops his argument, stating that 'People organise political alliances on a number of bases including personal interests, ideological commitments, interest in particular policy issues, and particularistic ties. . . . In Matsu township political alliances *can be formed only on the basis* of a "close" particularistic tie called a *kuan-hsi* [*guanxi*]' (Jacobs, 1979, p. 242, italics added). The reason why alliances can be formed only on the basis of close particularistic ties is because, although Matsu has substantial autonomy in choosing its own leaders, policy formulation takes place above the level of the township. Thus 'politics in Matsu centres on prestige and face rather than issues. The Matsu data cannot, therefore, support the development of a higher-order model integrating the roles of issues and particularistic ties in alliance formation, but it does isolate the political roles of particularistic ties from complicating factors and simplifies the development of this lower order model' (p. 242).

Jacobs has now introduced a number of additional subvariables and forces. *Guanxi* are separate from personal interests, interests in policies and ideology; and in Matsu, political alliances can only be formed on the basis of 'close' particularistic ties because individuals are primarily interested in 'face' rather than in 'issues'. In Matsu, then, political alliances depend upon particularistic ties that are 'close'. It is so because the assumptions built into the model and the philosophy upon which those assumptions are founded determine that it should be so. The discussion then begins its search for origins. Jacobs has argued that alliances depend upon the 'closeness' of *guanxi*. But upon what does the 'closeness' of *guanxi* depend? A new series of subvariables is then presented. First, there is the existence of a *guanxi* base (or shared identification) such as family, a common home town or place of work. Secondly, there is 'distance' of *guanxi*. And 'distance' depends upon the third subvariable—*ganqing* (emotion or sentiment).

The pattern of identification (and thus the typology of *guanxi* bases) is itself dependent upon a number of third-order variables. First, there is the level in the political system. For example, Jacobs argues that kinship will (for some unspecified reason) form a more important base in settlements classified as a village or township than in a higher-level centre. The second is 'whom the *guanxi* ties'. Thus, for example, a co-worker link is important between township leaders but unimportant between leaders and voters (Jacobs, 1979, p. 243). From this it would appear that Jacobs is suggesting that occupation and status can determine patterns of identification and therefore 'types' of *guanxi*. The third subvariable is the type of political system. Thus '*kuan-hsi* relevant to a territorial system with universal suffrage may be unimportant in a bureaucratic arena' (p. 244). Finally, there is the force of time: 'For example, co-worker (bureaucratic colleague) *kuan-hsi* appear more important than classmate *kuan-hsi* during the Sung dynasty while the reverse is true during the Ming and Ch'ing dynasties' (p. 244).

Consider, then, the third-order variables which affect the type of identification and thus the type of *guanxi* base: size and administrative hierarchy of settlements; occupation and status; types of political and administrative organisation; and time. The interaction of these subvariables, it is claimed, produces distinct types of *guanxi* bases: locality, kinship, co-worker, classmate, sworn-brotherhood, surname, teacher–student, economic and public.

What, then, of the other two second-order subvariables—distance and *ganqing*? Distance, Jacobs argues, depends upon *ganqing*, and *ganqing* depends upon social interaction, utilisation and helping. *Ganqing* cannot be present without social interaction and reliability. It usually occurs in hierarchical relationships (among those of different social status and ages) but it may also occur horizontally, provided the individuals concerned are non-kin.

Jacobs (1979) summarises his 'preliminary' model as follows:

' . . . [it] consists of three *variables*. The independent variable, the existence or absence of a *kuan-hsi* base, determines whether or not a *kuan-hsi* exists and is simply a nominal variable. If a *kuan-hsi* exists, its value, i.e. its "closeness" or "distance", also depends on an intervening variable, the element of affect or *kan-ch'ing* [*ganqing*]. Both the intervening variable *kan-ch'ing*, and the dependent variable, the value of a *kuan-hsi*, are ordinal variables. In other words, the quality of *kan-ch'ing* and the values of *kuan-hsi* in various relationships can be compared, ranked and thus measured' (p. 258).

Towards the end of his paper, Jacobs puts his argument in a slightly different way. The variables which comprise distinct political, economic and social structures, technology, religion and language become subsystems of the primary system—culture. Particularistic ties, one of the bases of political alliances, become a strand of this system. This strand runs through numerous cultural subsystems. The origins and subvariables of particularistic ties now become part of the foundation of culture and its subsystems. By the application of this concept to any culture, Jacobs argues, particularistic ties serve as an excellent entrée into the study of cultural influences upon politics (that is, the influence which each subsystem of a culture has upon the other) and, in addition, constitute a sound base for cultural comparisons. And the purpose of comparisons, so the cultural–scientific approach has predetermined, is to enable social scientists to identify those phenomena which are peculiarly 'Chinese', and those which are universal.

Consider now this strange phantasmagoria. It comprises systems and subsystems which can be broken down into variables and subvariables *ad infinitum*. These systems, subsystems, variables and subvariables are wielded by unspecified forces in a subtle and complex manner rather like themes of a fugue or symphonic movement, here twisted around each other, here separated, but always following a structure which is ultimately fathomable and predictable. For some, this is an admirable technique. But when all is said and done it portrays a dry, lifeless world. Individuals are apportioned into a particular cultural 'system'—the Chinese—and are ultimately subject to the machinations of the flat, unidimensional and clockwork world by which they are defined.

There seems to be an unconscious arrogance underlying the cultural approach: the world, it is implicitly assumed, *is* as I see it; these *are* the variables which determine the formation of alliances and *guanxi*; and although final confirmation from cross-cultural comparisons is required to determine which variables and subvariables are universal and which are peculiar to the Chinese, the variables and subvariables identified here not only explain, but also determine why Chinese political structures are as they are. These variables are part of a specific 'Chinese' culture and its subsystems which comprise and define the concept of 'the Chinese'.

M.M.H. Yang's (1989) analysis of reciprocity in China is, perhaps, a more sophisticated treatment (see also Yang, 1994). Taken together with a consideration of the interaction between trade and state procurement and distribution, it may be suggested that Yang's views on reciprocity in Communist China provide a clear illustration of the coexistence of, and interaction and conflict between, different forms of exchange and associated wealth values, institutional patterns and moral imperatives. But Yang's analysis is still, in many respects, unidimensional. In particular (and aside from the obvious point that Yang pays little attention to the crucial political, economic and social implications of trade), only the tension between reciprocity and state procurement and distribution is examined. Moreover, although Yang recognises that in certain instances the tactics of reciprocity can be used to bolster government strength, reciprocity is presented as an 'opposition economy in itself rather than for itself'. These views would suggest that Yang believes reciprocity, by its very nature, will act to subvert state procurement and distribution. In these pages, and elsewhere, however, it has been argued that while different forms of exchange may create tension, they may also complement each other, for different forms of exchange and their associated institutions and values possess many different aspects: individual choice, decisions, ideas, beliefs, values, desires and actions constitute exchange, its purpose and significance.

Yang's unidimensional analysis derives from a number of problems which also characterise Jacobs's scientific approach. Yang places greater emphasis upon the conscious manipulation of *guanxi*: the 'art' of *guanxi*, Yang (1989) maintains, 'lies in the skilful mobilization of moral and cultural imperatives such as obligation and reciprocity in pursuit of both diffuse social ends and calculated instrumental ends' (p. 35). However, as she builds up her arguments, theory and science begin to take over: individuals may manipulate each other in order to realise specific ends and desires, but their actions and the consequences of their actions are very much predetermined.

The operation of reciprocity, Yang believes, is built into a redistributive economy which relies upon a bureaucracy of state distributors to dispense livelihood and discipline. 'On the one hand, some officials take advantage of the system to promote their own class-status positions. On the underside of personal official power, however, dwells a repertory of the tactics of *guanxi* for the population. Not only does it challenge official power, the gift economy also subverts the dominant mode of economy. In other words, prefigured in the redistributive economy are the seeds and possibilities of challenges to its power' (Yang, 1989, p. 36).

Thus *guanxi*, so Yang argues, possesses a fixed symbolism which will inevitably cause it to subvert state procurement and distribution. In other words, it is assumed that reciprocity will mean the same thing to all men. Following Gregory (1982), Strathern (1983), Mauss (1985), Dumont (1970) and Bourdieu (1977), it is argued by Yang that a gift is regarded as inalienable from its owner; and since possession of the gift is contingent upon repayment, a recipient does not have full rights of ownership over the gift. Moreover, the absence of a disjunction between persons and things in gift exchange means that donor, gift and recipient share a common symbolic substance; it therefore follows that the donor has a hold on the gift in the form of a moral right to something in exchange for that gift.

Yang goes on to argue that there are five symbolic 'processes' (or symbolic 'mechanisms') at work in each transaction which produce reciprocity's oppositional effects to state procurement and distribution. The first is transformation or, in other words, the establishment of a basis of familiarity. By this is meant an attempt to transform the other into the familiar and therefore 'to bridge the gap between outside and inside' (Yang, 1989, p. 40). The second is incorporation—the symbolic destruction of boundaries among people—which represents a 'tactical incursion into the recipient's personal space' (p. 42). The third is moral subordination of status antagonism—a struggle for moral superiority between giver and taker in which it is possible for the social hierarchy to be temporarily reversed. The fourth is appropriation or possession. Yang explains this 'process' as follows: 'For the recipient, incorporating another's substance is to be appropriated or possessed by the other in oneself . . . the recipient becomes subject to the internalised will of the other . . .'(p. 44). The fifth is the conversion of value. This takes place when the recipient repays the debt owed in order to compensate for the loss sustained in the acceptance of a gift. The items or services repaid can then be converted into usable value. Gifts are converted into symbolic capital of indebtedness; symbolic capital is then converted (by repayment) into official or political capital, which might then be converted into material capital.

The predetermined oppositional nature of *guanxi* is reinforced by selected elements of Chinese culture, for these elements coincide with reciprocity's fixed symbolism. For example, the 'fit' between incorporation—the second 'process' or symbolic 'mechanism'—and Chinese culture is described by Yang as follows:

> 'Heart is the central symbol for person or self in Chinese culture. Whereas in the Enlightenment West, the mind is the key to the self ("I think, therefore I am"), in the Chinese context, the heart and its feelings signify the person. Even such a modern Western invention as psychology, which examines the

individual mind, is translated as "the study of the principles of the heart" (*xinlixue*). When a gift is given, it is often accompanied by the phrase, *Zhe liwu shi wode yidian xinyi* "This gift is a small token of my regard". The word for "regard", *xinyi*, means literally "sentiments" or "spirit or the heart". Since heart represents person, the gift is really a token of the person of the donor. Personal efforts and labor are also called "heart and blood" (*xinxue*); therefore, when these are given in the form of gifts, favours, or banquets, they signify a transfer or incorporation of personal substance in another. And the term "heart and liver" (*xingan*) can designate either personal conscience or a beloved daughter or son, who are an extension of one's own substance. That is why, between Chinese friends, family members, and kin, obligations need not be activated by giving gifts, nor are debts carefully accounted . . . because there is already a sharing of hearts or a fusing of persons in these relationships' (Yang, 1989, pp. 41–42).

Yang's analysis seems very reasonable and probable. But consider carefully what lies beneath this scholarly prose. Reciprocity has been separated from individuals, individuals have been separated from distinct cultural traits, and a link between reciprocity and these cultural traits has been provided in the form of psychology. By separating reciprocity, cultural traits, and individuals from each other within the intellect, it then becomes possible to invoke deterministic chains of cause and effect. Yang has created something—reciprocity—which is characteristically 'Chinese' and, therefore, something to which all or most people who are Chinese will adhere. Individual actions can now be explained by the determinants encapsulated in this evolving concept of 'the Chinese'. In order to strengthen the links between the various components of her argument, Yang takes language as a clear and direct reflection of a psychology which is held to be peculiarly 'Chinese'. Consider, then, the notion that the mind is the key to the Westerner's self and that the heart is the key to the Chinese self; but first accept Yang's assumption that a relationship exists between the lexicon chosen and the block psychology of a whole society. It is true that in English and other European languages 'mind' can represent 'intellect' as opposed to 'emotions'. But that is only one aspect of 'heart' and 'mind'. 'Next their heart' is an archaic phrase in which heart refers to stomach, but this meaning is retained in the contemporary phrase 'heart in their mouth'. The heart can also imply the seat of understanding and thought as well as feeling, and in its wider sense is equivalent to 'mind'. Heart, then, may carry the meanings of feeling, volition and intellect; it can imply duplicity and insincerity; it is the seat of the soul; it is will and purpose; and it indicates the seat of mental or intellectual faculties—understanding, intellect, mind and memory (as in the phrase 'learn by heart'). The etymology of 'mind' is complex and appears to have derivatives in Old English, Gothic, Sanskrit and Greek. 'Mind' signifies memory, thought, comment, intention; and its Greek derivatives are the strong emotions of rage and yearning. 'Mind' can signify sentiments and feelings (as in the phrase 'speak your mind'), purpose, intention, wishes, desires and the satiation of those desires; and it signifies a state of thought and feeling such as dejection, cheerfulness, fortitude, fearfulness, firmness and irresoluteness.

Yang follows a similar vein of argument in support of a 'fit' between status antagonism and another element of Chinese culture—'face':

> 'Face is especially important in Chinese culture because it is not only a matter of prestige, but an emblem for personal identity, for the autonomy and integrity of personhood. . . . Face and identity are linked because, whereas Western identity tends to depend on conformity to abstract norms and group categories, Chinese identity depends more on internalising the approbation of others in the context of particular relationships. Threats to one's face constitute threats to one's identity which, in Chinese culture, is constructed relationally by internalising the judgement of others in oneself. . . . the Chinese relational constructs of personhood represented by the importance of face provides the mechanism for the art of *guanxi* to constrain the actions of a gift recipient' (Yang, 1989, p. 42).

These arguments concerning the nature of face and its connections with identity and its implications not only for reciprocity, but also for the Chinese company and organisational behaviour, to be examined in the following chapter, find general and popular support in studies conducted by social psychologists (see, for example, Bond and Lee, 1984). The legitimacy of these arguments will be considered later. At this stage it is enough to consider Yang's point that selected terms are a window into the psyche of 'the Chinese':

> 'The [Chinese] language is rich in portraying the things that can happen to face. Besides "wanting face" (*yao mianzi*), "losing face" *(diou mianzi)* and "having face" (*you mianzi*), one can also "borrow face" (*jie mianzi*), "give face" (*gei mianzi*), "increase face" (*zengjia mianzi*), "contest face" (*zheng mianzi*), "save face" (*liou mianzi*), and "compare face", as in the phrase "his face is greater than others" (*tade mianzi bi bieren da)'* (Yang, 1989, p. 43).

Again, the presentation is scholarly: the broad, sweeping sentiments possess a certain *gravitas*, strengthening the kudos attached to the study of all things Chinese. But are not prestige, identity, face and the opinions of others important among English, Americans or Italians? And is not English 'rich in portraying the things that can happen to face'? An individual can have 'two faces', 'open his face', 'shut his face', 'laugh on the other side of his face', 'look someone right in the face', 'show his face', 'set his face against', 'run his face', 'have a face', 'show a false face', 'face down' or 'face out'. The association of 'face' with reputation, name, prestige and identity is both implicit and explicit in these phrases and in the common meanings of 'face': to look big, to brag, and to swagger. The association is also evident in the Latin roots of the word *facies*—look, beauty, imposing appearance, dignified look, impressiveness, to appear.

Once the rolling, scholarly prose is swept aside, Yang's mixture of implicit forces and psychology, which allows words to dominate and determine thought, is all that is left. The impression given is that of an attempt to justify the beliefs and preconceptions which underlie the wider arguments concerning the nature and implications of reciprocity among Chinese. Observations have first been interpreted according to the beliefs preserved within Yang's conceptual framework, and then these same interpretations have been presented as scientific evidence in support of the conceptual framework. In addition, Yang has assumed a link between selected Chinese words and a block psychology which is presumed to exist and which is held to be peculiarly Chinese. The significance of these words is interpreted in line with the conceptual framework adopted and these interpretations are then presented, so it would seem, as evidence to support the validity of that same conceptual framework.

The form of exchange created by Yang's analysis possesses its own symbolism which coincides with (or perhaps arises out of) Chinese culture. Together, this fixed symbolism and the selected elements of Chinese culture determine the individual's actions. Once again the discussion returns to the sociologist's recursive structures, institutions and values which arise out of a people and then revisit, determine and explain those individuals and their actions.

Smart (1993) incorporates several of Yang's points into his analysis of reciprocity in China: different forms of exchange are often polarised by the observer; and reciprocity, like other forms of exchange, is a performance, not a 'thing'. Smart also recognises that there has been too much emphasis on 'attempts to discover the essential nature of the gift and the obligation that it imposes' (p. 397). Moreover, he directly tackles the questions surrounding the morality or immorality of manipulation. Yet although Smart appeals to common sense in expressing these views, in practice he treats the analysis of reciprocity in a rather different manner by implicitly assigning a pivotal role to certain rules or forces, to chains of cause and effect, and to reciprocity as a real and distinct entity. Smart also maintains that a realisation of the immorality of manipulation is inherent in reciprocity and that this realisation is the reason for reciprocity's effectiveness.

To explain *guanxi* Smart falls back upon the much-used notions of *ganqing*, face and trust. If it is concluded, writes Smart, that 'gift exchange and *guanxi* are nothing more but manipulative tactics used to gain profits and other desired goals, we fail to understand the very basis of their effectiveness. In *guanxi*, immediate instrumental purposes are subordinated to the greater aim of developing relationships that may serve as resources for solving problems over long periods of time. . . . The cynicism of academic analysis that construes all social interaction as investment undertaken to increase social or symbolic capital is not possible as a universal form of social interaction in everyday life. Manipulative strategic interaction relies on the normal interpersonal trust that it exploits. Were all to approach the relationship in the same manner, trust could not exist' (Smart, 1993, p. 403).

'Trust', then, is presented by Smart as a moral absolute (and, indeed, almost as a currency which can be devalued by counterfeit) and the foundation without which reciprocity cannot operate. This moral absolute is also the basis for one of the rules of reciprocity fixed by Smart: 'The obligation created by gift presentation must be implicit, because once it is made explicit, it no longer follows the forms of the gift and becomes something else (a bribe or a failed gift performance)' (Smart, 1993, p. 393). Attention to the constraints which social etiquette imposes upon the performance of reciprocity is also important because it 'provides a way to avoid the Scylla of assuming that gift exchanges and market exchanges are completely different types of relationship and the Charybdis of dissolving the distinctions in a unifying theoretical practice of explaining all actions as the outcome of the strategic pursuit of the advantage of the agent' (p. 389).

The doubts surrounding the explanatory powers of *ganqing* and face are discussed again in Chapters 4 and 6. The immediate question which must be raised here about Smart's analysis is this: by what law is trust a moral absolute and the foundation of reciprocity? The suspicion that Smart's views are merely a justification of subjective assessments about human nature must be raised by the distinction made between immediate short-term gain and the solution of 'problems' which may appear in the long term; by the separation of manipulation from the realisation of genuine affection and loyalty; and by the division between the strategic pursuit of goals and the values by which success and individual worth are judged. These views may also, perhaps, indicate a reluctance to ascribe to others the deliberate and manipulative use of relationships (even with immediate agnates). This is behaviour which Smart finds distasteful, and the ascription of such behaviour to Chinese may tarnish both the observer and his intellectual framework with a moral stance that would make them less acceptable to the academic community. Rather than take the invidious position that other people may not hold the same values as those in academia, the scientific method is used to marginalise the motivations and desires of individual Chinese. A form of exchange, certain fixed norms of behaviour, unspecified forces, and moral absolutes generally acceptable to social scientists are brought in to explain the actions of Chinese individuals. By arguing that Chinese recognise trust as a moral absolute and manipulation as an immoral act, and that this recognition comprises the very nature of reciprocity, Smart distances his analysis from any implicit judgemental criticisms of Chinese: he uses a generally acceptable morality to legitimise the marginalisation of individual motivations, desires, choices, decisions and goals, and thereby creates a palatable concept—'the Chinese'. The suspicion that Smart's arguments are a justification of his own feelings about human nature, and an expression of the politics and techniques of the social scientist, is strengthened by his insistence that the code governing the performance of reciprocity must be observed if reciprocity is to work: yet to say that manipulation must not be apparent merely implies that everyone knows the performance to be manipulation.

Another matter which underlies Smart's analysis is his concern with the details of theory. By reducing the distinction between trade (market exchange) and reciprocity, the notion of capital conversion—central to Bourdieu's conceptual framework adopted by Smart—is made plausible, and the main reason for Smart's apparently commonsensical objection to the polarisation of different forms of exchange becomes clear. On the other hand, the distinction between trade and reciprocity must be maintained, for if the rules governing and defining reciprocity are dispensed with, Smart's adopted framework becomes redundant; it also becomes impossible to construct 'scientific' generalisations about events in China.

The manner in which Smart's objections to the polarisation of forms of exchange lays the groundwork for the support and propagation of his adopted conceptual framework is similar to the way in which his criticisms of essentialism (the term is his) are used. Smart's criticisms derive not from genuine doubts about the social scientist's obsession with origins, but rather from the need for a device which allows the existence of rules and laws implicit in his adopted framework. The moral absolutes which he believes underlie reciprocity, and the rules of behaviour defining and governing reciprocity, are simply 'there'. Trust, the implicit obligations, the pursuit of solutions to future problems, and the choreography for the performance all exist, so it seems, for each other, and, by helping to distinguish one form of exchange from another, enable the conversion of social capital into economic capital and vice versa. Consequently, reciprocity, its moral absolutes and its rules of conduct possess only that one aspect which enables the conceptual framework to operate with an internal consistency. Individuals become actors following a script. It is a drama ruled by a series of intellectual and moral dictates. There are the rules of *ganqing* and face; the rule that the strategic pursuit of goals is inconsistent with the different values by which individual worth is judged; the rule that genuine affection and loyalty are inconsistent with manipulation; and the rule that the pursuit of solutions to future problems indicates genuine, rather than manipulative, relationships. There is also the rule that obligations must remain implicit; the rule that the pursuit of short-term goals indicates a manipulative relationship; the rule that these codes of performance must be adhered to if reciprocity is to take place effectively; and the rule that sufficient distinctions must remain between forms of exchange to enable the conversion of social capital into economic capital and vice versa. To argue otherwise offends these rules and the absolute morality of trust, and thereby taints the critic, both morally and intellectually. This technique is a familiar one: to create, develop or adopt a conceptual framework and its entourage of rules and forces capable of governing and explaining individual actions, and of endowing that framework with an internal logic and consistency; and to reinforce that framework with values which appeal to those readers whose acceptance is desired.

There are three important matters arising out of this discussion. The first concerns distinctions between forms of exchange. It has been suggested in this book that exchange is merely an act and that neither the assumed origins of exchange nor its associated institutional patterns and moral values defines exchange. Yet it has also been argued here that there is an association between particular wealth values and particular forms of exchange, and that while institutional patterns are not predetermined, a particular form of exchange requires certain logistical imperatives to be met. This would seem to suggest that exchange is conditioned and shaped into a particular 'form' by wealth values and logistical imperatives—an argument different in expression but similar in meaning and principle to Smart's view that a form of exchange is defined by certain values and rules of behaviour. This line of reasoning could be taken a stage further: that the distinction between different forms of exchange is entirely artificial. Exchange is, after all, only an act; it is the institutions, the values, the logistics of organisation and of economic activity that create apparent distinctions in types of exchange. The 'form' of exchange is imaginary and immaterial. Indeed, sympathy with these criticisms is shared in these pages. Exchange is merely an act and stated in these terms the emphasis on forms of exchange is clearly artificial. However, as suggested in Chapter 1, the significance of 'forms' of exchange lies not in assumed origins of that act, nor in its associated institutional patterns and moral values: rather, it lies in the *intention* of that act. The intention to make profit (trade) is very different from the intention to construct social networks (reciprocity), and both are different from the intention to ensure state control (state procurement and distribution). These intentions can exist independently of moral values, institutional patterns, wealth values, and exchange. What is of interest is how, and the extent to which, exchange conducted with certain intentions and imbued with certain meanings becomes institutionalised, regularised and allowed to dominate or predominate, and to what ends it is directed. Thus, for example, social networks founded upon reciprocity and its soft-framed institutions may be constructed as an end in themselves, but they may also be directed towards the extension and institutionalisation of trade.

The notion of conversion advocated by Smart implies that individuals must obey a large number of very specific rules and procedures in order to engage in a certain form of exchange, and that if economic or social capital are to be converted, the individual must 'hop' from one set of prescriptions to another. Individuals are dictated to by the structures and values which, presumably, they or their forefathers created. Once again the debate returns to the convenient methods and philosophies of recursive social structures, unidimensional phenomena and chains of cause and effect. On the other hand, the notion of wealth values, forms of exchange, institutional patterns and moral values which comprise a loose association of phenomena (expressions of actions, desires, choices, goals and decisions which possess many different aspects, the significance of which alters with the different and changing contexts thereby created) suggests that social networks and trust are not 'converted' by switching to a new set of rules, laws, procedures or etiquette. Social networks and trust are used, manipulated, directed towards, and even incorporated within, the institutional patterns which extend, institutionalise, legitimise and safeguard trade.

The second important matter is that Smart's insistence that reciprocal relationships are not manipulative and are therefore essentially moral requires him to separate reciprocity from bribery and corruption. But to argue that reciprocity becomes manipulative and therefore either a failed gift performance or a bribe when the ends rather than the relationship are given pre-eminence, is to draw a very fine line between moral and corrupt behaviour. Yet values attached to the ideal of fairness and justice in professional life spring from, or rationalise, actions and behaviour that are very commonsensical—actions and behaviour which prevent or mitigate the centralisation of power in individuals and their whims, the formation of monopolies, the denial of opportunities and choice to some by the restrictive practices of others, and the diversion of goods, material and human energies to serve political goals. These actions may not only reduce and defuse tension: they may also avoid the centralisation of economic power. Acceptance of Smart's argument demands that his actors be very simple souls indeed with unidimensional minds, or leads very easily to the conclusion that effect is justified by intention or, worse still, by culture. Whatever his motivations, by tying reciprocity to an absolute standard of morality, Smart has placed himself in the position of justifying what he regards as immorality. The main question which Smart has failed to address is whether or not reciprocity works for or corrupts the prosecution of trade.

This last point leads on to the third important matter arising out of the discussion on Smart's paper. It has been suggested in these pages that he has built recognition of the morality of trust and the immorality of manipulation into his analysis of reciprocity, partly to distance himself from charges of judgemental criticism, partly to deflect the onus which would follow any argument that the Chinese are manipulative (and, therefore, that Smart's interpretation is both naïve and justifies behaviour that may often corrupt professional life), and partly to legitimise the marginalisation of individual choice and actions and thereby to help create a conceptual tool—'the Chinese'. This, at least, is one interpretation of Smart's determination to present trust as a moral absolute and as the foundation of reciprocity among Chinese.

In this connection it is perhaps instructive to turn to Firth's (1956) comments on the manipulation of emotions and feelings:

> When I went to the isolated island of Tikopia I was dependent, as every anthropologist is, on the co-operation of local people for information and for guidance. This they gave, freely in some respects, but with reservation in others, particularly on religious matters. Almost without exception, too, they showed themselves greedy for material goods such as knives, fish-hooks, calico, pipes and tobacco, and adept at many stratagems for obtaining them. In particular, they used the forms of friendship. They made me gifts in order to play upon the sense of obligation thus aroused in me. They lured me to their houses by generous hospitality which it was difficult to refuse, and then paraded their poverty before me. The result of a month or two of this was that I became irritated and weary. My stocks of goods were not unlimited, and I did not wish to exhaust them in this casual doling out to people from whom I got no special anthropological return. . . . Moreover, I came to the conclusion that there was no such thing as friendship or kindliness for its own sake among these people. Everything they did for me seemed to be in expectation of some return. What was worse, they were apt to ask for such return at the time, or even in advance of their service.
>
> Only in one man did I think I saw disinterestedness. But even he, having treated me with much kindness, at last showed the cloven hoof. Before we set out on a trip one day, he said quite frankly that we were going to pass by his sister's house and he wanted for her a quantity of fish-hooks and other small goods. I was disgusted. He, too, shared the general acquisitive attitude.

Then I began to reflect. What was this disinterested friendship and kindness which I expected to find? Why, indeed, should these people do many services for me, a perfect stranger, without return? . . . In our European society how far can we say disinterestedness goes? How far do we not use this term for what is really one imponderable item in a whole series of interconnected services and obligations? A Tikopia, like anyone else, will help to pick a person up if he slips to the ground, bring him a drink, or do many other small things without any mention of reciprocation. But many other services which involve him in time and trouble he regards as creating an obligation. . . . Was my moral indignation at his self-seeking justified?

So I revised my procedure . . . I abandoned the pretence at disinterested friendliness. When a gift was made to me or a service done, I went at once to my stores, opened them, and made the giver a present roughly commensurate to the value of that received. Technically, this was a great advantage. By counter-giving at once, I usually satisfied the recipient. His service was not left hanging in the air. And while he was happy to take what I gave him then, I often forestalled by this means a request for something bigger when he would have had time to ponder on and exaggerate the value of what he had done.

But more important than the change in my procedure was the change in my moral attitudes. I was no longer indignant at the behaviour of these calculating savages, to whom friendship seemed to be expressed only in material terms. It was pleasant and simple to adopt their method. If one was content to drop the search for "pure" or "genuine" sentiments and accept the fact that to people of another culture, especially when they had not known one for long, the most obvious foundation of friendship was material reciprocity, the difficulties disappeared . . . (Firth, 1956, pp. 190–192).

Leaving aside the question of whether or not morality derives from 'social structure', morality is certainly used and adapted. A writer may feel disgust at the 'calculating savage', and he may feel reticence at drawing attention to their raw manipulation. But, as Firth points out, why should values not be different? An expression of disgust or a reluctance to identify manipulation as such is interesting, not so much because it reveals a writer's sensibilities and his parochial concern that he may lay himself open to charges of ethnocentrism, but rather because it implies that other individuals are as they are *by default*. In other words, these 'calculating savages' would have been like the scholar had it not been for the values which the scholar possesses but which they lack. The guilt then surfaces as justifications which are inevitably patronising and, by implication, judgemental. The refusal to recognise manipulation for what it is also reveals a selective naïvety about the way people are. Considering the profession and status of academics, this naïvety is particularly odd. The adoption of certain manners and etiquette; the development of a particular scope of conversation; a knowledge of literature, music, art and architecture; the adoption of particular shades of political opinion and ideology; the use of certain words, style of dress and demeanour: all these interests and modes of behaviour possess many aspects. But they are also important social skills adopted by the individual in order to achieve certain ends, even if that simply be 'fitting in' and becoming 'one of us', and thereby smoothing his relations with others, and building the impression of mutual trust and reliability. Moreover, are not academics especially prone to the adoption and manipulation of fashionable opinions and values, and of techniques and theories which are so often founded upon or strongly allied with particular moral views? Morality is used to legitimise theory or technique; and technique is used to endow a moral position with objectivity. In this way technique and theory can be slipped easily into general circulation.

The refusal to acknowledge manipulative actions of individuals also indicates a determination to ignore the many different aspects of those actions and emotions. Manipulation does not negate compassion, and certainly within the Filipino community examined in Chapter 6 compassion does not easily degenerate into sentimentality. Extremes of emotion are felt and vividly demonstrated, and sentimentality is often affected. The emotions displayed are sometimes exactly what they appear to be. They may also be calculated ploys: expected reactions in given circumstances that are quickly defused by the appropriate and expected responses. It is then a society which is highly calculating and practised in the use of emotions—the soft-framed institutions of reciprocity. It is for many a world of suffocating emotional blackmail, and yet genuine feelings, emotions and manipulation exist side by side without contradiction.

But if the different aspects of reciprocity, its soft-framed institutions and their manipulation are recognised, should this recognition imply that no judgements should be made about this complex and murky world with its retinue of shades? Should the observer merely accept the manner in which emotions, values and beliefs are deployed at will with apparent ease by individuals who are able to retain a vast knowledge of relationships, each relationship perhaps connected with some story, some act of deceit or loyalty, some foible, some feeling of affection or distrust—a knowledge which can, if skilfully handled, enable those individuals to achieve their own ends? Perhaps the answer is that while these actions should be accepted for what they are, the practical outcomes of these actions should not be ignored, for this would be to ignore the different values, beliefs, judgements and opinions that exist within that society. In other words, there is often no clear boundary between right and wrong in the manipulation of relationships among the individuals concerned. To accept, without acknowledgement, that judgements are made, would be to misunderstand the nature of those relationships. If the observer is prepared to make judgements, not according to intellectualised moral absolutes formed in another part of the world and at another time, but according to the circumstances and his own experiences and instincts, then it is perhaps easier to appreciate those relationships more fully. Ignoring the outcomes of these relationships leads to rationalisations and warped interpretations just as easily as the academic's parochial concerns with charges of ethnocentrism. A recognition and judgement of the many different aspects and manipulation of the varying and changing relationships, beliefs, values, desires, institutions and practices in a notional group or society, transform these relationships from awkward, wooden, scripted performances, into an unpredictable and often unfathomable swell of feelings and desires.

Conclusions

Many of the arguments which derive from attempts at rigorous scientific analysis of *guanxi* are difficult to accept. But the objections raised in these pages lie not so much with the information presented and ideas generated, but rather with the obsessive adherence to a science which restricts and marginalises the significance of many observations and ideas, and so creates a dull, lifeless world in which human actions, decisions, choices, values and desires become incidentals.

Arguably one of the most outstanding features of these analyses is the recognition that exchange, emotions, values and relationships are deliberately manipulated by individuals as they attempt to achieve their own ends. Yet the significance of this tends to be submerged in the paraphernalia of scientific rigour, an obsession with theory, by a parochial and moralistic timidity, and by cultural sycophancy. Deliberate and calculated manipulation, even of family and kin, is either ignored or explained away by the operation of impersonal cultural determinants. Fried's view that *ganqing* includes a 'recognisable degree of exploitation' is regarded by Jacobs as 'harsh', for it implies that emotions and sentiments are manipulated solely in order to achieve selfish ends. Jacobs's answer is to place men at the behest of variables. *Ganqing* occurs when people work together and cooperate, for working together requires *ganqing*; *ganqing* is deepened by continued use and by helping each other; yet the Hobbesian view of the world and a rather hostile society means that help tends to occur only when *ganqing* is present. Strong emotional attachments are produced, but clearly there are also strong elements of self-interest in all this, for it must be remembered that, according to

Jacobs, individuals deliberately nurture *ganqing* and 'close' *guanxi* in order to form political alliances. Jacobs has now deliberately reversed the line of his argument. Social interaction is necessary for *ganqing*, and the 'use' and 'help' of *ganqing* is necessary for the development of 'close' *ganqing* and, therefore, for close *guanxi*. And close *guanxi* is necessary for political alliances. The actions and decisions and choices of individuals have become determined by the phenomenon which Jacobs is trying to explain. A clear chain of cause and effect is thereby established. By tracing a chain of cause and effect backwards in a search for origins in a dispassionate, objective scientific fashion, the variables which, according to the model, bring about political alliances are set out. The chain is then traced forward and each variable explained ultimately by the formation of alliances. Teleology is used to give credence to the model. The apparent choices, decisions and actions of individuals forming alliances are, it turns out, predetermined by the paths and the forces implicit in the model. Individuals have become convenient, depersonalised links between variables. Indeed, because particularistic ties can now be used to explain all prior actions, there is no longer any need for vestigial beings: the subsequent discussion can now be confined to the notion of culture and its subsystems.

In Yang's analysis, too, a constantly repeating theme is the conscious manipulation of reciprocity. Moreover, different aspects of reciprocity are hinted at: it can be a means of social contact in a society where there is no state authority capable of imposing social order; and it may be used to bolster official power in a state dominated by procurement and distribution. Even so, Yang's analysis permits manipulation to occur only within the parameters set by what she identifies as *guanxi's* fixed symbolism, the nature of the Chinese psyche, and state procurement and distribution. Yang (1994), too, like Smart, presents a number of rules to ensure that *guanxi* is thought of, and analysed, as something quite distinct from corruption, cynical manipulation, and the pursuit of profit. There is the rule that friendship, trust and affection, though part of the formula of *guanxi*, are nevertheless separate from, and are more pure (ethically) than, '*guanxi*'; the rule that adherence to the particular moral values and personal qualities associated with friendship, trust, indebtedness, obligation, and affection is as important as the realisation of any particular goal or material benefit; the rule that *guanxi* self-consciously defines itself against impersonal money transactions; the rule that gifts are inextricably tied to the identity of those giving and taking; and the rule that it is *guanxi's* symbolic capital which allows one form of capital to be converted into another. If the actors do not follow the script written by Yang, then they are not practising *guanxi*.

Yang argues that the art of *guanxi* betrays a struggle for domination between interpersonal ethics and cynical tactics as the values of a socialist state begin to wane. But the parameters and rules set out by Yang determine that *guanxi* is essentially honourable and ethical, that it can remain untainted as trade is further extended, and that it can only work against state procurement and distribution. As with Jacob's and Smart's conceptual frameworks, the choices and decisions which individuals make, their actions, motivations and desires, their ideas and beliefs are already predetermined by theory. For Yang, the evolution of the state apparatus (*guanxi's* 'historicity'), in conjunction with those many features of Chinese culture which *guanxi* bears, create a strand of 'Chineseness' that is at once timeless and contemporary, specific to China and yet shared by the Overseas Chinese. Yang (1994) acknowledges that *guanxi* is interpreted in many different ways by 'its native practitioners and observers' (p. 173). But it is Yang who has penetrated these interpretations to their true source, and it is Yang who has thereby uncovered the phenomenon of *guanxi*. For it is Yang who claims to recognise and combine two important advantages. The first, the professional and epistemological standards of the anthropologist, is used to create a framework or bench-mark for the second—a willingness to enter into, and take an active part in, another world, and thus to perceive and interpret that world as a 'native' would do. To disagree with Yang's interpretation is therefore to misinterpret *guanxi* either as a pure or corrupt practice, or as a rationalisation of market exchange. To disagree is to see the world through the eyes of a 'Westerner'. It is Yang, then, who now becomes the medium between 'the West' and 'the Other'. 'This world is my world. It is as *I* see it'.

It has been argued in this chapter that reciprocity and its soft-framed institutions are neither peculiar to, nor a product of, 'Chineseness' or some wider 'Oriental' culture. Reciprocity, and the actions, emotions and values which comprise its soft-framed institutions, are multidimensional, and may be 'turned' towards the prosecution and support of trade. This argument represents an important theme in the discussion on the Chinese company and the Chinese association in the following chapters.

Notes

1. See also: Omohundro (1983, p. 73); Sit, Wong and Kiang (1979, p. 322); Fried (1953); Gosling (1983, p. 155); Silin (1972).
2. Skinner (1964–65) began his studies with the necessary assumption that Chinese marketing systems were not particularly remarkable. More

than 20 years later, with evidence from other parts of the world clearly out of step with his interpretation, markets in China had become 'exceptional' (Skinner, 1985b, p. 7).

4

The Chinese company: a nexus of reciprocity

Introduction: towards a model of a 'Chinese' company?

In this chapter it is suggested that the Chinese company, like the market, may be understood as an expression of the extension and institutionalisation of trade itself, or, in other words, as one logistical response to the desire or need for the direct conduct of trade. More specifically, the Chinese company may be understood as a visible and tangible junction or nexus of reciprocity and its soft-framed institutions created by individuals in order to facilitate the institutionalisation of trade.

However, it must be emphasised that all that is presented here is an imperfect generalisation—an interpretative description of companies set up, owned and run by Chinese. The company outlined here is not determined by culture, nor is it unique to the Chinese—it is merely an expression of the decisions, choices, desires, actions, ideas, beliefs and values which are diverse and in constant change, particularly since trade permits material progress and brings individuals into contact with new stimuli.

It was argued in Chapter 3 that reciprocity and the networks built up through, or strengthened by, reciprocity are a realisation, affirmation and symbol of the ability to manage other people. But these networks of relationships do not in themselves necessarily endow an individual with social status and power. Reciprocity is used to build up social networks; and in doing this the individual may draw upon the status derived from the realisation, affirmation and confirmation of trade's associated wealth values—the ability to innovate, create and to contribute. Infused with this status, social networks thereby become a powerful instrument which can be used to extend and to institutionalise trade and thereby to allow sustained material progress. But the importance and efficacy of these networks nevertheless remain dependent upon trade and the realisation of its associated wealth values.

It is often noted in the literature that networks of social relationships enable an individual to build up a reputation of trustworthiness, reliability, business acumen and ability, and are used to create complex flows of information about possible deals, suppliers, customers, funds, prices, labour and possible events and political decisions which may affect trading conditions. Reciprocity and its associated networks, then, are a way for the individual to manœuvre himself into a position which opens up the greatest opportunities for trade. Furthermore, by focusing social networks upon himself and a few key individuals, and thereby fostering dependency on the reputation and connections of one individual or a very small and select group, it becomes possible for the company head to exercise considerable direct and indirect control over personnel. This control is reinforced by direct reciprocal relations between the head and selected personnel—relationships formed upon actions ranging from the exchange of apparently insignificant favours to deep involvement in the welfare of an individual and his family. These acts, which are closely intermingled with the exercise of the 'common touch', may be taken simultaneously as acts of kindness and generosity. Key personnel become tightly bound into a network which comprises an extremely subtle manipulation of feelings, desires and goals, but which, if well handled, is nevertheless warm and very strong. Such control is sometimes developed and exercised most effectively over kin—especially within the immediate family—whose personalities, abilities, temperament and wishes can be shaped and directed with great care. Clearly, this represents a very pragmatic attitude towards family and friends, and for this reason kin may be deliberately excluded and marginalised and even jettisoned if it is thought they are more trouble than they are worth.

The success of a company depends in large part upon the manipulation of its members as a group. But the notion that this group of Chinese is submissive and comprises individuals without a clear sense of their own worth and individuality could not be further from the truth. The creation and maintenance of a close and successful collective are an achievement demonstrating considerable skill and subtlety precisely because of the individuality of the company's constituent members. It is also because of their individuality, and because of the means by which these individuals are held together and directed to a common purpose, that to claim the group is a temporary device for the pursuit of profit would be to present only one aspect of the company. The survival of the company depends upon the successful pursuit of profit, but it is towards this end that coercion, reason, emotion, appeals to self-interest, and the common touch must be exercised. Control, if achieved, is flexible but—as far as possible within human affairs—assured.

The use of reciprocity and associated networks is also crucial for the establishment, maintenance and growth of trading empires, both national and international. By focusing social networks upon himself and thereby gathering in many influential leaders, the individual provides himself with a flow of exclusive information; he greatly improves opportunities for making deals; he places himself in a position to influence those conditions—such as the ground rules for competition within a whole business sector—external to his directly owned companies; and he endows himself with greater credibility if it should prove necessary to draw high-level politicians into his social networks. Indeed, the use of reciprocity to draw in and to form extensive networks with politicians, administrators and politico-military officials is often essential in alien societies in which reciprocity and its associated wealth values are dominant or predominant. Although directed towards the extension and institutionalisation of trade, and thus towards the communication, affirmation and confirmation of different and even conflicting wealth values, networks of reciprocity possess a chameleonic quality which make it possible to infiltrate and permeate indigenous elites.

imp.

Splicing reciprocal networks with those of the indigenes is essential for the protection of companies; it enables the individual to sharpen the focus of networks upon himself and thereby fosters greater dependency and widens opportunities; and it may actively and directly facilitate trade. Connections with indigenous elites in more than one country may therefore prove crucial to the establishment, maintenance and growth of trading empires. Such connections provide direct and rapid entrées into political and business life; enhance an individual's political influence in any one country; further improve opportunities for making deals; and constitute an escape route if conditions in one country should deteriorate.

The dependency which his position at the nodes of social networks breeds among his kin and employees may, if skilfully handled, endow that individual with greater assurance of control over those companies he entirely or largely owns. Such dependency allows him to operate these businesses with considerable speed, flexibility and confidence, and with little monitoring; it thereby gives him greater freedom to direct his energies towards the expansion and diversification of his business interests. It may, then, often appear that Chinese companies and trading empires (especially those spread out over two or more countries) are disparate entities with little or no sense of direction. Yet it is precisely these characteristics which may be symptomatic of the steely control derived from reciprocal exchange and social networks focused upon select individuals. While the means and the opportunity to exercise strong, centralised control exists (a useful option for the orientation of an economy towards particular sectors of the international market), flexible and assured control by a few individuals over a large number of enterprises enables the atomisation of the economy and avoids the creation of damaging monolithic blocks which, in addition, may also attract unwelcome attention from the indigenes.

This particular understanding of Chinese companies is supported by numerous observations which identify features commonly held to be distinctively 'Chinese'. The importance of trust and personal networks is emphasised by most, if not all, analyses of Chinese business and Chinese business organisations (see, for example, D.E. Willmott, 1960; Silin, 1976; Barton, 1983). In his study of Hong Kong, Leeming (1977) observes of small businesses in Hong Kong that business contacts are often friends, and that international contacts usually focus on uncles and cousins living overseas; indeed, virtually all business is conducted through personal networks (p. 51). Sit et al. (1979) also note that the majority of small enterprises in Hong Kong secure their orders through recommendations and through friends: less than 20% of small enterprises advertised their products. Ward's (1972) study of a glass factory in Hong Kong suggests that the establishment of a company and the maintenance of power within it very much depends upon such personal networks. One managing director is identified by Ward as the key man responsible for setting up the company: his personal contacts and creditworthiness within the glass business were crucial, for through them he borrowed money, secured advance orders, acquired equipment cheaply, and recruited skilled workers. However, the focus of leadership later shifted to the first director's partner who had contacts essential for marketing: it was he who knew the buyers, who could obtain credit, and who was skilled at winning people over, including the workers within the factory.

Leadership, then, is highly centralised within the Chinese company (Mok, 1973)—too much so according to some (Suryadinata, 1988)—and didactic in style (Silin, 1976). As Redding and Wong (1986) point out:

'The implication is that the leader holds information and thus power, and doles it out in small pieces to his subordinates, who thus remain more or less dependent. . . . At the same time, the leader does not normally commit himself to a line of action, but rather keeps his options open, leaving the direction of his organisation or department to follow lines dictated by a somewhat nebulous but nevertheless powerful set of personal institutions. The latter are the responsibility of the leader, and his reputation, and to some extent his author-ity, rest on his capacity to intuit the right strategy . . . ' (p. 278).

Non-objective performance assessments are identified as another dis-tinctive feature of the Chinese company (Redding and Wong, 1986). By marginalising the evaluation of professional competence, the subordi-nate becomes further dependent upon his personal relationships with his supervisors, and thereby strengthens his superiors' use of personal relationships in their control over the company in their pursuit of profit:

'The lack of a structured system of performance measurement is inevitable with a lower structuring of organisational procedures, and leads to a lack of data on which individual performance can be assessed. At the same time, the definition of responsibilities is often loose and changeable, and formal organi-sational charts are rare. In these circumstances, a person's contribution to the organisation is viewed by his superiors more in terms of loyalty and "follow-ership" than in terms of his objectively measured achievements. That, of course, produces a high level of upwards dependence in the relationship, and removes power from the subordinate. . . '(Redding and Wong, 1986, pp. 281–282).

The pervasive influence of these reciprocal groupings and networks upon which the company rests also finds expression in cliques within and around the company. 'To the outsider the Chinese organisation may appear to be "one large family"; underneath it may be seething with hostility between cliques and tense interpersonal relations based on dif-ferences in obligation nets' (Redding and Wong, 1986, p. 280). These groupings may extend from, and are manipulated by, family members who have little compunction about manipulating each other ruthlessly (Wilson and Pusey, 1982). Silin (1976) also notes in his study of business organisations in Taiwan that collective groups are often understood as temporary expediencies. The operation of cliques in and around small businesses in Hong Kong—where they are often based on clan and regional affiliation—has also been commented on (Leeming, 1977). And in the Philippines, larger and more intensively competitive cliques have been noted:

'Cooperation usually follows definite circles outside of which there is frequent cutthroat competition. One will speak of various influential groups or cliques dominated by some powerful individual or family: the Sycip group or the Yu Khe-tai group or the Peter Lim group. Business cooperation does not usually follow clan lines but requires closer relationship which may or may not be that of kinship . . . it implies a very close personal connection arising from a long association in China or in the Philippines. Business alliances are frequently sealed by marriage. Among these "close" people, cooperation always exists in a general way in the sense that they will share useful information, make oppor-tunities available to them rather than to outsiders, invest in one another's enterprises, and so on. More directly, however, it happens frequently that related people—kinsmen, townmates, schoolmates, or just plain friends—will get together and pool their resources to start a business' (Amyot, 1973, p. 65).

The ability of Chinese to graft these networks on to those of the indigenous elites has also been observed widely throughout Southeast Asia.[1] In Malaysia, for example, Chinese firms have managed either to hold on to or improve their economic position despite the New Economic Policy.[2] This has been achieved partly by actively seeking out and establishing alliances with Malay allies in the bureaucracy, military and government, and among the new Malay entrepreneurs; and by strengthening links overseas—with Singapore, Thailand, Indonesia, Hong Kong, Japan, Australia and Britain—which serve as channels for possible expansion and as escape routes for capital during times of economic hardship or political hostility. Interestingly, it has been the family companies in alliance with powerful Malay patrons that have prospered the most. Tycoons and small businessmen appear

'to have chosen a two-pronged strategy: to work closely with Malay patrons to achieve financial success and social status in Malaysia, and at the same time build relations with non-Malaysian capital that can serve as a potential source of wealth should conditions in Malaysia deteriorate' (Heng, 1988, p. 142).

So, too, has the disparate and yet centrally controlled Chinese business organisation been commented upon widely. Skinner (1957) argues that while the conduct of business life among the Chinese in Thailand may have appeared disparate and diversified, it was in fact 'concentrated in the hands of a small elite' (p. 177). Social networks, the successful practice of trade, and the status derived from these achievements were all used to establish informal groupings; to achieve influence over members of formally organised business associations and leadership of numerous small enterprises (often comprising all the members of entire business associations) brought together under a single company or combination in which each constituent enterprise bought shares; to place kin in strategic positions; and to arrange marriages within or into elite business cliques. Reciprocity and the construction of social networks may, then, provide a way of centralising power and control without putting an end to intensive competition. Gosling (1983) has noted a practice of 'competitive cooperation' among Chinese crop dealers in Malaysia and Thailand. This may indicate informal centralised control, or simply another means of achieving the same ends—remaining competitive and ensuring mutual protection and cooperation without the creation of monolithic blocs that would be damaging economically and unnecessarily attract the attention of the indigenes. Gosling writes:

'This Chinese system of "competitive cooperation" is one which is designed to accommodate a large number of small-scale, under-capitalised dealers and merchants and to provide them collectively with economies of scale, access to capital, protection against risk, access to collective knowledge and advice, all without hindering competition. It also serves to make the entire group more competitive with non-Chinese, forcing everyone's prices to low levels, with modest profit margins. In contrast, indigenous merchants or traders faced with high costs of inventory or lack of inventory, cannot price as competitively.

All merchants handling the same commodity may be located in the same area of town, often in adjacent stores. They will share inventory – that is, if you seek to purchase an item in one shop which is not in stock, the merchant will simply send to one of the neighbouring shops for it; thus they maintain a large collective inventory, which provides every shop with a wide variety of goods to attract customers plus economies of scale in purchase and storage. The prices among all the shops are virtually standard, at a relatively low mark-up over cost. The customer's choice of one shop over another is based on service rather than price, primarily on differential terms of credit. External competition is not required to enforce modest price mark-ups; internal competition among the cooperating merchants will retain low prices and give good service even where there are no external competitors. Lower prices mean greater volume, more customers, more contacts and more security—all far more important than a quick profit' (Gosling, 1983, pp. 153–154).

In Hong Kong, too, it has been noted that even companies owned by the same family may be in competition with each other. The sons of Lo Fong-seong—an industrialist who established the popular chain of fast-food restaurants Café de Coral (or Dajiale – Everybody Happy) and whose family has run Vitasoy for more than fifty years—set up Fairwood Fastfoods which runs parallel to (for, apart from their names and logos, both are very similar) but in competition with Café de Coral (K. Wong, 1994).

Given this evidence, is it not possible to 'firm up' the rather nebulous understanding of the Chinese company proposed in this chapter by integrating these various characteristics into a harder generalisation or even a model? After all, there have been numerous attempts to do so. Ward (1972), in her detailed study of a small factory in Hong Kong, is primarily concerned with whether or not there is anything specifically 'Chinese' about socio-economic relations within the factory. In its organisation and in its operations Ward finds nothing peculiarly 'Chinese':

'Layout, workflow, job allocation, labor recruitment, even most aspects of the workers' previous experience and background did not appear to be specifically culture-linked. Even among the managing staff and directors there were no particularistic ties of a traditional nature. Only the system of rewards by categories seemed to be somewhat unusual, and even that is quite common in other parts of the Orient . . .' (Ward, 1972, p. 384).

However, Ward does argue that there are certain peculiarly 'Chinese' traits to be found:

They include beliefs about the value of education, the virtue of hard work, and the self-evident goal of economic self-betterment, shared by workers and management alike. This set of values was accompanied by an ingrained habit of giving "face", in my opinion closely linked with the strong preference for multiplex rather than single-stranded relationships . . . insofar as it may be possible to isolate any aspects of the internal organisation of this factory as uniquely "Chinese", they will not be found in the structural framework or the organisational plan, but in intangibles such as these, none of which alone is specific to Chinese culture but all of which in sum constitute a peculiarly Chinese style of running economic institutions' (Ward, 1972, pp. 384–385).

Deyo (1983), on the other hand, argues that it is difficult to identify work values that are uniquely 'Chinese' in Bangkok and Singapore, but suggests that a Chinese company can be identified by virtue of certain managerial practices—such as centralised decision-making, the emphasis on personal touches which enhance trust and control, and the importance attached to loyalty as a criterion in the selection, recruitment and promotion of personnel—which, Deyo argues, are derived from:

'. . .direct linkages . . . with culture-bearing Chinese community institutions, such as associations, family networks, and, more particularly, ethnic-based economic institutions (credit networks, guilds, trade associations and so on). Businessmen must and do contribute to the maintenance of these institutions in order to enhance business contacts, local reputations and cooperation from others. These institutions in turn become important influences on business and management practices' (Deyo, 1983, p. 228).

L. Lim (1983) presents a succinct description of those characteristics which are often taken to constitute the typical Chinese company:[3] it is usually owned by a family and exhibits a mix of wage-labour and family labour; its scale of operation is small and localised geographically; it is in a commercial sector of the economy (such as money-lending, import–export, and rural–urban retailing) that is dominated by a particular ethnic (dialect) group, and yet the sector remains competitive; and even if the company grows into a large, modern, corporate enterprise and is transformed into a joint-stock company, control and ownership still tend to rest with a single family or group of families related by blood, marriage, clan or dialect; and business activities remain concentrated in the traditional primary and tertiary sectors of the economy (p. 245). Other characteristics may be added: networks extend outwards from the family, permeating the company and reaching beyond it; the level of specialisation is low, job specifications are not detailed, there are few specialised departments, and individuals are responsible for a wide range of different activities; the style of leadership is authoritarian; goals are decided upon in much secrecy; lines of business are diverse; and, because of family tensions and sibling rivalry, the typical Chinese company rarely lasts for more than three generations.

Siu-lun Wong (1985, 1988a) builds this last characteristic into a model of the Chinese company. Wong argues that the concept of familism has different layers or meanings: paternalism (in which case the family serves as a model for the administration of personnel within the company); nepotism; and ownership—the family enterprise. Three development phases are then identified: the emergent phase which is characterised by a partnership that is inherently unstable; a period of centralisation and vigorous growth during which the majority of shares come under the ownership of a single entrepreneur and there is an intense urge to enlarge the family; and a segmented phase which describes the transfer of responsibility for the company to a younger leader who must now earn the respect that his new position demands if he is to be successful. As friction builds up among competing brothers, this third phase is followed by the disintegration of the company.

Wong further suggests that in Hong Kong the Chinese company's distinguishing features—a fluid economic hierarchy, a centrifugal force encouraged by the desire of family and employees to set up on their own, and a tendency to make decisions centrally—all stem from 'entrepreneurial familism' (the term describes the family as an impetus, and a support, for innovation and the necessary risks associated with innovation) which, in turn, derives from China's bureaucratic past. In Qing China, Wong (1988a) argues, there was a relative scarcity of strong organisations mediating between the imperial bureaucracy and local kin groups:

> 'This could have limited the range of organisational possibilities for the formation of Chinese industrial enterprises and led to the sharp bifurcation of public enterprises modelled on the imperial bureaucracy and private firms modelled on the family.

> The bifurcation of public bureaucratic enterprises and private family firms provides us with a clue to the understanding of another special feature of Chinese industrial competition . . . the relative absence of oligopolistic groupings similar to the *zaibatsu* in Japan and the cartels in Europe. . . . In the Chinese case, private enterprises are generally not durable because the family on which they are modelled has an inherent tendency towards segmentation. Just as each son in the family is a potential *pater*, each heir in the firm is a potential entrepreneur. The Chinese state, for its part, was particularly watchful of attempts by families and kin groups to form alliances, as it was feared that these would challenge its authority and upset the system of political integration. The strong hostility to private enterprise expressed by both the Guomindang and Communist administrations in recent times could represent the continuation of this attitude' (Siu-lun Wong, 1988a, p. 172).

Redding (1980), too, incorporating the ideas of a number of writers (Hall, 1976; Moore (ed.), 1946; Nakamura, 1964; Armstrong, 1973; Litterer, 1965), assigns great importance to Chinese culture and its historical roots in his explanation of the Chinese company. The way in which Chinese companies are organised and business is conducted, argues Redding, is determined by paradigms of behaviour which are, in turn, determined by culture of which the most important element is Confucianism. Culture affects the formation of 'maps' or paradigms—that is, systems of beliefs and understanding—within the mind. It is these maps which are the main guide to behaviour. And if one type of behaviour rather than another brings rewards, then behaviour is soon confirmed by values. Redding selects five categories of variables which he uses to describe Chinese mind-maps and which therefore influence the nature and operation of Chinese business organisations.

The first is causation. The Chinese mind, it is argued, is intuitive. By this it is meant that Chinese utilise concrete rather than abstract concepts. Things are as they are; things are immediately comprehended; and the Chinese mind believes that, in order to understand, one must contemplate a particular phenomenon within its context—that is, its relationship to the other phenomena which surround it. The second category is probability. The Chinese mind, Redding argues, is fatalistic:

> '. . . the world is perceived in sets of mutually balanced interconnections and one of the outcomes of such a view is a sense of fatalism. This rests on the inevitability of the balance being fully restored and gives an almost preordained view of the future. In this context the future is not seen as being "for calculation". One does not make one's own destiny. Even success is taken as having been predetermined' (Redding, 1980, p. 144).

The third category is time. The Chinese mind perceives time as cyclical—as something associated with events. The fourth category is self. The Chinese, in common with most 'Orientals', exist in a 'high-context culture' and thus the individual is embedded in social networks—webs of obligations, cliques and patronage. The 'Oriental' therefore has a less pure sense of consciousness, and any innate desire for achievement will be suppressed or dampened by society. The final category is morality. The Chinese are concerned to avoid shame derived from the infringement of social norms which are 'situational'.

There are two important effects which the interaction of culture and paradigms has upon the organisation of the Chinese company and the conduct of business. The first leads to the absence of a formal planning framework. According to Redding, planning requires sophisticated abstract concepts which, in turn, require a linear, sequential logic to link them together. Yet apparently neither abstract concepts nor linear logic are 'the normal mental approach of the Chinese. Their vision is more of large numbers of interconnected specifics' (Redding, 1980, p. 143): the Chinese world comprises sets of mutually balanced interconnections. Furthermore, planning requires a sense of control over the future and a linear sense of time. An 'Oriental' culture, however, 'lacks the necessary foundation' (p. 144). For this reason, then, planning in Chinese business is rudimentary, and detailed organisation into the future is not highly developed (p. 144).

The second effect which, according to Redding, culture and Chinese mind-maps have upon the Chinese company is to produce an informal organisational structure. In particular, the Chinese company is heavily reliant upon 'interpersonal' relationships and upon decisions made centrally for control and coordination—a feature derived partly from the social sensitivity of Confucianism. The reason for this, Redding maintains, is that the Chinese mind does not seize upon abstract concepts (such as 'marketing functions', 'financial control' and 'coordinating mechanisms') which are essential for the differentiation and integration of practices which enable an organisation to grow in size. The Chinese mind's rather fuzzy perception of time is another trait which contributes towards the informal organisation of the Chinese company: several things are perceived to happen at once; there is only a general aim of completion without scheduling; individuals are controlled by the minutiae of what they do, rather than by when it is to be done. Again, these characteristics limit the size of the Chinese company.

Confucianism also forms a main plank of Mackie's (1992b) study of Overseas Chinese entrepreneurship. This is one of the very few attempts to examine the reasons for the economic success of the Overseas Chinese in Southeast Asia, and it is one of the most comprehensive and subtle analyses. Yet Mackie begins almost immediately by creating 'the Chinese'. Although it is recognised that 'the Chinese' embraces considerably diverse groups, they are nevertheless thought of and treated as a single block:

'Using the term "Overseas Chinese" to refer to all people of Chinese descent outside China, regardless of their nationality, status or length of sojourn abroad, poses problems of definition to which there are no simple answers (Wang Gungwu, 1981: 249–60). It can be objected that the differences between, say, the Sino-Thai and Sino-Indonesians are now at least as significant as their common ethnic origins, hence that the racial element should be disregarded. But that is a doctrine of perfection which would make it impossible to discuss most of the issues under investigation here' (Mackie, 1992b, p. 41, footnotes).

Mackie thus makes 'the Chinese' a distinct variable which forms a link in a chain of cause and effect: 'the Chinese', entrepreneurship and economic success. The task of the social scientist is to uncover the pattern of these phenomena and the links between them and therefore to provide a scientific explanation of Chinese economic success. The science used to explain the 'processes' involved is rudimentary, but the approach is nevertheless 'scientific'. The other main variables and subvariables are: traditional Chinese 'cultural' values; the Chinese family; Chinese networks; the firm; political connections; the wider institutions, strategies and cultural milieux in which the Chinese operate; and the circumstances in which they find themselves. For much of the paper, Mackie is content to suggest lines of inquiry for further research. But despite the scholarly understatement, there are hard assumptions which are taken as hard fact rather than as uncertain possibilities.

Mackie takes it as axiomatic that certain elements (or subvariables) of Chinese culture and Chinese values are an important part of an explanation of entrepreneurial ability and business success. Citing Hicks and Redding (1983), Redding (1988, 1990), Berger (1988) and, one might add, Wang Gungwu (1981), Mackie (1992b) rightly points out that:

> 'The significance of the neo-Confucian ethic in the economic miracle of the "Four Little Dragons" is widely acknowledged' (p. 43).

Indeed, Mackie goes on to argue that:

> 'It can hardly be doubted that those factors [Chinese values, family structures and family firms] were very important in the early years of Chinese migration to Southeast Asia, when family firms were the classic form of Chinese economic organisation and family labour, along with the discipline fathers could exert over their sons and daughters, were crucial to their success' (p. 53).

Another crucial subvariable is networks of personal connections infused with trust. Indeed, together with traditional Chinese cultural values, personal connections and trust are regarded by Mackie as sufficient explanation for the success of Chinese prior to the rise of large Chinese corporate conglomerates in the 1970s.

> 'The success of [this] earlier group can be adequately explained along the lines of Myers' view that traditional Chinese cultural values and preference for personal connections and trust, rather than impersonal structures, "nested comfortably with the proclivity of businessmen to rely upon the market to cover transaction costs", at the level of small family enterprises; i.e. to externalise them in the market rather than to internalise them within the firm' (Myers, personal communication, 1991) (Mackie, 1992b, p. 59).

However, while Mackie takes it as axiomatic that such variables are essential for success, he accepts that they must be considered within the wider socio-economic and political context—the situational logic of the Overseas Chinese, and the nation state and its political framework. The influence of these variables can be clearly seen in the pattern of small and large firms in the West Pacific Rim: relations between small and large firms have been different in Hong Kong, Taiwan, South Korea and Japan:

'Even though all these societies have been lumped together under one cultural label as neo-Confucianist in some sense, the differences between them in their business structures are significant (and possibly also of family and socialisation patterns), for they indicate that other factors besides culture must have played some part. South Korea has had nothing like the plethora of SMEs that Taiwan has had, because of the dominance of the twelve or so great *chaebol* since the 1950s to 1960s. . . . And while Taiwan has also begun to generate a few large-scale conglomerates since the 1970s, they appear to be very different types of organisation from the *chaebol*, or Japan's *zaibatsu/keiretsu*. Hong Kong has had both, but is much closer to the Taiwan pattern than the Korean. Yet in all three cases, the large enterprises as well as the small are essentially family firms . . . the reasons why each of these countries has its own distinctive pattern in this respect cannot be reduced to any one or two structural determinants. They have to be traced out in historical and socio-cultural terms, involving family types, family enterprises and . . . "structure and strategy" issues . . .' (Mackie, 1992b, p. 55).

Mackie also considers more recent changes in Chinese firms, and he wonders how relevant Chinese values still are, as 'more and more small Chinese family firms have grown into larger, more complex corporate enterprises, utilising modern managerial techniques and employing graduates from overseas' (p. 53). Mackie goes on to point out that 'nearly all the larger Chinese conglomerates, as well as the smaller enterprises in Southeast Asia are still family-owned firms' (p. 53), and suggests that it is likely the traditional Chinese business practices and attributes—the old, distinctively 'Chinese' characteristics deriving from the small family company—will persist among the Overseas Chinese. Nevertheless, he also suggests that things may be changing faster in Southeast Asia than in China, Hong Kong and Taiwan. This is particularly true of personal relations:

'The traditional Chinese preference for personal relationships rather than reliance on impersonal market mechanisms may therefore prove to be weakening there. There is abundance evidence, of course, that personal relationships are still very strong, but there is also evidence of gradual change to what I have elsewhere called corporatization and reliance on more impersonal managerial practices and market mechanisms' (Mackie, 1992b, pp. 57–58).

*   *   *   *   *

Clearly, many of these observations could be built into the interpretation given at the beginning of this chapter, and another variant model could be produced. Yet such an attempt only serves to mislead. Purportedly distinguishing characteristics are shaped into a typical Chinese company; the circumstances, events or phenomena which are thought to bring this company into being are traced; chains of cause and effect begin to determine the significance of phenomena and to limit their aspects; the discussion moves inexorably into China's past and Chinese culture (and Chinese social psychology); the scientific method, and the writer's initial assumption that culture is indeed the idol of origins, begin to complement each other; all observations are interpreted accordingly; and these interpretations are now presented as fact. Wong's or Redding's, Mackie's, Ward's or Deyo's scholarly prose lulls and soothes, for their explanations of the Chinese company are solidly founded upon historical determinism, upon the assumptions which saturate the cultural determinists' and sociologists' notions of 'recursive' human actions and social structures, and upon the cosy surety which is generated by a mutually supporting relationship between a belief in the importance of culture and the scientific method. Culture is treated and thought of, not as a word describing the residual and coagulated expressions of changing values, beliefs, desires, choices, decisions, actions, circumstances, institutions and forms of exchange, but rather as a causal entity or force.

It is argued here that there is no such thing as 'the Chinese company' in any theoretical sense: that is, there is no model or conceptual framework for analysing a company which is either the product of, or which adds yet another layer to, the cultural mantle of 'the Chinese'. Reciprocity and its soft-framed institutions are simply part of a mix of loosely associated sets of wealth values, forms of exchange, institutional patterns and moral imperatives—all with many different aspects. The apparent characteristics of the Chinese company represent particular aspects of reciprocity and its soft-framed institutions, and of other expressions of the extension and institutionalisation of trade, which have been selectively extracted and interpreted according to the dictates of unidimensional scientific theories and assumed chains of cause and effect thought to be rooted in culture.

The use of reciprocity and its soft-framed institutions is simply one response taken in conjunction with others to the imperatives for extending and institutionalising trade. Together, these actions, choices, decisions and desires may find expression as nebulous webs of credit or as complex business organisations. In short, the conduct of reciprocity, and the hard procedures and modes of organisation which comprise the company are not determined by any formulae or by any dictate of 'culture'; and neither the practice of reciprocity, nor its soft-framed institutions, nor their use to extend and institutionalise trade, are peculiar to Chinese. While similarities among 'Chinese' companies may be produced by the transmission of ideas, by the decisions of governments which may, directly or indirectly, require companies to enforce certain practices, by similar logistical problems, and by the similar perceptions and judgements of individuals who comprise companies, it is also precisely because of the transmission of ideas, of the decisions of different governments, of different logistical problems, and of individual perceptions and judgements that there is diversity and constant change. This change is likely to be rapid since trade, by definition, brings individuals into contact with different ideas, practices and stimuli, and because the material progress which trade permits creates changing conditions, and new opportunities and new methods of operating.

The ideas presented at the beginning of this chapter evidently conflict quite sharply with the cultural explanation—the popularity of which has recently been strengthened further by growing commercial interest in the West Pacific Rim and, most especially, in China. Scholars have not been slow to create ready-made cultural explanations of 'Chinese' behaviour and forms of organisation which the proponents of these explanations hope will find a market among business managers, and among those academics who are content to accept convenient, immediately communicable, fashionable explanations based upon common preconceptions girded with the respectability of 'science'. In view of their influence, these explanations are now considered in rather more detail.

The holy trinity: a critique

Variation and dimension

One criticism of the cultural explanation and its application of scientific techniques to analyses of Chinese companies is that the different and changing aspects of institutions, values and exchange and, consequently, the possible significance of variations in observations (many of them conflicting), are ignored.

Nonini (1983) in his study of the Chinese truck transport industry in a market town in Malaysia notes that while kinship is a major 'principle' of organisation within companies, many of the 'typical' features of Chinese business are absent: individuals in one company do not marry into another company; interlocking ownership by individuals who are partners in more than one company is uncommon; and very few of the people employed by the core kin group in each company are related. Moreover, Nonini's study reveals a much earthier, perhaps more realistic, side of business in which the rather ethereal notions of *guanxi*, *ganqing*, *jia* and the other fine niceties presented by Sinologists are either out of place or are harshly manipulated.

Silin (1976) and Redding and Wong (1986) observe that the rejection of 'the affective component in creating solidarities' is a striking characteristic of leadership in Chinese companies. One consequence of the absence of warmth is that the leader must be constantly on his guard against his subordinates who may withdraw their support at a crucial moment. Ward (1972), on the other hand, argues that although social distance between directors and staff is accepted on both sides as right and proper, the impact is greatly modified by genuine friendliness in conversation, and by paternalism coupled normally with great care in giving face (p. 378).

To some writers the Chinese exhibit a very strong tendency towards cooperation; other writers note that the Chinese are intensely individualistic (Purcell, 1965, p. 284) and that their ability to cooperate is poor (D.E. Willmott, 1960); some writers suggest that the Chinese are no more materialistic, ambitious and aggressive in their economic pursuits than the Malays (Deyo, 1983; Stough, 1983; cf. Winzeler, 1983); while many others argue that hard work (or at least the emphasis on the claim of hard work) is part of a distinctively Chinese company.

Assessments of the practice, and effect, of nepotism in Chinese companies also vary considerably (see, for example, Levy, 1949; Levy and Shih, 1949; Chao, 1977; Pelzel, 1970; Ward, 1972; Sit et al., 1979; D.E. Willmott, 1960; Ryan, 1961; Wu, 1982; King and Leung, 1975). While the owners of one company unashamedly put their children and close kin into key positions, the owners of other companies deliberately exclude their relations. Nor does there appear to be any consensus about the existence of a correlation between the Chinese family and the size of Chinese companies. While some writers argue that the reliance on the family restricts the size of the Chinese company and makes the company unstable, others regard the whole matter as an open question, and still others see no inherent disadvantage in a family company. Indeed, Heng (1992) argues that, in Malaysia, although 'the Chinese family-oriented closed corporation, based on an individual tycoon and his family has heretofore often been thought to limit Chinese capacity for capital mobilization and organisational expansion, it is precisely such firms which, in alliance with powerful Malay patrons, have prospered the most' (p. 127).

It may be suggested that all such variations are an expression of different, but conscious, choices, decisions, desires, beliefs and actions, and, therefore, an expression of the different aspects of forms of exchange, institutions and moral values: there is no formula which defines or explains 'the Chinese company', for in a theoretical sense there is no such thing as a 'Chinese' company. The view that the family and economic development are antipathetic, Mackie (1992b) argues, is no longer tenable and, as he rightly points out, the conventional wisdom is now almost the opposite: 'today's writers invoke Confucianism and the particularistic family to explain Chinese economic behaviour and commercial success' (p. 53). But the possibility that the family can work, either directly or indirectly, both for and against the extension and institutionalisation of trade, depending upon the desires, goals, beliefs, values, decisions and actions of the individuals concerned, is never considered. Indeed, later in this chapter it is noted that in those cases where Chinese do not appear to fit into the mould prescribed by a writer's scientific explanation, they are removed from the discussion: they become a 'different kind' of Chinese.

Similarly, the belief that reciprocity works against achievement and helps explain the purported collectivist or group mentality of the Chinese and their preoccupation with 'face', reveals thought plagued by deterministic chains of cause and effect, and considers only one aspect of reciprocity and its soft-framed institutions. The core of reciprocity is to serve yourself while serving (or appearing to serve) others. There is no true group or collectivist mentality here: it is an individualistic philosophy guided by self-interest. The 'sameness' desired of the group is a product, not of its public justification for unity and harmony, but of a crude philosophy of control—a philosophy and practice often confused with leadership—and of a desire to hide deficiencies. Rather than achieve competence or proficiency, individuals may work to avoid mistakes, to ensure vindictive mediocrity, and to develop an image (the 'businessman' who sleeps for but a few hours each night at the factory, the 'scholar, sage and mentor' who shapes 'the studious student' with his wisdom, 'the artist' who is at once self-consciously unconventional in his behaviour and yet conventional in his desire to serve his community as the guardian of its culture, and 'the worker' who is 'the salt of the earth') which conforms with the others' expectations, and, above all, to improve their own status as judged by an ability to handle social relations. This, at least, is one aspect of reciprocity, but it does not deny the simultaneous presence, expression and motivation of compassion and affection.

The attainment of status and prestige by the successful management of social relations may well attract criticisms and stimulate resentment in some quarters as part of the cut and thrust of the attainment of that status. And it may also be the case that reciprocity can work against achievement defined according to different wealth values. Such achievement will reflect badly upon others, questioning the foundations of their own sense of worth; or it may simply be regarded as invalid—a selfish, mean and immoral act which could not have been accomplished except at the expense of others. Those who achieve according to those wealth values associated with trade may be excluded, chastised or made the target for active retribution. Reciprocity and its associated wealth values may also act to console: it is a substitute expression of worth. To say that an individual achieved because he is selfish and did not consider others may reflect a different sense of worth, but it may be little more than a justification of one's own failings as judged according to other competing wealth values. The formation of networks of like-minded souls confirms those individuals' own sense of worth and provides a base from which to denigrate alternative conceptions of achievement and success.

But it is clear that reciprocity may be 'turned', or directed towards achievement according to different wealth values. The disciplines imposed by the need and desire for profit and material progress make it essential that individuals brush aside their exceptions to material success, particularly when efficient, reliable and impartial legal frameworks and procedural checks and balances capable of reining in the excesses of politicking are absent. Reciprocity and its associated networks no longer remain an introspective obsession and an end in itself—it becomes a tool for the prosecution of trade. None of this implies that the techniques of reciprocity are absent from trade—the friendship and compassion, and the cut-throat deception, dishonesty and duplicity are as much a part of business life as of life within bureaucracies. But this cold manipulation of emotions is directed at the realisation of a clear goal, rather than at the suppression of an alternative understanding of worth and achievement. Nor does the tension which such achievement brings disappear. Reciprocity for its own sake may exist simultaneously, especially in those circumstances and institutions where individuals are free from the disciplines imposed by profit. Overseas Chinese societies are not a single block: they possess different mixes of competing, complementary and contradictory wealth values, forms of exchange, institutions and moral values. Individuals not only have to deal with the cut and thrust of business; they must also steel themselves against criticisms, legitimise and safeguard trade, and remove or circumvent any obstacles to the realisation of that end.

The different aspects of reciprocity also help towards an understanding of its associated soft-framed institutions, such as 'collectivism' and 'face' which have attracted much comment. Some writers make a distinction between individualism and self-reliance (Ryan, 1961), the latter concept being compatible with collectivism, and between different types of face—most especially between ascribed moral worth (*lian*) and the reputation (*mianzi*) drawn from an individual's own achievements (Hu,

1944). So fine are these distinctions, however, that they appear contrived, particularly when the distinction between individualism and self-reliance is used to explain the proliferation of Chinese family companies, and when the distinction between *lian* and *mianzi* is used to explain why Chinese are less sensitive to face in some circumstances than in others. Collectivism and face are not a product or cause of reciprocity, nor do they possess only one aspect. Joining a group and acting together as a group may do no more than demonstrate a lack of confidence to make decisions and take responsibility; or it may reflect defensiveness or a simple desire to avoid loneliness, or it may constitute a way of behaviour that individuals find useful in the conduct of reciprocity and its support of trade.

Face is similarly presented as a very wooden concept according to which stiff and lifeless dolls are subservient:

'. . . in order to generalise [research on "face-work"] to an Oriental culture like that of the Chinese, the hierarchical structure of society with its permanency of [status] should be taken into consideration. Stover (1974) took the cocktail party as an example for illustrating the American game of one-upmanship, and compared it with the Chinese game of face. In the typical American cocktail party, everybody plays the game of one-upmanship using the polite boasting and free-floating expressions of sentiment that go with elevating oneself as a means of establishing a positive image. They joke about this and that, jostling for position while gradually revealing something of their personality and feelings. At the Chinese dinner party, on the other hand, rank is fixed by the seating plan. Everybody invited knows his standing relative to everybody else. They are all expected to follow *li*, doing the proper things with the right people, bowing and gesturing in verbal ritual, and paying respect to others' (Bond and Hwang, 1986, pp. 244–245).

When so much depends upon relationships, appearing to put oneself or, perhaps more importantly, somebody else in a bad light unintentionally and through carelessness will create embarrassment. Much will also depend upon context. In those circumstances where something more concrete depends upon personal esteem, exception may be taken to a casual remark or a careless action, though again much will depend upon individual personalities and judgements. Those less confident, more bombastic, more tense and unable to exercise that all-important 'common touch' may be more sensitive to the offhanded and careless remark or action. Otherwise anyone in the right circumstances is fair game for lively repartee, and a target for raw and unsavoury jokes. In those circumstances, most especially when relationships are being established and strengthened, taking exception would be regarded by many as unwarranted grandeur, and stiffness would be perceived as an obstacle to the development of trust and confidence. At one gathering (an office barbecue) in Hong Kong a very wealthy Fujianese industrialist—who owned among his many concerns a factory which made health drinks—was recounting in vivid biological detail an illness he had inflicted upon himself some years previously by drinking spirits in large amounts too frequently. A junior remarked with heavy sarcasm, 'Oh, so that's why you want the rest of the world to buy your drinks?', at which everyone giggled. The humour was perhaps rather weak but circumstances made the joke permissible. If circumstances had been such that something more than self-importance rested upon the industrialist's self-esteem, then the junior's remark would have brought about a very different response. An individual who is Chinese is no more concerned with 'face' than someone who is not Chinese; and a Chinese individual is no less open to the acceptance and apportionment of criticism and ridicule, although the circumstances and manner in which liberties are taken with ego and esteem may vary from one person to the next.

It is in a flat, unidimensional world that theory can be propagated most effectively. The search for chains of cause and effect and the idol of origins is made so much easier if the world is fragmented into variables and subvariables, if those variables are thought to possess only one aspect, and if that aspect's only significance is its function as a link between the idol of origins and the phenomena to be explained. Yet variations, it is suggested here, cannot be marginalised for the convenience of a model and a 'scientific' train of thought, for often these variations illustrate different aspects of forms of exchange, institutions and values being manipulated in different ways to achieve particular goals.

Cultural links

A second criticism of the cultural explanation of 'the Chinese' company is that, despite the creation of unidimensional variables, links between those variables and their presumed cultural origins and the specific features of a Chinese company are simply assumed. If an interpretation of 'Chinese' thought or organisational traits seems to offer a plausible explanation for the posited characteristics of the Chinese company, then that interpretation is accepted as evidence of causal links: in other words, the very fact that the identification of a variable is claimed may often be taken as sufficient proof that the assumptions with which the writer began are valid and that the connections between the variables, their cultural origins and the company are demonstrated.

For example, the links between the cultural elements (and often the cultural elements themselves) which Redding (1980) assumes profoundly influence the paradigms for behaviour (the Chinese mind-maps), if specified at all, are at best loose and vague. Redding (following Chan, 1967, Hall, 1959, 1976, Hallowell, 1955 and numerous other writers) gives no explanation of the Chinese 'cyclical', 'polychronic' or 'loose' perception of time: their perception of time is as it is because they are 'Oriental'. As for the Chinese perception of self, Redding maintains that its foundations lie in Confucianism. Indeed, Confucianism is regarded by Redding as the source of most other variables, or traits, including Chinese morality. The link between Confucianism and these variables is 'socialisation'. Following Benedict (1946) and M. Ng (1977), Redding argues that the Chinese 'shame culture' relies on externalised sanctions for good behaviour (see also M.M.H. Yang, 1989):

> '. . . the Chinese socialisation process . . . [is] one in which a tradition-based and lofty model is prescribed which requires unsparing effort for its attainment. At the same time the individual is trained into developing a highly sensitive pride and it is the wounding of this pride which controls conformity. The latter is a largely social force. . . . A manifestation of this in social behaviour is 'face' (Redding, 1980, p. 138).

Shame is taken as the medium of the Chinese 'socialisation process': children are trained to develop a pride which is highly sensitive, and conformity is achieved by every individual threatening to shame every other individual.

In a later work, Redding and Wong (1986) are more definite about those Confucianist characteristics which determine Chinese organisational behaviour. These characteristics are: a tendency to help the group; a sense of hierarchy and complementarity in relationships; and socialisation within the family unit or, in other words, the perpetuation of values surrounding education, sobriety, the acquisition of skills, and a strong sense of responsibility and obligation. And, indeed, observe Redding and Wong, most Overseas Chinese companies are family businesses, for even the largest public companies are still heavily influenced, if not dominated by, the family, and management control is achieved through nepotism, paternalism, obligation networks and non-objective performance assessment. Moreover, the power to make decisions is commonly centralised in a single dominant owner whose leadership is strongly authoritarian; there is little specialisation and, consequently, little standardisation of activities, few ancillary departments, and ownership overlaps with control; the establishment of goals is a family matter; and there is much secrecy about performance.

Again the links between Confucian values and these organisational characteristics are explained by elements of Chinese psychology and by socialisation—that ever-useful cement for theories of cultural determinism:

> 'Because of the importance of the family, and the tendency in a "collectivist" society for the attachment of the individual to the group, relationships at all levels of Chinese business organisation tend to be modelled on familial relationships' (Redding and Wong, 1986, p. 280).

Functionalism (the use of kinship, ethnicity or culture, to obtain material objects, funds, services or power) and networks which centre on the family and ramify throughout the company, produce a bevy of groups— each a cohesive unit but each in conflict with the other: 'The power of such a system to survive is based partly on ancestor worship and partly on the perception of wealth as family wealth, to which all contribute and from which all may benefit' (Redding and Wong, 1986, p. 284).

Following Hofstede (1980), Landa (1983) and Fei (1948), Redding and Wong (1986) argue that the sense of belonging to wider and different, but overlapping, networks is an expression of the Chinese people's high level of collectivism which centres on the family and from which, conceptually, concentric rings of successively weaker ties and obligations radiate: . . . individuals do not perceive themselves as related to each other on some fundamentally equal basis. Instead the individual is at the centre of a series of concentric circles, the closest having the strongest blood ties' (p. 284). As a result, obligations and rights differ according to the relative position of individuals within these circles and, therefore, different standards of morality operate. Building on Redding's earlier work (Redding, 1980) Redding and Wong (1986) argue:

> 'Duty to the community at large, based on some universally agreed moral principle, is replaced by ethics based on taking each situation and interpreting it according to loosely formulated principles of *jen* (or human-heartedness) . . . the application of multiple standards in dealing with people is not morally repugnant where the standards are set to [their] relative positions in the concentric circles' (pp. 284–285).

The maintenance of social harmony is achieved partly by the 'mechanism' or 'lubricant' of face. Again, Redding and Wong (1986), developing a point raised in Redding's earlier paper, argue that: 'a great deal of ritual behaviour takes place in a Chinese organisation in an effort to maintain the social harmony which is so highly prized . . . there is . . . an almost visible tendency to be wary of others' sensitivity . . . ' (p. 286). The development of 'face sensitivity', they maintain, is achieved by socialisation; and the development of 'face sensitivity' means that individuals will 'invest' in groups via the face they claim. The emphasis on shaming techniques and group consciousness in child-rearing reinforces compliance: 'The search for interpersonal harmony thus produces a person psychologically attuned to deference, compliance and cooperation within delimited groups' (p. 287).

The link between Confucian values, the individual, a sense of hierarchy and its incorporation within the company is also explained by socialisation:

'. . . Chinese socialisation teaches a person acceptance of hierarchy. It is noticeable, for instance, that in the *Wu Lun*, or Five Cardinal Relations on which centuries of impressive social stability have been built, four relationships are explicitly vertical, and training in accepting them creates an individual who understands the two-way obligation which they entail. . . . The exchange of obligations is a matter of Confucian duty and the expression of such duty serves to stabilise the family unit which is the framework for all relationships' (Redding and Wong, 1986, p. 287).

Relations between a superior and a subordinate in a company remain formal, aloof and reserved, 'as if to acknowledge tacitly that true loyalties are likely to be directed elsewhere. The remaining bonds which do keep the relationship operating are the traditional Confucian respect for authority and the utilitarian perception that it is rational for authority to be exercised for the common good. The tone is thus more calculating than emotional, when seen from below' (Redding and Wong, 1986, p. 288).

Materialism is another important element of Chinese psychology which, Redding and Wong believe, link Confucianism with the Chinese company. In combination with familism, materialism produces both the need for entrepreneurial skill (a skill which is consequently held in high regard) and the drive to start one's own business.

More will be said of Chinese social psychology later in this chapter. At this point it is enough to reiterate that Redding and Wong's analysis (and many other studies like it) amounts to little more than their own particular interpretation of a unidimensional world of their own creation. Selected observations are interpreted in a way consistent with the initial hypothesis. Indeed, this is made quite explicit. Redding and Wong follow the 'post-Confucian' hypothesis which 'suggests that the economic success of Japan, South Korea, Taiwan, Hong Kong and Singapore is due in large part to certain key traits common to cultures which, although they may now vary in their expression of it, derive their main social ideals from Confucian tradition. Recent research from sociological viewpoints has provided substantial support for Kahn's idea . . . ' (Redding and Wong, 1986, pp. 269–270).

This hypothesis and all its assumptions are the terms of reference for their analysis, and all observations are interpreted accordingly. Redding and Wong begin with the assumption that Confucianism is the idol of origins, that Chinese society is strongly 'collectivist', and that there are elements of Chinese psychology, rooted in Confucianism, which are peculiarly Chinese. Once particular characteristics of the Chinese company have been selected, it is then just a matter of linking these components together with the help of a few well-chosen tools such as 'socialisation' and 'materialism'. Not surprisingly, the result is that observations are transformed into evidence which supports the authors' initial hypothesis.

For Mackie (1992b), too, the stress which Confucianism places upon the family, and the observation that in many countries most Chinese companies are based upon the family, are taken *ipso facto* as proof that the Chinese company is a consequence of Confucianist values. To strengthen the assumed link between Confucian values, the Chinese company and material success, the ubiquitous notion of 'socialisation' is once again brought into the analysis. Since the family is also regarded by Mackie as the prime agent of socialisation, the notion of socialisation is particularly useful to his argument. The effect of this chain of thought, however, is to substitute the assumed link of socialisation with a tautology: the family transmits Confucianist values which emphasise the crucial importance of the family; and since the family finds expression in the Chinese company, then the Chinese company is perpetuated and transmitted from one generation to the next by the family. In short, the Chinese company is based upon the family because the family is the basis of the Chinese company. As Mackie (1992b) puts it:

'Virtually all accounts of the economic behaviour of Overseas Chinese emphasise the versatility and entrepreneurial qualities of their family firms, as well as the particular characteristics of Chinese family structure, as the primary source of socialisation and of the transmission of values conducive to business success. Other forms of socialisation may also play a part, such as schools, clans and trade associations, newspapers etc., but the primary role of the family seems to be unquestionable' (p. 53).

Another point of view with which Mackie sympathises is that put forward by Myers:

'In dealing with the links between Confucian values and economic development in China, Hong Kong and Taiwan, Ramon Myers (1989: 13–14) also touches on the connection with the family firm. . . . He stresses the consequences for personality development of the sibling rivalry between younger and older brothers, arguing that Chinese children everywhere are more "autocentric" and likely to behave according to well-defined rules developed in the course of their upbringing and socialisation than the more "heterocentric" children of the West (i.e. inclined to look to others as their authorities for behaviour patterns). "Is this characteristic related to Confucian values?" he asks. The answer is yes, because their family instruction, education and community training have strongly indoctrinated them to follow the authority of their parents, school or immediate community, since they will be shamed if they fail to do so. This kind of motivation, argues Myers, inclines them strongly towards achieving the goal of both individual and family welfare by which they gauge success' (Mackie, 1992b, pp. 55–56).

A number of comments need to be made about this sort of argument. The cultural determinists' choice of values, beliefs and ideas which they claim are 'socialised' is highly selective. If one cannot accept a belief in the force of socialisation, then it is only common sense to accept that children may learn something from their parents and schools which may or may not serve them well, and which they may or may not choose to adopt, adapt or reject. Since the late nineteenth century, in parts of Southeast Asia under British rule, Chinese merchants have often educated their children at British schools. Moreover, the views expressed by Myers, and by many other writers, and supported by Mackie, are somewhat confused. It is claimed that Chinese children are autocentric because they have been indoctrinated by the family and the rest of the community into behaving according to well-defined rules, whereas 'Western' children look to others as their authority for behaviour patterns. It is further claimed that as a result of their upbringing Chinese children will be more concerned with individual and family welfare than are 'Western' children. If there is any distinction between 'Chinese' and 'Western' children to be found in this odd set of assertions, it is one of convenience. Interestingly, M.M.H. Yang (1989) and Redding (1980)— both proponents of the cultural explanation—hold views on 'Chinese' socialisation which appear to be almost exactly the opposite to those of Myers and Mackie. Redding and Yang suggest that Chinese behaviour is very much conditioned by what others think, and that 'Chinese' morality depends not upon 'internalised' moral abstractions (as is the case among 'Westerners') but upon context, situation and the use of shame. Yang (1994) goes on to argue that 'internalising' the approbation of others in the context of relationships (a concept which lies at the core of *mianzi*) is, in China, more common than *lian* (another version of 'face' which represents an individual's ultimate moral worth measured according to society's fundamental code of ethics) in constraining '*guanxi* reciprocity' (p. 140). And yet she also places considerable emphasis on the crucial importance and peculiarity of 'relational' or 'situational' ethics to Chinese society in China and Overseas. However, these apparently contradictory assessments of the 'differences' between 'Chinese' and 'Westerners', and between different categories of 'face', serve the same end. By manufacturing false distinctions between 'internalised' and 'situational' values (and thereby distorting the commonsensical knowledge that individuals create, alter, adapt, reject or accept values, beliefs and practices), these values can be presented as a force in their own right subject to the operation of other forces—such as shame and socialisation—which together are capable of producing understandable and predictable beings.

In his analysis of the inculcation of economic values in Taipei business families, Olsen (1972) does not assign the 'socialised' individual with a passive disposition. Society and the individual, Olsen suggests, participate in a probabilistic rather than in a deterministic 'process'. Nevertheless, Olsen argues that:

> 'there exists in urban Taiwan a differentiated business value-system, and that a significant part of this value-system is transmitted from one generation to the next within business families. This is a process I call economic socialisation. . . . Differentiation of the business value-system is on a fairly high level of generality; the evidence suggests that it is shared in large measure by business families representing the whole range of entrepreneurial activities . . . ' (p. 294).

In Olsen's study, then, socialisation takes on a purely statistical form. The values and aspirations held by children are first determined. The values which the children appear to hold (or do not hold) are linked to their parents and their parents' occupations. In those instances where links appear strong, socialisation is particularly strong.

Again, however, all this is pure supposition. The credibility of Olsen's argument lies only in the common-sense observation that children may learn and adopt something from their parents, and that if the children and the parents concerned, and the circumstances prevailing, are amenable, then children may choose to absorb techniques, knowledge, values and a particular philosophy or approach to work which they believe may serve them well. This common-sense understanding has been distorted into a self-propelling force which turns individuals into vehicles for the transmission of values from one generation to the next— values which give rise to the nature and the form of the Chinese company, and which are capable of producing distinct groups. This is a convenient tool, but its use requires acceptance of a great many assumptions which have only faith as their foundation.

'Socialisation' would appear to be little more than a technical expression for the belief in the sociologists' force which determines that every individual behaves in a certain way because every other individual behaves in that way. It is a useful device which creates for the social scientist a bundle of drab, lifeless toys—individuals who are unable to change, who are incapable of independent action, independent thought, independent decisions and independent emotions, but who do behave exactly as the scholar dictates and predicts.

The utilitarian family and different 'kinds' of Chinese   When Chinese simply do not behave in a way that is compatible with the links which are assumed to exist between 'culture' and 'the Chinese' company, departures from the behaviour, decisions, values, choices and actions predicted are explained either by an alien force or by the introduction of an entirely different type of Chinese. The assumptions upon which scientific analysis and the cultural explanation are based are not questioned; the different aspects of phenomena are not recognised; rather, the traditional Chinese and their institutions, which it is assumed can be explained by culture, are replaced by Chinese less constrained by that culture and who, incidentally, appear to be much more like the modern scholar. For 'traditional', read 'explicable by culture'.

The concept of the utilitarian family, or more accurately the line of thought upon which this concept is based, has proved a useful contribution to the cultural explanation of the Chinese company, especially when used in combination with the notion of 'socialisation'. Lau (1978, 1982) argues that in Hong Kong the family has become increasingly utilitarian as it moves away from its traditional form. The traditional family is more or less self-sufficient, hierarchically structured and authoritarian; each member is governed by a set of prescribed norms of behaviour which vary according to the individual's status and role within the family; relationships between individuals are never conceived solely to be instrumental; and membership is defined by fixed criteria such as genealogy, creating a clear boundary around the group. The primary concern of each individual within this exclusive group is with the family's reputation which each individual will do his utmost to preserve and promote. Reputation is derived partly from the correct observance of prescribed behaviour within the family, and partly from the fulfilment of graduated duties and responsibilities which serve as links between the family and the wider society. If the family possess a filial son or a daughter-in-law who observes her duties with particular rigour, and if a family member has passed official examinations to become an instructor or a bureaucrat and has fulfilled those responsibilities correctly, then the reputation of the family, and the stability of society, is assured.

The Chinese immigrants, Lau argues, have been uprooted from their natural setting and placed under a colonial government which cares little for social cohesion, and has thus left the immigrants with little choice but to engage in commerce. In order to facilitate and protect their material interests, individuals have been forced to restructure ties both within the family and between the family and the rest of society:

'In essence, utilitarianistic familism can be defined as a normative and behavioural tendency of an individual Chinese to place his familial interests above the interests of society as well as its constituent individuals and groups, and to structure his relationships with other individuals and groups in such a fashion that the furtherance of his familial interest is the primary consideration. . . . Moreover, materialistic interests take priority over all other non-materialistic interests' (Lau, 1978, p. 4).

The enhancement of family pride, then, is no longer the primary motivation of individual behaviour. The group becomes less exclusive and relationships within the family are more elastic and egalitarian. Indeed, to such an extent is authority downgraded that children (no longer closely monitored and controlled, and now allowed to make decisions for themselves) are thereby given further experience of 'learning the world through their own unimpaired manipulation of it' (Mitchell and Lo, 1968, p. 315). Even within the family social, emotional and ritual ties are marginalised in favour of economic matters, for while the family is an 'affectively charged unit', and while its members will still help each other, free from utilitarian motives, the utilitarian characteristics of the family have assumed enormous importance. 'Though services and help can be extended to those members who cannot reciprocate, in many cases this will be considered as a long-term investment with the hope that it will pay off in the future' (Lau, 1978, p. 7).

In sum, the traditional family is based upon a clear set of values, ethics and ascriptive relationships which clearly prescribe the roles, status, behaviour and duties of individuals. It is an exclusive group with a corporate personality which is 'not artificially constructed to serve some limited purposes' (Lau, 1978, p. 23). The traditional family reflects a philosophy of life which merges the individual with the family, and the family with society. The utilitarian family, on the other hand, is 'the normative order for the regulation of the operation of a group deliberately organised for some specific purposes' (p. 34).

Clearly, this concept of the utilitarian family ignores the many different aspects of both individuals and the family which they constitute. The argument that the traditional family was not utilitarian requires acceptance of a very limited understanding of 'manipulation'. Moreover, if the 'traditional family' is considered within the wider unit of the lineage, then the family was integrated within an organisation essentially political in nature. The 'traditional family' was not manipulated and directed purely or even largely towards the extension and institutionalisation of trade (see Chapter 2), but manipulated it was.

Nevertheless, the concept of the utilitarian family remains attractive, for even though it is suggested that traditional values and the detailed structures of traditional institutions have been marginalised and distorted, the notions of a traditional family and a utilitarian family, and the line of thought upon which these concepts are based, provide a logical superstructure around which a link between 'Chinese' cultural roots and 'the Chinese' company may be constructed and explained. More importantly, the utilitarian family provides a convenient transitional stage between the traditional Chinese and an entirely different kind of Chinese.[4]

Although Mackie makes no reference to Lau, his line of thought is the same and takes the argument one step further by changing traditional Chinese into something new. One reason why the distinctive characteristics of the Chinese company and Chinese entrepreneurship may be retained, Mackie argues, is the importance of the family in the socialisation of values, attitudes and practices conducive for success: so far there is no evidence that traditional values, the family and personal relationships will not persist, but it remains an open question. Indeed, Mackie leans more towards the view that things have already changed and will continue to do so: '. . . the characteristics that seem to explain their initial success do not necessarily fit so well in explanations of why some of them have been able to make the transition to much riskier levels of large-scale enterprise, although thousands more have not. In trying to explain these processes we need to treat them as quite distinct analytical issues' (Mackie, 1992b, p. 59). Following Myers, Mackie suggests that:

> '. . . the latter-day Chinese who are willing and able to operate large-scale enterprises along modern lines, relying less than previously on personal connections as a basis for trust in their business dealings, should be regarded as in many respects a *quite different kind of Chinese* from their predecessors. They are *more capable of transcending the constraints imposed by the traditional values and cultural predisposition* that made it difficult to trust strangers. In many cases, the impact of better education, exposure to the modern world and the processes of acculturation to Southeast Asian values, have eroded the influence of traditional Chinese values' (p. 59, italics added).

In a later article, Hicks and Mackie (1994) appear to increase the differences between the 'new' and the 'traditional' Chinese still further: today's Overseas Chinese, they observe, are more like the indigenes than Chinese. It is in this way, by creating the 'new' Chinese of today, that the 'traditional' Chinese of yesterday can be used to demonstrate the validity of the cultural argument. The 'new' Chinese, it would seem, are part of an attempt to use common sense to establish a false contrast to legitimise the cultural argument, not only intellectually, but also morally; for it is a contrast which helps distance cultural determinists from present-day racial tensions. A very similar line of thought has been pursued by Murphey (1977) and King (1987) who suggest that new kinds of Chinese, pragmatic and rational in their use of traditional Confucian values, have arisen in commercial and cosmopolitan Southeast Asia.

A more resilient and ingrained brand of 'Chineseness'—but one which helps emphasise the essential distinction between 'traditional' and 'new' as understood by cultural determinists—is portrayed by Redding (1980) who, it was noted earlier, maintains that the purported fundamental differences in cognition between 'Orientals' and 'Westerners' are a consequence of different influences, the most important of which is culture. It is culture which affects the mind-maps—the systems of belief and understanding—that guide behaviour. In order to examine the contrast between the cognition of 'Orientals' and 'Westerners', artificial categories of behaviour which are said to have specific implications for business organisations are first selected. These categories are: explanations of events, the perception of time, the individual's place within social networks, influence on behaviour, and the assessment of 'rights' and 'wrongs'. From these five categories, another five are refined and have already been referred to in this chapter: causation, probability, time, self and morality. Redding claims that whereas the Chinese mind is intuitive, and phenomena are contemplated within a wider context, 'Western' thought is characterised by logical sequential connections and emphasises cause; while the Chinese mind is fatalistic, the Westerners' logical, sequential 'thought processes' and their emphasis on cause leads easily to extrapolation and prediction; while the Chinese mind perceives time as cyclical, the Westerner has an accurate sense of time; and while the Chinese mind's sense of individual consciousness is less pure, the Westerner regards the individual as isolated and important in his own right. Indeed, Redding (1980), drawing on Hall (1976) as his precedent, argues: 'The individualistic Westerner finds it difficult to appreciate the networks of influences and relationships typical of a high-context society, and the idea that the person is not an individual, in the Western sense, is almost untransmittable . . .' (p. 135). And whereas the Chinese is concerned to avoid shame derived from the infringement of social norms which are 'situational', the Westerner acts to avoid guilt derived from absolute moral principles.[5]

According to Redding (1980), three major differences between the Chinese and Western company stem from these differences in cognition. Formal planning frameworks, which require the use of abstract categories and linear logic and therefore come naturally to the Western mind, are absent in the Chinese company. While the Westerner makes decisions in the light of facts and scientific extrapolations, the world of the Chinese mind comprises sets of mutually balanced interconnections. And whereas the Westerner's linear sense of time allows him to conduct programmes and scheduling, Oriental culture 'lacks the necessary foundations' (p. 144).

A formal organisational structure requires the apprehension of abstract concepts such as 'marketing functions', 'financial control' and 'coordinating mechanism' which comes easily to the Western mind. The Chinese mind, however, is unable to seize upon these constructs which are essential for the specialisation of activities and for their integration. The Chinese mind's rather fuzzy perception of time is another trait which contributes towards the informal organisation of the Chinese company. The Western company, Redding argues, is heavily reliant upon a common perception of time: it is essential that 'everyone understands the significance of a deadline, that the necessary sequencing of events is understood, that the time-frame of budgetary control is accepted' (Redding, 1980, p. 145). Reliance on this accurate, linear perception of time means that one thing is done at a time; there can be an emphasis on scheduling and promptness; there can be extensive delegation and a long hierarchy; and goals can be scheduled, but the job's minutiae can be left to the individual.

Redding develops this last point, arguing that in the Chinese case:

'. . . growth along Western lines, which usually is by growing the corporate body itself, appears to be resisted. There are large Chinese companies, it is true, but they appear still to be run in the same way as small Chinese companies. They remain in family control. Rational/legal authority is not adopted. Size is often achieved by collecting together a set of small businesses and leaving them uncoordinated except at the financial level. More complex forms of large-scale enterprise have not developed: there are no Chinese multinationals.

It is possible to argue that the Western bureaucracy developed because the rationality of the Western mind fostered it. In the Far East, the rise of the corporation may not be a matter of time waiting for such a natural development to take place. Growth in Chinese business may continue by using its present units—and simply more of them' (Redding, 1980, pp. 147–148).

One striking difference between 'the Western' mind and 'the Chinese' mind arises out of Redding's analysis. The acultural Westerner is obviously in control of his actions and in the designation and pursuit of his goals, and he attempts to exercise control over those circumstances in which he finds himself. The Chinese, however, is at his culture's beck and call: the Chinese mind has no clear consciousness of cause and effect, of extrapolation of goals, nor even of its individuality; and its moral perceptions shift with circumstances. The characteristics of 'the Chinese' company, then, are derived largely in default of 'Western' characteristics. Formal planning is either absent or partially developed because abstract categories, linear logic and a linear sense of time are not characteristics of the Chinese mind. There is no formal organisational structure for the same reason: the Chinese mind cannot seize upon abstract concepts and has an imprecise perception of time. And because the Chinese mind has no clear sense of individuality, the company is reliant on widespread networks of personal relationships.

There is no implication in this kind of argument that 'the Western' mind is superior to 'the Chinese' mind; but Redding reveals far more about the psychology and techniques of the social scientist than about the Chinese. Redding's paper is an example of a clear, concise exposition of the scientific approach. The creation of a people whose actions and decisions are determined by their cultural mantle provides neat, well-defined subjects who will fit easily into neat, ready-made 'scientific' explanations.

Comparison

The conversion of the Chinese from one kind to another brings the discussion to a third set of criticisms of the cultural explanation of the Chinese company and of Overseas Chinese economic success. In some instances comparisons are used to remove from the discussion those Chinese who do not fit in with the cultural explanation by highlighting their similarities with Western industrialists who are, by definition, not constrained by culture. In other instances, comparisons are either ignored altogether or used selectively to isolate 'Chineseness', or, if this proves impossible, to broaden the cultural explanation by identifying similarities between 'the Chinese' and other cultures (usually described as Confucianist) in the West Pacific Rim and thereby to create a distinctively Oriental culture in which 'Chineseness' is one of several specific regional variations.

The reason for this selective use, marginalisation or dismissal of comparisons is clear. In the absence of a proven and deterministic causal link between those traits arbitrarily designated by a writer as components of Chinese culture and those phenomena which the writer is attempting to explain, it cannot be assumed that 'culture' is the cause. Unguarded comparisons only emphasise the quandary into which the social scientist has fallen. It might be argued that the same or similar phenomena may exist for different reasons and are therefore no less Chinese. However, such an argument is very weak without an established causal link, for it merely presupposes that the initial assumption of a link between culture and those phenomena is valid. Moreover, it is an argument which can be turned around: the existence of the same or similar phenomena in different contexts illustrates that institutions, values and forms of exchange possess many different and changing aspects and, therefore, that no causal relationship between these sets of phenomena can be assumed.

Comparison, then, is used selectively in cultural explanations of the Chinese company to create or highlight differences which it has already been decided are cultural determinants. Alternative interpretations are ignored and opposing evidence is dismissed or presented as a consequence of the alien influence of 'modernisation' and 'industrialisation' which shifts individuals away from 'traditional' values (an argument which either requires acceptance of mystical forces, or a determination to ignore tautology, for 'modernisation' is no more than an expression of deliberate human action and choices). Yet all this intellectual twisting and turning is merely to weave further baseless assumptions into the analysis. It is to assert that the assumptions are correct; it is to use these assumptions as a framework for the selection and interpretation of evidence; and it is to present those interpretations as fact. 'The world *is* as I see it.'

The checklist of differences and similarities with which the social scientist looks for a 'deeper' culture is a distraction—dancing reflections of circumstances, choices, judgements which tease and mesmerise the writer as he tries to grab them and shape them into a tangible form of 'the Chinese' company which conforms with his preconceptions. A harmless distraction perhaps, but also a crafted, scholarly denial of the uncomfortable, imprecise, unstructured, untidy and difficult world of the humanities. If it is not assumed that the company is a product—or at least an imperfect reflection—of Chinese culture, then all that is left is the play of circumstances, chance events, individual choices, judgements, the transmission and acceptance or rejection of ideas and practices, and the spontaneity and multiple intelligence of the human mind—a complexity which comparison helps to emphasise.

Chinese thought, Chinese morality and the Chinese company
Redding (1980) summarises the differences between Chinese and Western cognition as follows:

'Western cognition: Logical, sequential connections. Use of abstract notions of reality which represent universals. Emphasis on cause.

Chinese cognition: Intuitive perception and more reliance on sense data. Non-abstract. Non-logical (in the Cartesian sense). Emphasis on the particular rather than the universal. High sensitivity to context and relationships' (p. 132–133).

In support of his views, Redding cites Needham who argues that

'We are driven to the conclusion that there are two ways of advancing from primitive truth. One was the way taken by some of the Greeks: to refine the ideas of causation in such a way that one ended up with a mechanical explanation of the universe, just as Democritus did with his atoms. The other way is to systematize the universe of things and events into a structural pattern which conditioned all the natural influences of its different parts. On the Greek world view, if a particle of matter occupied a particular place at a particular time, it was because another particle had pushed it there. On the other view, the particle's behaviour is governed by the fact that it was taking its place in a "field of force" alongside other particles that are similarly responsive: causation here is not "responsive" but "environmental"' (Needham, 1978, p. 13).

Northrup's views are also used by Redding to present a selective interpretation of 'Chinese' thought:

'Formal reasoning and deductive science are not necessary if only concepts by intuition are used in a given culture. If what science and philosophy attempt to designate is immediately apprehended, then obviously all that one has to do in order to know is to observe and contemplate it. The methods of intuition and contemplation become the sole trustworthy mode of inquiry. It is precisely this which the East affirms and precisely why science has never progressed for long beyond the initial natural historical stage of development to which concepts by intuition restrict use' (Northrop, 1946, cited in Redding, 1980, p. 132).

But then compare Redding's summary of the differences between Western and Chinese cognition with another passage by Needham:

'The key word in Chinese thought is Order and above all Pattern (and, if I may whisper it for the first time, Organism). . . . Things behaved in particular ways not necessarily because of prior actions or impulsion of other things, but because their position in the ever-moving cyclical universe was such that they were endowed with intrinsic natures which made that behaviour inevitable for them. . . . They were thus parts in existential dependence upon the whole world-organism. And they reacted upon one another not so much by mechanical impulsion or causation as by a kind of mysterious resonance' (Needham, 1956, p. 281).

However questionable Needham's views may seem (see Qian Wenyuan, 1985), his interpretation hardly indicates concrete, non-abstract thought which emphasises the particular. Indeed, according to Needham, this 'Organistic' thought of the Chinese is a proto-science which could provide the impetus for the science of the future. In contrast to both

Needham and Northrop, Qian Wenyuan (1985) argues that science did not develop in China because of China's political and ideological rigidity. From time to time there appeared isolated geniuses, but:

'China was not truly powerful, because its people were on the whole neither economically nor intellectually powerful. European people were comparatively powerful; and that is why no single power, military, political, or spiritual, has ever succeeded in ruling them all. (p. 27) . . . What traditional China unfortunately lacked were a series of totally new intellectual elements: new attitudes, new ways of thinking, new fields of interests, new epistemological standards, and most importantly a propitious politico-social condition to produce, sustain, and promote them' (pp. 31–32).

Qian Wenyuan illustrates China's political, economic and intellectual stagnation by a comparison with Japan—a nation that has been at least since the twelfth century very different from her neighbours (Reischauer, 1977):

'. . . none of the numerous similarities between China and Japan such as racial identity, geographical proximity, Confucianism, Buddhism, rice agriculture, the self-sufficient agrarian economy, the writing system, the many mutual cultural borrowings . . . were decisive in the two nations' respective social metamorphosis; but in the different politico-social structures of the two societies lay the potentials for their respective future development, potentials which were realised in the nineteenth century when both countries faced the threat of Western imperialism. . . . In Japan, modernisation was carried out in both political institutions and economic constructions, while in China the Manchu regime looked to the West for only superficial technological innovations with the sole purpose of maintaining its outmoded political establishment' (Qian Wenyuan, 1985, p. 28).

If 'the Chinese' mind cannot be characterised by concrete and intuitive thought, then neither can the 'Western' mind be characterised by abstract, logical and sequential thought which emphasises chains of cause and effect. Redding has simply ignored the way in which, for example, top decision-makers in Britain operate, and long-standing debates by Western intellectuals such as Dilthey, Rickert and Weber on the distinctions between the natural sciences, the social sciences and the humanities. It is also interesting to note that Needham's use of the term 'existential' to describe the dependence of things (in 'Chinese' thought) upon 'the whole world-organism' betrays the influence of the debate (or of ideas which also find expression in the debate) between Kierkegaard and the Idealism of Hegel—entire schools of Western thought which Redding has also chosen to ignore. Comparison, then, has been used to create 'Chinese' and 'Western' minds that are conveniently unidimensional.

Similar criticisms can also be made of Redding's assessment of the differences in morality and self between 'the Chinese' mind and 'the Western' mind. It was noted in Chapter 3 that the notion of chains of cause and effect between 'social structures' (or culture) and morality is open to more than passing doubt. Redding (in common with many proponents of the cultural explanation) argues that within a 'high-context' society (the basis of which is Confucianism) individuals merge with each other. The lack of consistency in moral standards is regarded as a characteristic of Confucianism and therefore lies at the heart of his exposition of Chinese morality. Yet does this not raise the possibility that Confucianism has many different aspects? For if morality is used according to circumstances—or, more specifically, varies according to whom the individual is dealing with—then clearly morality is being manipulated, unless one is prepared to believe that the conscious manipulation of morality by the individual is determined by Confucianism. Both Christianity and Confucianism preach with great common sense that you should not do unto others what you would not have done unto you—a code of conduct which clearly recognises the manipulation of morality. Are we to believe that adherence to this code of conduct is a matter of choice for all Christians but pre-programmed into all Chinese? Although Redding acknowledges the use of Confucianism in maintaining social order (1990, pp. 45–46), he presents Confucianism as a set of ethics which arose from men but which now revisits and shapes them. To such an extent is this view emphasised, that in his hands Confucianism becomes a powerful force in its own right, capable of directing and determining individual actions. But if Confucianism were to be taken as little more than the codification of values, prejudices and emotions by which the status quo was maintained (see Graham, 1988, Franke, 1972, and Pan, 1989, 1990)—a collection of ideas, values and feelings which could be bent to the will of individuals—then the different aspects of Confucianism which appear soon betray the operation of individual consciousness and choice.

Furthermore, if it were to be accepted that there is, as Redding believes, a connection between social structure or culture and moral values, then there is no reason to think that Chinese moral values and norms are more, or less, dependent upon context than those of the 'West'. There is a contradiction in Redding's argument: if it is accepted that there is a causal link between culture and moral values, then it is difficult to appreciate how moral values in one society can be absolute while in another they are 'situational'.[5] More importantly, Redding has again ignored relevant 'Western' debates: this time an entire branch of philosophy—ethics—and the notion of cultural relativism expounded by Herskovits.

There is something a little ironic about an attempt to use the rigour and precision of the cultural determinists' self-styled 'scientific' techniques to examine, explain and understand a people whose thought, it is claimed, is intuitive (cf. Wilson and Wilson, 1965, p. 69). To argue that the Chinese are characterised by a particular type of thought is to create a unidimensional mind—a mind with only one aspect; to suggest that this assumed characteristic explains the failure of 'traditional' China to develop modern, 'Western' science is to build yet one more assumption upon another; and to trace the origins of this assumed characteristic from traditional China forwards to the Chinese company is an act of faith. The Overseas Chinese escaped—or at least found themselves free from—the values by which individual worth was judged, the imperatives of political status and of loyalty to the state, and the necessity to suppress trade or to redirect it towards the maintenance of a status quo legitimised by Confucianism. It was this freedom, most especially under the European powers, that allowed those Chinese who so wished to redirect their values and institutions towards the support of the extension and institutionalisation of trade.

Whereas Redding has ignored similarities, Mackie uses similarities to create a template of unique Chinese characteristics: by bringing together analyses of companies from different cultures, those characteristics which are specific to the Chinese can be identified. Mackie (1992b) acknowledges that family enterprises play an important part in business in many different countries and cultures: 'they are not a uniquely Chinese phenomenon, nor dependent solely on the value systems or tight family structure of the Chinese' (p. 54). Nevertheless the significance of such evidence is already predetermined by Mackie's initial assumption that Chinese values and 'culture' (Confucianism) largely explain Chinese companies and Chinese economic success. Evidence from other parts of the world, then, does not lead Mackie to question his assumptions. On the contrary, for Mackie (in common with Jacobs, 1979, 1980, 1982) the purpose of comparisons and the identification of similarities is to enable the social scientist to pin-point those unique Chinese characteristics and the interplay of unidimensional variables: 'the more we can learn about the general structural characteristics or morphology of such firms, the better for our purpose of trying to understand what is distinctive about the Chinese variety, as against what is common to them all' (Mackie, 1992b, p. 54).

Rather than bring out the similarities which demonstrate the many different and changing aspects of family organisation and the lack of relationship between organisation and function, Mackie chooses to marginalise comparisons and at the same time to give emphasis to the task of isolating those variables and their pathways of interaction that will give higher definition to 'Chineseness'. Comparison, then, is a device for selectively extracting observations which seem to confirm 'Chinese' characteristics and which appear to legitimise the assumption that causal links exist between these characteristics and commercial success.

Consequently, while Mackie regards as untenable Weber's argument that the Chinese extended family is not conducive to economic success, Mackie neglects Weber's argument that, for the sake of effective large-scale organisations, making decisions should be separate from their execution. Decisions made informally among individuals who know each other well and who may intuitively know and understand what the other is thinking or may do in given circumstances, is crucial for quick and effective action. This, it is suggested, is a matter of common sense and instinct. In small companies, the need for separation does not arise, but in larger organisations making decisions may well have to be separated from their implementation. However, this separation does not necessarily mean that reciprocal networks do not underlie the executive channels of action: arguably they may be mutually supportive and protective. Administrative procedures can be bypassed or referred to or reinforced as seems appropriate, for a precise distinction between decision-making and executive action may not always be possible. These considerations are particularly relevant to the discussion, for it is frequently noted in the literature that Chinese firms do grow and that although they remain based upon the family they show a mix of purportedly 'Western' and 'Chinese' practices.

In their analysis of top decision-makers in Britain, Lupton and Wilson (1959) observe that evidence from proceedings of a tribunal appointed to inquire into allegations that information about the raising of bank rates was improperly disclosed, revealed:

> 'some important decisions were taken and others accepted because colleagues knew about, and relied upon, each other's beliefs and special aptitudes. Lengthy analyses were not a necessary prelude to decision-making. This is not surprising. When decisions have to be made quickly most persons have to act according to precedent and "hunch" and not in the light of detailed analysis of the current situation' (Lupton and Wilson, 1959, p. 31).

And indeed, highly complex networks of relationships—other than those which arose out of business—exist among Cabinet and junior ministers, directors of the Bank of England, directors of the 'Big Five' banks, directors of City firms, directors of insurance companies, and senior civil servants. In the late 1950s, certainly, between around one-quarter and one-third of all the individuals (except senior civil servants) included in Lupton and Wilson's study were educated at Eton; two-thirds of the directors of the Bank of England and half of all ministers had been educated at one of six public schools; and nearly half of all individuals (including senior civil servants) had been educated at one or other of these six schools—Eton, Winchester, Harrow, Rugby, Charterhouse and Marlborough. Of all ministers, senior civil servants, directors of the Bank of England and directors of the 'Big Five' banks, between one-half and three-quarters had been educated at Oxford or Cambridge. Two-fifths of senior civil servants were members of the Reform and the University Clubs; more than half of all ministers were members of the Carlton Club; nearly half of the directors of the 'Big Five' were members of the Carlton Club, Brookes', and the Yacht Club; around one-quarter of directors of City firms were members of Brookes' and Whites; and nearly half of the directors of insurance companies were members of White's, the MCC and Brookes'. Links among these individuals also existed in the form of multiple directorships and through extraordinarily complex kinship ties of blood and marriage that also drew in the landed gentry, the judiciary and merchant banks. Top positions in some firms were occupied by adjacent generations of the same family. For example, at the time Lupton and Wilson were writing, the Chairman of Lazard Brothers and the director of the Bank of England were both occupied by Lord Kindersley, and, before him, by his father.

The individuals who comprised these relationships were acutely aware of the role conflicts which could arise when friendship, kinship and association ties exist within modern organisations for, as already noted, a clear distinction between making and executing decisions is not always possible. Thus the organisations concerned seemed to incorporate both 'traditional' and bureaucratic structures. And, indeed, if one sees human affairs as fluid, uncertain and imprecise, it is not difficult to appreciate why the advantages of these arrangements should have been thought to outweigh the disadvantages, especially given the nature of decisions which have to be made and implemented quickly.

Clearly, then, the notion that reliance on, and use of, complex webs of personal relationships is an 'Oriental', let alone a specifically 'Chinese', characteristic is easily disabused by even a cursory glance at the literature on British, American and African societies, on Jewish communities, on illegal underground economies, or even on academia (Bottomore, 1964; Ferris, 1960; Sampson, 1962; Parry, 1969; Chapman, 1969; Cohen, 1974; Hennessy, 1990; Eitzen, 1968; Sommers, 1993).

Mackie (1992b) is aware of the possible variety and significance of comparisons, but chooses to ignore or marginalise them:

'Two related issues must also be mentioned, but cannot be explored at length here for lack of space. One is the question of why indigenous entrepreneurs and their business firms have lagged so far behind the Chinese in modern Southeast Asia, *even though claims have been made* that certain ethnic groups such as the Minangkabau, the Bugis, the Mon and some Filipino peoples have demonstrated strongly entrepreneurial characteristics in appropriate circumstances. . . . No general reasons can be advanced in explanation of this phenomenon, but the following are relevant considerations in some times and places. Most have simply not been as strongly motivated to engage in commercial pursuits as the Chinese, having found an adequate livelihood traditionally in agriculture or government, or aspired towards prestige in non-commercial fields. . . . Rarely have they had access to comparable networks of credit, market information and support in times of adversity that many Chinese have generally been able to call upon. . . . But the particular reasons why some indigenous peoples have shown more business enterprise than others, or why they have done so in particular times and circumstances but not others, are too complex to go into here' (pp. 44–45, italics added).

There are two points to be made here. First, Mackie's use of the word 'claims' to describe observations that some ethnic groups have demonstrated entrepreneurial characteristics is a little surprising given the literature which he himself cites, the economic restructuring of the Malay-dominated states of Southeast Asia over the last two generations, and the obvious material success of these 'ethnic' groups such as the thriving Filipino business communities in the Philippines and overseas (Clifford, 1994; Tiglao, 1994a, c, d; Silverman, 1994; Jayasankaran, 1994; and see Chapter 6). It must be assumed that Mackie's terminology is either careless or is designed to play down observations which cannot easily be dismissed and which would otherwise require a discussion and an interpretation of the similarities between the Chinese and other ethnic groups.

Secondly, Mackie has ignored the different aspects of reciprocal networks and, consequently, nowhere does he mention that a preference for personal ties (in business or in other fields) is not a unique feature of the Chinese. In common with many other writers, Mackie has bypassed the questions arising from a strong resemblance between the values and institutions of Chinese and those of the indigenes. These similarities, it is suggested in this book, illustrate the different aspects of institutions and values, and therefore indicate how similar institutions may be used for different purposes, and how different sets of wealth values, forms of exchange, institutional patterns and moral imperatives may interact.

Underlying this whole debate on personal networks within and around the Chinese company is one nagging question: why should it have ever been assumed that the use of reciprocity in business is so characteristic of, if not unique to, the Chinese? The answer seems to be that the obsession with Chinese culture has meant that writers have ignored the more pertinent questions. In what circumstances (if any) is reciprocity and its associated social networks not of some significance in the conduct of trade? And why is it that there arise more impersonal, formalised procedures and dealings frequently enshrined in law? It may be that formalised procedures and dealings presuppose an environment generally favourable to trade: trade is acceptable or desirable, the rule of law can be relied upon, and thus individuals can be expected to operate effectively according to more impersonal, formal practices, thereby opening up a much wider field of potential deals and customers. As Dewey (1962) has suggested, trust, based on non-legal sanctions, is of crucial importance to trade in a society that lacks a well-developed and well-enforced civil code. This is not to suggest, however, that in an environment generally favourable to trade, reciprocity therefore becomes redundant. This would be to ignore the other aspects of reciprocity and its associated social networks: these include facilitating the rapid formation and implementation of decisions which are necessarily intuitive, and ensuring cohesion and speed of operation within and between organisations which would otherwise be impossible to realise and enforce through pedantic bureaucratic procedures and sanctions covering every nuance of action and behaviour. Seen in this light, there is nothing peculiarly 'Chinese' about the Chinese company.

The Chinese company and social psychology

It is clear from the material presented in this chapter that many analyses of the Chinese company draw support from the field of social psychology. Faced with a choice between ethereal forces and genetic determinism, it is perhaps inevitable that the social scientist should turn to psychology as a means to dig out the ultimate origin or in order to provide the link between economic success, the Chinese company and the prime cause.

The belief that the observed behaviour of selected individuals can be extrapolated to Chinese throughout Southeast and East Asia is itself founded upon two other beliefs: that an individual's behaviour is rooted in Confucianism; and, that after centuries of socialisation, most Chinese will behave in the same or a similar way. The social psychologist, then, attempts to demonstrate objectively that the idol of origins is Confucianism while at the same time building this very assumption into a conceptual framework that is used to interpret observations. In other words, all observations are interpreted according to initial preconceptions which the social psychologist simultaneously claims are scientifically tested hypotheses.

It is interesting to note that China (because of its relative poverty) is usually excepted from these generalisations by calling upon a different set of variables stitched together with the same logic and reliance on forces which, it is held, explain the relationship between Confucianism and Chinese economic organisation. Huo and Steers (1993) put these arguments most succinctly:

> 'Although the economic/political structure is often imposed upon the people as a result of revolution in a country, over time the values imbedded in that structure may be taken for granted, implying that they gradually become part of the culture . . . [the Taiwanese] were shocked to find that the gifts and money they brought back [to China] were received by their relatives with far less appreciation than they expected. While the older people tended to express deep gratitude, many youngsters who had been educated in the socialist system simply interpreted their generosity as a way to redistribute the national wealth for the sake of egalitarianism' (p. 77).

Because it is possible to interpret behaviour according to the social psychologist's initial preconceptions and assumptions, it is therefore believed (not only by social psychologists, but also by many specialists on the Overseas Chinese and by writers from a range of disciplines) that behaviour so interpreted confirms the validity of the initial assumptions. The inclusion of 'external factors'—such as government policy and the influence of 'Westernisation', 'industrialisation', history and geography—which are said to affect the details of the final outcome, are presented as refinements to the basic assumptions which, it is believed, have therefore been shown to be sophisticated enough to take account of a complex whole.

Stated in these terms it is perhaps surprising that this circle of mutually reinforcing assumptions has gained so much credibility, particularly when its empirical material comprises subjective assessments of very limited aspects of a few individuals' observed behaviour in specific instances. However, the concept of 'the Chinese' is a convenient tool which, provided its basic assumptions are accepted, possesses its own internal consistency; and internal consistency is a feature which is taken by many social scientists as proof positive that their intellectual constructs exhibit a true likeness to the world outside their own imaginings.

These crude assumptions are a scholarly version of a 'national character' which appeals to the expectation that other peoples are a little less in control of their decisions and actions, and a little less conscious of their individuality and rationality. 'The Chinese', rather like 'the Malays' or 'the Indians' or 'the Africans' or 'the Americans' or 'the English', is an intellectual device which creates a people difficult to empathise with, but easier to understand intellectually. And yet, paradoxically, its scientific gloss also provides an emotionally satisfying reaction to the belief that 'native' cultures are unable to cope with the demands of material progress—a belief that is now associated with the outdated, old-fashioned, judgemental and patronising colonialist.

Bond and Hwang (1986) begin with the assumption (or hypothesis) that Chinese social behaviour—which they characterise as harmony-with-hierarchy—is determined by culture or, more specifically, by particular elements of Confucianism. These elements may be described as the hierarchically structured relationships by which an individual lives, by which he is defined, and by which social order is maintained, provided every individual abides by the requirements of his own role within this hierarchy. Confucianism, they argue, explains a large variety of categories of Chinese social behaviour: the Chinese assess an individual's conscientiousness and good nature by the responsibility he shows towards others and by his cooperation during interaction; the constructs Chinese use to perceive and judge others emphasise conscientiousness and good nature, morality, kindness, modesty and reliability; self is played down in favour of the collective; the concern which an individual attaches to the establishment, maintenance and improvement of 'interpersonal relations' is regarded as a desirable trait; an individual's 'face' is connected with others, and the protection and enhancement of face are disciplined by a concern with hierarchical order; and the Chinese regard authoritarianism married with compassion as the defining quality of good leadership. Indeed, so powerful is Confucianism that it may even explain why certain issues have not been analysed sufficiently by Chinese psychologists. Research by the Chinese into 'the self' has been lacking, so Bond and Hwang (1986) maintain, 'perhaps because of the unhealthy alliance between self and individualism in the mind of the Chinese' (p. 236).

Bond and Hwang (1986) make the charge that most of what has been written about Chinese social behaviour 'is episodic and unsystematic, lacking the synthesis and power to predict that are the hallmarks of scientific theorizing' (pp. 216–217). Yet it is precisely their application of the scientific approach to the uncertain world of the humanities with its many different aspects and its complex individuals that produces the weakness in Bond and Hwang's own analysis. Observations are interpreted according to a hypothesis which is taken to be true, and the conclusions reached are the very assumptions with which the analysis began. As part of their scientific approach, comparisons are manipulated to highlight differences and to produce explanations more in keeping with the guiding assumptions.

For example, it is argued that while the Chinese use constructs such as effort, ability, difficulty, luck, stimuli, circumstances and the individual's nature to explain individual behaviour, they also consider the influence of relationships more frequently than do Americans. Given the demands of building relationships in collective cultures, Bond and Hwang maintain, this finding is not surprising. Moreover, in Hong Kong it was found that the Chinese prefer individuals who are self-effacing in their explanations for their own performance, and 'group-serving' in explanations for their group's performance. Yet in their summary of research on 'attribution processes', Bond and Hwang (1986) point out that 'the bulk of research to date has been generally supportive of hypotheses derived from Western models. As researchers move beyond replications and incorporate more social variables into the context of data collection, distinctive cultural processes are certain to be unearthed' (p. 233).

Deterministic cultural traits, it would seem, have been assumed, results interpreted accordingly, and more sensitive techniques, it is already decided, will merely confirm what is already known. But where are the links between Confucianism and the ways in which tested individuals perceived and judged others? It is also worth noting that when observations can be interpreted in line with the guiding assumptions, no doubts are raised about the methodology used. When these observations do not highlight differences between the Chinese and Westerners, the methodological techniques must be improved and widened to unearth the 'cultural processes' which, it is known, exist.

Precisely the same criticisms can be levelled at Bond and Hwang's (1986) interpretation of 'implicit personality traits'. In tests designed to pin-point the constructs used by individuals to assess others, it was found that American subjects were primarily concerned with constructs (words) such as extroversion versus introversion, good nature versus ill temper, conscientiousness versus unreliability, emotional stability versus neuroses, and cultured versus boorish. Chinese subjects, on the other hand, showed disproportionately more concern with constructs (words) such as kindness, modesty and reliability, which fit into the category of 'social morality': these factors 'seem important in maintaining the strength and harmony of interpersonal relationships against the fragmentation of individual abrasiveness and self-seeking . . . these integrative characteristics are of signal importance in highly collectivist cultures' (p. 235).

Again, however, the link with Confucianism is merely assumed. Results have been interpreted according to the hypothesis. If the Chinese had shown disproportionate concern with 'emotional stability' or 'cultured', would this not also fit the hypothesis? Certainly among the British middle class, relationships would be difficult to build without demonstrating both skill and scope in conversation, without the possession of a reasonably broad knowledge of literature, art, music and contemporary issues, and without exhibiting similar values, beliefs, interests and, indeed, a fair degree of emotional restraint and stability. And could not all five categories demonstrated by American subjects also be classified under the heading of 'social morality'? Are not the traits exhibited by Chinese subjects—modesty, kindness, conscientiousness, reliability and emotional stability—those traits which any nursery or primary school, generally regarded as 'good' in Britain, Canada, Australia or the USA, would wish to see in their children? Indeed, as Bond and Hwang show, both Americans and Chinese share a concern with the constructs of 'conscientiousness', 'reliability' and 'good natured'.

Bond and Hwang (1986), it has already been noted, argue that the unhealthy alliance between 'self' and 'individualism' formed within the Chinese mind accounts for the lack of research by Chinese into 'self'. They go on to argue: 'The topic of the self becomes doubly fascinating in a collectivist culture, however, precisely because concern with the self is often derided and played down in favour of group considerations' (p. 236).

Once again the hypothesis has been accepted as fact, and observations are interpreted accordingly. Tests on Hong Kong and Taiwanese Chinese show that they are more likely to endorse 'group-oriented self-concepts' than the Americans: this, it is maintained, is consistent with what one might predict from a knowledge of Chinese collectivism. Indeed, even when the results of the tests are contrary to expectations, observations may still be shaped by the initial hypothesis. In another exercise carried out by Bond and Cheung (1983), it was found that 'the Chinese students in Hong Kong were similar to Americans in the ways they talked about themselves'. This similarity is explained away by the fact that Chinese in Hong Kong 'generally study curriculum materials written in English and by British authors' (Bond and Hwang, 1986, p. 236). It is therefore important, Bond and Hwang insist, 'to replicate this result in Chinese societies less influenced by a Western intellectual output' (p. 236). When the tests and observations do not conform with expectations and predictions, the desire to replicate experiments and to introduce more sensitive and complex social 'variables' becomes far more intense and the influence of 'the West', of 'industrialisation' or of a new 'kind' of Chinese is then brought into the discussion. K.S. Yang (1986), for example, similarly argues that 'modernisation' is moving Chinese students away from authoritarian attitudes and towards a higher level of internal control. It must be asked whether or not these considerations would have been regarded as necessary or pertinent had the outcome of tests been as expected, for when results are as predicted, the fact that the subjects are Hong Kong Chinese ceases to be of any consequence.

Indeed, it is not being too cynical to suggest that when discrepancies arise, the social psychologist's initial assumptions are usually powerful enough to explain observations and descriptions in favour of those assumptions. In a comparison of managerial traits in Canada and Hong Kong, Okechuku and Yee (1991) examine assessments of managers by their supervisors, and the managers' own assessments of themselves:

'A surprising finding was that while overall managerial effectiveness, as judged by the manager's superior, was positively correlated with the manager's scores on supervisory ability, achievement motivation, self-actualisation, self-assurance and decisiveness in the case of Hong Kong, it was not correlated with any of the traits scores in the case of Canada' (p. 234).

A possible explanation for this finding, argue Okechuku and Yee, is that 'Canadian superiors do not understand the innate abilities, personality and motivational characteristics of their subordinates to the same extent as Hong Kong superiors. Perhaps cultural and organisational differences dictate greater arm's length interaction between the Canadian superior and his or her subordinate' (p. 230). However, such material could be interpreted in a number of ways. It could be suggested, for example, that whereas Hong Kong managers appear to demonstrate traits which are usually regarded as North American traits (an emphasis on ability, achievement, self-actualisation, self-assurance and decisiveness),[6] the Canadians appear to be exhibiting traits normally regarded as Chinese— marked 'power-distance' in an organisational hierarchy.[7] Nevertheless, Okechuku and Yee (1991), following Redding (1980), maintain that this result:

'appears to support other studies which reported greater subjectivity in performance evaluation in the typical small Chinese company. Since job descriptions were not typically given and areas of responsibility not clearly defined, there was a lack of objective performance measurement. Rather, evaluation was influenced by personal relations between the superior and the subordinate. . . . The Canadian superior, on the other hand, appears to base his or her evaluation of the subordinates primarily on demonstrated on-the-job abilities' (p. 234).

Schermerhorn and Bond (1991) demonstrate similar flexibility in inter-pretation. Their analysis of Chinese and Americans also seems to show a reversal of expected traits. Although the Hong Kong Chinese were more likely than Americans (who belong to a moderate 'power-distance' cul-ture) to prefer 'the assertiveness tactic of managerial influence' and less likely to ingratiate themselves with their superiors (behaviour which implies an attempt to breach 'power-distance'),[8] the Americans and Hong Kong Chinese shared similar techniques of upward and down-ward influence.

> 'The results showing *direction of influence attempt differences* are consistent with prior research indicating that people are prone to pursue different tactics when attempting to influence superiors and subordinates in a work situation. . . . *In upward-influence situations* both Hong Kong Chinese and American sub-jects preferred rationality and coalition building as managerial tactics. *In downward-influence situations* both Hong Kong Chinese and American subjects preferred ingratiating themselves with superiors, assertiveness, blocking, exchange, upward appeal, and sanctions as managerial tactics' (p. 154, origi-nal italics).

In this instance Schermerhorn and Bond (1991) solve the apparent rever-sal of traits in different situations, and therefore the similarities between the Chinese and Americans, by turning to what is held to be a universal trait associated with hierarchy (claims of a regional culture in this case are obviously impossible). Their explanation of these unexpected find-ings is softened with an appropriate degree of obfuscation:

> 'Such direction effects may be explained in the expectancy-value framework by the fact that superior status in a hierarchy legitimises and increases the likelihood of compliance with certain influence tactics, while subordinate sta-tus does not. Such a "process universal" (Lonner, 1980) may be built into the social logic of hierarchy and thus stand relatively "culture free". As a result, organisational superiors in any culture may be more likely than subordinates to prefer directive, confrontative, authoritative tactics because both the proba-bility of compliance and the capacity to punish non-complying subordinates are relatively higher' (p. 154).

This is yet another soothing argument which relies upon the use of words to build up dream-like impressions which demand little thought. It is also an argument which runs counter to Bond and Hwang's view that Chinese society is strongly hierarchical because it is Confucianist: Schermerhorn and Bond now appear to be suggesting that it is not Confucianism but the logistics of hierarchy that determine managerial behaviour within the Chinese company. Or is it to be assumed that hier-archy may be a consequence of many things (including Confucianism) but that the behaviour it produces is universal? But then surely it is not implicitly being suggested that a Chinese company is hierarchical because of Confucianism? Bond and Schermerhorn are now acknowledging that there is something other than culture at work, and in doing so raise for themselves some awkward questions. In particular, why is no comparison made between different types of institutions? What similarities might be found between a Chinese firm and government departments; between a Chinese firm and a British Army regiment? Wilks (1990), somewhat dubious about anthropological interpretations of culture, suggests a political theory of choice: individual organisations have their own 'culture', and since this 'culture' is manufactured, it can be changed.

Ralston *et al.* (1993b) adopt a line of argument very similar to that of Bond and Schermerhorn in their analysis of the importance of managerial values in decision-making. In this case, the notion that Chinese and Americans are converging is used to explain similarities. Ralston *et al.* begin with all the usual assumptions and reveries of the cultural determinist and social scientist: '. . . it is the antecedent value system of any group that collectively influences behaviour. . . . A group's culture subsequently becomes the dominant variable determining an individual member's value framework' (pp. 22–23). Moreover, the fact that culture influences individual values, and collectively shape the actions and reactions of a social group 'has been determined through research on learning theory and on how value sets are passed from generation to generation through cumulative group biases, beliefs and attitudes' (p. 22).

According to their results, students in both Hong Kong and America seemed to be more willing to use social power and influence to accomplish their objectives. In America, Ralston *et al.*, point out, longitudinal studies on differences between generations would seem to indicate that Americans are becoming more manipulative in their behaviour and 'more concerned with material issues than with developing a meaningful philosophy of life. . . . In Hong Kong, educators also perceive a change in student attitudes. Hong Kong students are apparently becoming more interested in power and influence than in theoretical and social commitment' (p. 31).

The same manipulation of comparisons and of preconditioned interpretations is adopted by Bond and Hwang (1986) in their examination of 'conformity'. The hierarchy of Confucian social structure, they argue, suggests that 'when higher-status Chinese targets are influenced by lower-status sources, they would show less, rather than more, conformity than would be shown by targets from less authoritarian cultures' (p. 254). Moreover,

'Submission to control from a superior would thus be reinforced more strongly in cultures like the Chinese with high power distance. Consequently, there should be greater responsiveness to this status variable among Chinese samples than among samples from cultures characterised by lower power distance. . . . Surprisingly, however, Huang and Harris (1973) found that, contrary to prediction, Chinese in Taiwan were no more sensitive in general to their status manipulation than Americans' (p. 255).

They also found that Chinese were influenced more than Americans by a low-status model. The first of these discrepancies is explained away as follows:

'In fact, a closer inspection of the results would have shown that their conclusions held only when the model explicitly disclaimed any competence about the target issue. When the model claimed competence about this target issue, the predicted interaction between culture and relative status was shown in the means, although not at significant levels' (Bond and Hwang, 1986, p. 255).

The second of the discrepancies is explained by invoking another strand of Confucianism:

'In public situations involving strangers thrown together, we believe that the core concern of the Chinese to avoid conflict becomes a prepotent issue. As the interactants have no probable future together, the agenda becomes one of avoiding potentially disruptive differences. Imitation of the other's responses is the obvious solution. . . .

We believe this desire to avoid open conflict lies behind the results of studies comparing the response of Chinese and Americans to peer-group pressure. In these situations, the status variable is not engaged. Instead, the collectivist concern of the Chinese to avoid interpersonal disharmony . . . becomes salient' (Bond and Hwang, 1986, p. 256).

In their analysis of 'interpersonal attraction' Bond and Hwang (1986) simply marginalise similarities and emphasise differences in order to support the presumed links between their observations and their preferred hypothesis. Although it is generally found that 'most models of interpersonal attraction proposed by Western psychologists are also applicable to Chinese people' (p. 241), Bond and Hwang stress that:

'. . . the results of research [conducted in Taipei] studying the relationship between affiliation tendency and interpersonal attraction in Chinese society contradict those found in the West . . . affiliation tendency was positively and linearly related to sociometric status as perceived by others (interpersonal attraction) for boys but not for girls. For both sexes affiliation tendency was positively related to self-perceived sociometric status and to self-evaluation. The results are very different from findings of research conducted in the West. The latter generally revealed a significant negative relationship between the affiliative motive and interpersonal attraction. . . .

The difference between East and West can be interpreted in terms of their divergence on cultural values surrounding affiliation. In Chinese culture, where collectivism and interpersonal "dependency" are highly valued, an individual's concerns about establishing, maintaining, and improving inter-personal relationships can be viewed as desirable traits, which may make him attractive to others or to himself. On the contrary, in Western culture, which sets high values on individualism and independence, a strong affiliative tendency can be perceived as a weakness which consequently decreases an individual's interpersonal attractiveness' (Bond and Hwang, 1986, p. 241).

The assumption of links between observed behaviour and Confucianism, and the selective manipulation of comparisons, are criti-cisms which are equally applicable to analyses of other categories of Chinese social behaviour. In their analysis of 'face' for example, Bond and Hwang recognise that 'face' is not peculiar to the Chinese; and yet, it is argued, within Chinese culture—with its hierarchical social struc-ture—face takes on a peculiar 'Chineseness': it becomes an expression of concern with one's own place within that hierarchy. For the social scien-tist this interpretation is fact: the world is as I see it. In their analysis of leadership, Bond and Hwang are content to assert that Chinese leader-ship is authoritarian: they make no attempt to marshal an explanation. Their interpretation relies upon the assumptions with which the authors began: 'It is generally assumed that the "typical" Chinese concepts of leadership evolved mainly from the traditional ways by which the head of a family manages his household' (Bond and Hwang, 1986, p. 252).

These analyses by social psychologists amount to little more than inter-pretations of behaviour—interpretations founded upon the very assumptions with which the study in any event set out to prove. Interpretation is imposed upon description, and yet the whole exercise is presented as scientific explanation. Depending upon the internal logic of the arguments (a logic dictated by the initial assumptions) and upon the nature of potentially awkward evidence which cannot be left unmen-tioned, differences are emphasised, and similarities between 'the Chinese' and the rest of humanity are either ignored, marginalised, transformed into a regional culture or designated as 'culture-free' behav-iour. If variations appear, or if observations do not conform with those expected or demanded by the initial assumptions from which analyses and arguments proceed, then declarations are issued: the subjects were influenced by Western culture, Western books, industrialisation and lan-guage; the model is faulty; the researchers concerned made a mistake; or other elements of Confucianism have interfered with the results of the tests.

Many social psychologists seem unwilling to consider the extraordinarily complex mosaic of aspects which actions, beliefs, values, institutions and forms of exchange create; the spontaneity of the human mind and its multi-intelligence; the desires and choices of individuals; the changes which occur; the different and changing contexts the shifting aspects create; and the conflicts and tensions which exist within society and within the individual. All these are as nothing. These social psychologists, for the sake of their science, are determined to present chains of cause and effect in a unidimensional world of black and white. It is a much easier world to interpret; it is a world in which analysis and the expression and communication of ideas are less taxing; and it is a world of comforting certainty. But if the reader does not share their faith, then the point of the psychologists' tests is left in doubt. It is no longer clear what the tests are supposed to reveal; nor is it clear by what means and according to what beliefs observations should be interpreted; nor is it clear upon what grounds scientific generalisations are so easily made.

Conclusions

It is interesting to note that many of those writers who support cultural explanations are careful to stress the influence of non-cultural 'variables' and to explicitly distance themselves from cultural determinism and from a mechanistic scientific approach. Of Hicks and Redding (1983), Mackie (1988) comments that they 'have given particular attention to the interplay of cultural and other factors relevant to "the sources of East Asian economic growth", while avoiding the temptation to assign too much causal determination to culture itself' (p. 253). Mackie (1988) also praises L. Lim for presenting important correctives to cultural determinism (p. 231).

Redding (1990) argues that the word 'cause' raises fundamental problems in attempts to explain economic success:

'Social science is not physics, and even so, physics is not what it was in the simple days of cause-and-effect. What the laymen thinks of as cause-and-effect must give way to a whole complex of chains of factors, each one, if it is separated out for analysis, being determined by others, and being influenced in return by its results' (p. 6).

Redding illustrates this argument with the notion of family ownership of business which, he suggests, results from six factors: the Confucian political philosophy of designing a state based on the stable family; a set of values instilled in consequence over centuries of socialisation; facts of Chinese history which have made it sensible for the Chinese family to act as the main source of identity and succour; psychological characteristics of dependence and loyalty; patrilineal and patrilocal kinship structures; and more recent politico-economic environments which have encouraged family business units. The Chinese family ownership of business is itself a determinant of high levels of identity, loyalty and commitment in key organisational roles; paternalistic organisational climates and organisational dependence; a narrowly defined set of stakeholders in the organisation; nepotism, together with its accompanying tension; and common economic outcomes of increasing family wealth. These results, in turn, affect Chinese family ownership of business. More specifically, these results tend to reinforce the established pattern by releasing motivation in key organisational areas and thereby promoting efficient performance; and by perpetuating the welfare of key stakeholders, both practically and psychologically.

> 'To understand fully this stream of connections would also require some incorporation of ideas from what might be seen as parallel streams of explanation; looking at, for instance, political factors, institutional structures, social psychology, history, economics, etc.
>
> It is necessary then to abandon any notion of "mechanical" cause, that one thing results from another in a simple linear way like one billiard ball pushing another. Instead, we are faced with a vast net of connected elements, none of which is especially dominant, and all of which have somehow to be acknowledged, even though they may be only seen with peripheral vision' (Redding, 1990, p.7).

Redding argues that there is no single chain of cause and effect—there are several. But this is billiard-ball physics writ large. The factors, results and interactions which Redding believes explain the Chinese family company contain all the hallmarks of the scientific approach: they are expedient devices that enable the construction of scientific explanations. The challenge which Redding presents himself with is akin to untangling the intricate pathways of a well-tried chemical experiment, the outcome and cause of which is already known. For Redding, as for many other writers on the Chinese company, the idol of origins in his world is Confucianism. For these writers Confucianism is not a crafted, manipulatory doctrine. Rather, Confucianism is a spirit philosophy which arose out of a people, surrounded them, and over the millennia, by the grace of socialisation, infused their psyche and permeated their social 'structures', and now determines their behaviour, actions, desires, choices, values and beliefs:

> 'For there to be an affinity between business behaviour and the spirit which infuses it, assuming consistency of behaviour requires also consistency of the notionally independent spirit. It will be argued that Overseas Chinese businessmen think sufficiently alike, and differently from others, for such consistency to be proposed. Internal consistency is a matter of the parts of the belief system making sense as contributing components to a sensible overall framework. This will be argued to be the case, and will strengthen the case for the distinctiveness of the belief system. The role of Confucianism as the bedrock set of beliefs is crucial in this regard' (Redding, 1990, p. 11).

Like so many social scientists, Redding practises his mechanistic science while explicitly rejecting it. What, then, is the nature of their faith in the scientific approach?

Despite the social scientists' claims to the contrary, the crucial importance of culture in so many explanations of Overseas Chinese economic success cannot be hidden. Other factors may be considered, but they are taken to be marginal influences on the details of the final outcome: culture is the prime cause and the key variable to be analysed. Moreover, there is obvious sensitivity to alternative non-cultural explanations, as Mackie's (1992b) criticisms of L. Lim (1983) demonstrate.

Mackie (1992b) comments: 'Lim says little on the question of Chinese entrepreneurial skills (she seems to be striving hard to reject any stress on ethnic distinctions)' (p. 51), but the questions she addresses are relevant to the matter of Chinese entrepreneurial skills 'insofar as they throw light on the economic roles of the Chinese' (p. 51). Why do Chinese firms predominate in certain occupations? How do they relate to indigenous businessmen? How are Chinese communities divided within themselves? And have ethnic differences in economic activities been increasing?

Mackie continues:

'Lim's answers to the first question do not add much to what has frequently been said before. Kin and social networks give Chinese one important advantage over other ethnic groups: access to labour, credit, information, market outlets and security. But "while the form of these networks may be distinctively Chinese, their function is not", Lim observes; and at the highest levels of business they have to compete with other forms of interlocking business and social interests. . . .

Lim's most interesting and challenging ideas emerge in answer to her final question: Have ethnic differences in business activity been widening or narrowing? Here she argues strongly for what might be called "the convergence thesis"—that as industrialisation and modernisation proceed, traditional cultural traits become less significant, both among factory workers and their managers and bosses, be they Chinese, Malay or any other. Chinese culture is highly adaptive, she asserts, and may be de-emphasised where necessary or retained where useful. . . .

A case for this view can no doubt be made, but it remains to be seen whether ethnicity will become entirely irrelevant as development occurs. It has not yet become so in the USA, or Europe, and there is a long way to go before it will do so in Southeast Asia, even though the overlap between ethnic differences and economic roles may gradually become blurred' (Mackie, 1992b, pp. 51–52).[9]

Mackie has misrepresented Lim, for Lim's statement that 'while the form of these networks may be distinctively Chinese, their function is not' is taken out of a much wider context:

'Chinese kin and social networks clearly give them *one* important economic advantage over other ethnic groups, in access to labour, credit, information, market outlets and security. But while the form of these networks may be distinctively Chinese their functions are not. These are replicated by modern personalistic networks which are familiar even in the Western corporate business world which has long recognised the value of "the old school tie", the "old boy network", and the "mentor" system. Where personalistic networks do not exist, they are created, for example, through the establishment of service organisations of business and professional people such as the Rotary, Lions and Kiwanis Clubs . . . ' (Lim, 1983, pp. 4–5).

Moreover, Lim (1983) goes on to argue that ethnic business traditions may be preserved only so long as they help or at least are not detrimental to business success. As she later notes,

'the flourishing or disintegration of traditional Chinese cultural practices does not seem to be correlated with economic growth or decline. Indeed, far from being destroyed by economic decline, Chinese ethnic identity may well be reinforced by it. . . .

If any generalisation is to be drawn from these diverse studies about the relationship of economic activity to ethnicity—and it is arguable that it should not be—it must be this. The culture (broadly defined) of Chinese in Southeast Asia is pragmatic, flexible, adaptive, dynamic and evolutionary. It is highly responsive to the economic environment, and particularly to market forces. Where it is economically useful to emphasise Chinese ethnic identification, cultural characteristics, and social organisation for business purposes, this is done. . . . Where Chinese economic activity requires close interaction with indigenous populations, "Chineseness" may be selectively de-emphasised and indigenous cultural and social elements incorporated . . . ' (Lim, 1983, pp. 19–20).

Arguably, Lim has been misrepresented because she notes comparisons and behaviour which bring into question the cultural explanation. Lim is not arguing for convergence, but that 'culture' is more an expression of pragmatic manipulation than a deterministic spirit philosophy.

In an earlier work, Mackie (1988), a strong advocate of the skilled manipulation of separate variables and their 'systematic' analysis, is a little more strident in his criticism of Lim:

'I do not find her theoretical framework entirely convincing. . . . Lim seems to be trying to deny or play down *the saliency of ethnic elements as a key to economic behaviour or success*. She attributes the latter far more to historical circumstances of time and place. "Chinese social organisation is neither necessary nor sufficient as an explanation of the economic dominance or monopoly (by the Chinese) of a particular line of business in Southeast Asia" (Lim, 1983, pp. 5–6). This is overstating the case, in my view; what we need to understand better is how far (and how) these social and cultural characteristics come into the picture. Lim's approach is another useful corrective to excessively "cultural" explanations of Chinese business success, but she has carried an almost economic determinist type of analysis to the extreme of repudiating socio-cultural factors altogether. She seems to regard them as little more than unfortunate, undesirable legacies . . . ' (Mackie, 1988, p. 231, italics added).

Mackie, it would seem, finds it difficult to entertain Lim's views because she does not begin with the assumption that culture is a key to economic behaviour or success; and because she does not begin with this assumption, her views are not rigorously 'scientific'. In the context of the lengthy quotes from Lim given above, Mackie's criticisms, especially with regard to accusations of determinism, are (to mix metaphors) a case of the pot building a straw kettle and calling it black. Again, Lim's comments have been taken out of a broader context which clearly raises pertinent questions about the credence which Mackie and others give to the cultural explanation:

'. . . while successful businessmen tend to stick to Chinese kinsmen and fictive kinsmen as business associates early in their careers, as they become more successful they include more and more "outsiders", Filipino fictive kinsmen and friends, and even foreigners. . . .

. . . the fact that successful businessmen have more links with both insiders (kinsmen) and outsiders suggests that these links reflect rather than determine their business success and class position: those who have more kinsmen to resort to for capital are more likely to be from rich and established business families, which also gives them more links to the outside world. It is the unsuccessful businessmen, lacking such fortuitous personal contacts, who must resort to Chinese lineage and home-village associations, and to outsiders early in their career, in the attempt to forge such links.

In other words, Chinese social organisation is neither necessary nor sufficient as an explanation of Chinese economic dominance or monopoly of particular lines of business in Southeast Asia. It simply defines the form in which Chinese businessmen acquire some working capital and reduce the risks and costs of credit transactions. These benefits can be and are obtained through other social networks involving non-Chinese as well . . .' (Lim, 1983, pp. 5–6).

*   *   *   *   *

It may be suggested that the defensiveness and sensitivity of proponents of the cultural explanation find expression as concern over charges of determinism, the morality of manipulation, and the intellectual strength of 'utilitarian reductionism', and as attempts to devise a form of words which explicitly denies the existence of natural forces, while simultaneously allowing presumed forces to operate within theory.

In these pages criticism has been directed at the methodological trinity of the scientific approach to the study of the Chinese company—a trinity which comprises: the identification of 'variables' which are regarded or treated as intimately related but nevertheless distinct entities separate from each other and from men; a belief in the existence of cultural links and a faith in the search for the idol of origins; and the comparison of these variables in order to identify and isolate uniqueness (which may also necessitate the creation of a universal or regional culture). More specifically, criticism has been directed at the view (or philosophy) which implicitly or explicitly maintains that reciprocity exists as a distinct entity separate from man, and that if a particular group of individuals possess certain traits then it is likely or inevitable that those individuals will engage in reciprocity. The intellectual separation of exchange from these traits (including values, beliefs, ideas and behaviour), and from men, allows these particular traits and this specific form of exchange to define 'the Chinese'. For if traits, exchange and individuals can be separated, then the possibility that individual actions are determined by these traits can be invoked quietly and the assumption of a chain of cause and effect can be established. In other words, 'culture' comprises chains of variables which can be used to construct scientific explanations of human society.

The logic of the scientific approach and the willingness of many social scientists to let intellectual devices—which, it is frequently claimed, are merely technical expedients—separate what are held to be intimately connected parts of a presumed totality, lead to the creation of distinct variables which 'interact' and together influence man. Mere words become distinct variables which are thought of and treated as distinct entities. According to current fashion, this endows analysis, in its initial stages, with neatness, clarity and scholarly rectitude; and it adds further strands and weight to the notion of 'the Chinese'. Once the variables and the assumptions of links between them are established, the logic of the approach then requires the writer to trace pathways of cause and effect into the gyres of a search for the idol of origins.

For many, the logic of the scientific approach cannot seriously be questioned, for it is more than a technique applied to the study of the Chinese—it is a fundamental philosophy. The cultural mantle of 'the Chinese' which has been built up gradually is too useful to be discarded: it is a convenient set of given 'facts', subvariables and subsystems which serve as a reliable base for the examination of other phenomena, and from which to issue advice and immediately communicable and acceptable analyses. 'Culture' becomes a particularly useful intellectual construct when it is defined by its ability to determine or condition behaviour. Following Kroebner and Parsons (1958), Redding and Wong (1986) argue that the notion of culture can be limited 'to the transmitted and created content of values, ideas, and other symbolic meaningful systems which are factors in the shaping of "human behaviour"' (p. 271). The term 'social system', they suggest, can be used to designate 'the specifically relational system of interaction among individuals and collectivities' (Kroebner and

Parsons, 1958, p. 583). This distinction between 'culture' and 'social system', argue Redding and Wong, 'allows for the recognition that social actions and relationships can be conditioned by "culture", the latter being the set of normative and preferential conditions for action, as well as by non-cultural variables such as technology and economic factors' (1986, p. 271). Harrell (1982) presents culture as one of two possible absolutes. Culture is either purely derivative of something and can therefore be explained. Or it possesses a life of its own. Whichever is the case, Harrell believes, energies should be focused on how culture may be used to explain observed phenomena. The observer has no choice because the scientific approach has determined that this is so. Hofstede (1980) defines culture with a certainty which is even more stark, for while Hofstede believes that no two individuals are programmed in exactly the same way, and that the link between individuals, collectives and universal programmes cannot be delimited with ease, culture is nothing less than 'the collective programming of the mind which distinguishes the members of one human group from another' (p. 21). Huo and Steers (1993) in their examination of cultural influences upon the design of incentive systems similarly argue that culture predisposes individuals and organisations to see problems in a certain way: 'Simply put, different cultures see problems (and solutions) quite differently, and these differences affect subsequent behaviour' (p. 73).

Explanations of 'Chinese' social psychology and, indeed, the Chinese company, are generally founded upon three beliefs. The first, that observed behaviour of selected individuals may in each instance be extrapolated into extensive generalisations for all Chinese, is itself founded upon the second belief that an individual's behaviour is rooted in Confucianism, and upon the third belief that after centuries of socialisation, most Chinese behave in the same or similar way. These assumptions depend for their empirical evidence upon subjective assessments of observed behaviour of individuals in particular circumstances—assessments which are interpreted according to the initial assumptions. Thus, observations are taken as factual evidence because they are interpreted according to the assumptions which this evidence now shows to be valid. This circle of mutually reinforcing assumptions, strengthened by its manufactured evidence, is now made impenetrable by defining culture as values, ideas and other symbolic and 'meaningful systems' which determine or condition human behaviour. Around and around the gyres of his imagination runs the social scientist: the world *is* as I see it.

Notes

1. See for example Mackie (1988) and Robison (1986) on the *zhugong*; D.E. Willmott (1960) on the *langganan* system; M.H. Lim (1981, 1983); and Skinner (1957).

2. This policy discriminated in favour of Malays until they owned 30% of the nation's corporate wealth. Between 1970 and 1985, Malay ownership of corporate wealth grew from 2.4 to 17.8%; and foreign ownership fell from 63 to 25.5%; and non-Malay (largely Chinese) ownership rose from 34.3 to 56.7%. Not surprisingly, Chinese businessmen disputed these figures, arguing that certain statistical categories such as locally owned subsidiaries of foreign-based corporations (in which Malay ownership is considerable) have been included in estimates of non-Malay ownership, while Malay ownership of corporate plantations has been excluded from estimates of the Malay share of corporate wealth.

3. L. Lim herself does not attempt to create a model or typical company. Indeed, Lim is somewhat unusual in that she is very suspicious of the emphasis which many writers on the Overseas Chinese give to culture in their explanations of Chinese economic and social institutions and of Chinese economic success. Chinese companies, Lim argues, lose whatever ethnicity they may have had when they enter modern technological industries oriented to the world market. It is because of her suspicions that her views have been subject to a number of strictures (see the Conclusion of this chapter). Lim's arguments might also suggest comparisons with the decline of family businesses (which tended to dissipate when heirs took over) and the rise of more complex business organisations in England from the end of the seventeenth century (see Grassby, 1970).

4. A similar logic has been applied to the analysis of Koreans (see Chang Yunshik, 1991).

5. A point which takes us back to the apparent confusion over whether Chinese operate according to 'internalised' or 'externalised' values.

6. Ralston *et al.* (1993a), for example, argue that Americans believe that strategies which make oneself stand out as an excellent performer and therefore visible to the Boss is the way to progress. The Hong Kong Chinese, on the other hand, see low-key approaches which make the Boss look good as the path to success. Thus, 'An American manager in Hong Kong might misread the true capabilities and motivations of his or her staff, believing them to be passive and unmotivated . . . a Chinese manager in America might see his or her employees as too self-centred, outwardly aggressive, and with too little loyalty' (p. 171).

7. Following Hofstede (1980), Ralston *et al.* (1993a) in fact begin their analysis with the assumption that 'The United States is a prime example

of an individualistic culture, while Hong Kong is much more collective, as are most Asian cultures' (p. 158), and possess a 'high power-distance'.
8. A finding which appears to be contradicted by Ralston *et al.* (1993a, b).
9. See, also, Mackie (1988), especially pp. 250–251.

5

The Chinese association

Introduction: the association and its aspects

It has been argued in these pages that institutions should not be considered or treated as discrete phenomena with specific 'roles' and 'functions'. Rather, they should be understood as expressions of the actions of individuals who accept intact, or accommodate, alter, adapt or reject, or who generate and evolve their own, values, beliefs and objectives. Change within institutions and in their external relationships, then, is inherent. Change may occur slowly and pass almost unnoticed, or it may be accelerated by external circumstances and chance events or by a will imposed. It is for all these reasons that institutions, which may appear to form constant and stable 'structures', are, on closer inspection, better understood to comprise many different aspects. Apparent order and pattern may alter or dissolve into chaos over time, or as the observer alters his own values, beliefs, desires and objectives, and so changes his own perspective.

It is with these ideas in mind that the Overseas Chinese associations are now considered. It is suggested that the associations have been stimulated partly by friction among different Chinese groups, and between these groups and the indigenes and colonial authorities. As a symbol and focus of loyalty and mutual protection, the associations institutionalised those conditions which, in some instances, may have provided the initial stimulus and opportunity to trade. But associations are best understood as organisations which have arisen in response to the Chinese immigrants' desire and perceived need to protect, legitimise and extend trade in alien territories that are frequently hostile.

These qualities are most clearly expressed by the business association which, in addition to those actions (such as disseminating market information, raising capital, arranging credit, and organising group purchases and boycotts) which directly facilitated trade, accumulated a bevy of other responsibilities, including the provision of welfare for its members and liaising with government. Trade's institutionalisation in the form of these associations is perhaps most strikingly illustrated by those trade associations dominated by a particular dialect group, and by dialect associations which represented a dialect group dominant in a particular line of business.

Yet it must also be recognised that the formation of associations has often been a response to considerations of defence, cooperation, the establishment of law and order, welfare, education, health, and communications with government and Mainland China—matters in which trade is of little or no concern. This is as true of a single association as it is of any collection of associations. Even many of the business associations have most regard for their role as benevolent societies or protection agencies. The importance of these considerations is clearly evident in the wide range of activities in which associations are engaged: the provision of scholarships and legal services, and of housing, employment and financial relief for the needy; liaison with government officials, legislative bodies and labour organisations; the mediation of disputes within the Chinese community; the maintenance of schools and hospitals; the payment, and arrangements, for a funeral cortège; the organisation of immigration; the repatriation of the elderly to China; and, in some instances, the establishment of political movements and parties. How, then, may these different aspects—often exhibited by a single association—be reconciled with the argument that associations may also be a response to the need to protect, legitimise, extend and institutionalise trade, without resorting to an understanding of trade's extension and institutionalisation so broad that the phrase becomes almost meaningless, and without regressing to laws which demand that all activities are motivated and directed towards trade?

Although friction, trade, the immigrants' perceptions of their own needs, and the uncertain nature of the territories in which the Overseas Chinese had chosen to reside may have stimulated the appearance of associations, these stimuli did not cause or determine the appearance of associations, nor do they account for the associations' multiplicity of aspects. Except in the instance of a colonial or indigenous government—as in the Philippines, Vietnam and Singapore—to establish associations in order to facilitate the state's control over the Chinese, it is perhaps impossible to give any precise and definite reason why the Overseas Chinese should have reacted to their circumstances, desires and needs by creating associations. It is true that similar institutions were formed in China and that these institutions, like the associations overseas, were organised according to locality, descent and language—a characteristic which has been taken by many writers to indicate a shared origin and the existence of causal links between the lineage, the associations found in China, and the association overseas. For example, many of the 'hometown' associations examined by See (1981) in the Philippines are not miniature clans, but true lineages with definite agnatic links among their members:

> 'Each of them knows the *fang* or family branch he belongs to, and he often also bears the genealogical name . . . to indicate his kinship rank. When a village organisation is wealthy enough to build its own clubhouse with an ancestral hall to worship its founding ancestor, the agnatic corporate descent group in rural China is virtually transferred to the overseas community' (p. 231).

Indeed, there are many writers who maintain that the origin of the association is rooted in certain traits and principles of behaviour which are peculiarly 'Chinese'. This is an argument considered in more detail later in this chapter. Certainly there can be little doubt that the Chinese immigrants brought with them ideas and practices from China: credit associations, burial associations, guilds, secret societies and lineage organisations were all 'transferred' from south China (Gosling, 1983). But it cannot be assumed that an institution or value possesses only one aspect, and, therefore, that their appearance elsewhere indicates a shared origin, or the existence of causal links, or their 'transfer'. The rotating credit associations of China, for example, were also indigenous in Japan, India and West Africa (from where they spread via the slave trade to the Caribbean and possibly to the southern United States), and among the Malay groups in Southeast Asia, where they operated, initially at least, more as social institutions and as ritualistic symbols of solidarity (Geertz, 1962). Other examples are locality consciousness, segmentary organisation and organisation according to descent: none of these phenomena have ever been the exclusive preserve of the Chinese. Moreover, ideas and practices 'transferred' from south China may themselves have been a response to similar conditions in China. For example, it may be suggested that the eighteenth and early nineteenth century financial guilds of Ningpo (Jones, 1972, 1974), which made use of occupational and *tongxiang* (or, very loosely, hometown) ties, were a logical response in a state where the attitude of government towards trade and merchants was ambiguous, and, consequently, where the absence of clear commercial regulations, and an inability to enforce those which did exist, meant competition was effectively lawless.

If ideas and practices were borrowed, then either it was because the challenges and difficulties which the Overseas Chinese had to contend with in Southeast Asia were in many respects similar to those which they had faced in China, or it was because those ideas and practices were useful in different circumstances or could be adapted to suit new conditions. The associations, then, were and remain a response to the desires, needs and ambitions of the Chinese immigrants and to circumstances rather than a manifestation of underlying 'Chinese' principles of social organisation. This argument is supported by a number of observations.

Firstly, not all the Overseas Chinese wished to participate in trade so extensively or single-mindedly. The *peranakan* and *babas* adopted (or retained) values closer to those of the indigenes; and the rural Hakka communities of Sarawak were based on families linked to each other through networks of clan and affinal ties. Trade links were confined largely to the outside world.

Secondly, not all the Overseas Chinese formed associations, nor do they all—not even the majority—join associations. After all, if the most important objective of an individual is economic success, the association cannot guarantee that success. It cannot impart reliability, judgement, drive, the ability or luck to arrive at the right decisions at the right time, nor effective means of organisation and operation. It was noted in Chapter 4 that in Malaysia under the New Economic Policy, the narrowly focused Chinese institutions like the Chinese associations and the chambers of commerce have not fared particularly well; rather it has been the individuals and their companies, determined to forge links with the Malay political elites and to look for opportunities overseas which have survived and prospered (Heng, 1992). As Lim (1983) points out, it may be the weaker members of the business community who attempt to use membership of an association to bolster their contacts, status and credit standing. A similar view is held by some businessmen in Hong Kong. Membership of associations or clubs is regarded as a sign of, rather than a means of, success: to join an association or club with the intention of extending business links implies an inability to establish strong enough connections and to build oneself a solid reputation. Unless a club or an association boasts a particularly good restaurant or golf course, other businessmen will have little or nothing to do with such organisations for they are perceived as little more than rather tiresome forums for those individuals whose first and last intention is to expand their own social and political status, and who, to that end, use all kinds of pretexts to gather round themselves a string of acolytes. As Freedman (1957) observed of the clan associations in Singapore:

' . . . there is a world of difference between those who actively participate in association affairs and those who remain passively on its rolls. The difference is essentially that between the status seekers and the ordinary members. . . . To the status seekers . . . the clan association, in common with the whole reticulation of voluntary bodies, is an apparatus for the advancement of his position' (p. 94).

Thirdly, the putative sociological principles upon which the associations are said to be based are quite consciously manipulated, or even fabricated. A useful rationale for extending membership and for cooperation is provided by a common genealogical or geographical origin, and the promotion of fictive kinship to the same status as cosanguineal ties. This rationale was reinforced by the logistics of migration which, although it often followed social channels, largely comprised males from many different parts of south and southeast China, and which required effective organisation, not least because of the demands of the indigenes or colonial authorities. If there was no formal or factual genealogy to be found, then one could be manufactured which, depending on whether or not the purpose was to create an agnatic group to incorporate property, could stress lines of descent or honorific materials (Meskill, 1970; See, 1981; J. Hsieh, 1977; and see Chapter 6).

Fourthly, and most importantly, associations are not peculiar to the Chinese. The significance of this observation is pursued later in this chapter: associations, both religious and economic, were, for example, established by the immigrant and slave communities in ancient Rome; they were closely bound up with the extension and institutionalisation of trade in parts of Africa during the eighteenth century; they are created by the latter-day Creoles of Sierra Leone, West Africa, and by elite groups in Britain.

The associations of the Overseas Chinese in Southeast Asia, then, may be understood as a pragmatic response derived from the multiple-intelligence and spontaneity of the human mind. It is precisely for these reasons that the associations possess many different aspects. On the one hand, the motivation behind the establishment of associations is the deliberate extension and institutionalisation of trade. This is true even of those associations publicly concerned first and foremost with the provision of welfare. In Vietnam, for example, businessmen openly conceded that the main purpose of the various charitable and welfare associations—including hospitals, school committees, cultural societies and surname associations—was 'to promote the business fortunes of their members by facilitating communications and promoting mutual trust and cooperation' (Barton, 1983, p. 57). The fabrication or manipulation of religious beliefs and practices in order to support trade is illustrated by Tan's (1985) study of the *dejiao* associations in Malaysia and Singapore.

These associations were established by Teochow businessmen, and the spread of these associations coincides with the activities of these business-men. While genuine religious piety (or firm moral attitudes) should not be discounted entirely as a motivation, 'we should note that the formation of the *dejiao* association is advantageous to the establishment of business links and creates a meeting ground for common capitalist interests. Once an association is established, the business connections further help to make the association successful and prosperous' (pp. 19–20). Although religious in sentiment, the merchants employ no priest nor any full-time religious specialist; the associations are conveniently syncretic; planchette divination is used to legitimise decisions, particularly those concerning the establishment of new branches of the association; and the associations' most important activities—largely charitable—are reinforced by a range of myths, legends and moral virtues. The association, then, also serves to legitimise material success.

On the other hand, deliberate manipulation does not exclude the possibili-ty that in many instances the conduct of trade and the various other dif-ferent aspects of the Overseas Chinese association exist without mutual concern. But if individuals desire or feel constrained to engage in trade, this desire or necessity may be permitted to some extent to condition their other activities. Even if it is accepted that an association is established or operated by individuals motivated by pure altruism, the context created by trade endows the various aspects of the association with a different sig-nificance, and this may require adjustments in behaviour. Those individu-als who may establish and comprise the association, even if motivated by a wish to provide welfare, would be engaged in trade or would have to take into consideration the realities imposed by this desire or necessity for trade. The associations need financing; the men with prestige are mer-chants; the provision of welfare is dependent upon these merchants (a fact which in itself acts to legitimise trade); and these individuals who influence the provision of welfare would (even if harbouring no ulterior motive) justify and legitimise their success by their very actions. Even if born out of genuine altruism, the association would have to elicit support from the materially successful and, therefore, by definition, the influential. Deliberately or unintentionally, associations have grown up around trade and trade's underlying wealth values. The two worked well together, for as Cheng (1985) remarks of Raffles's policy of divide and rule in Singapore,

'. . . as the various dialect groups settled in different quarters [Hodder, B.W., 1953], the provision of social, religious and economic services and political con-trol within each dialect group could only be met by those who were economi-cally strong and public spirited. This laid the conditions for the emergence of leading merchants as dialect leaders who could through the formation of asso-ciations cater for the needs of their fellow dialect members' (pp. 29–30).

These other aspects of the association, then, may institutionalise trade unintentionally, or may be 'turned' deliberately, but only subsequently, towards the extension and institutionalisation of trade. In that an association cannot survive without financial support or without the support of the influential, the deliberate or unintentional symbiosis of the different aspects of the association constitutes a response to the desire or need for trade. These different and seemingly contradictory aspects of the association are perhaps easier to reconcile if the association is understood as a nexus of reciprocity, and as a symbol of elite designation.

The association as a nexus of reciprocity

It may be suggested that associations are merely formalised expressions of reciprocal groupings, although in some instances associations may be established precisely in order to encourage the formation of such groupings (Amyot, 1973; Cheng, 1985; Lim, 1983). In either case, the association and its different aspects are expressions of the various changing desires, needs and goals of individuals who engage in and construct networks of reciprocal exchange, either as an end in itself or as a means to realise other goals, including the prosecution of trade.

It was noted earlier that membership is often ascribed by social connections comprising individuals who, if they choose, may ignore the more commonly manipulated shared attributes such as descent, place of origin and language, or find some kind of symbolic justification for the inclusion of other individuals who do not share the same surname or come from the same part of China—the element of trust being 'more essential than . . . surname, dialect or a common place of origin in China' (Gosling, 1983, p. 152). This may also have been true in China. Fried (1953) observes that some members of the Commercial Guild in Chuhsien were not related to each other, nor did they come from the same part of China. Moreover, individual relationships within this organisation and with guilds in other places were cemented individually on the basis of *kan-ch'ing* (sentiment or affection) and friendship. Indeed, the merchant 'who is looking for a scarce commodity may be helped by the guild system but this is hardly likely, unless, in addition to his guild membership, he has ties of kinship, friendship or *kan-ch'ing* with persons in other guilds or he works through an intermediary' (p. 151).

It was also mentioned earlier in this chapter that it may be the weaker merchants who join associations with the hope of expanding their existing trade networks. Merchants would undoubtedly be brought into contact with others whom they might not otherwise meet. But connections do not exist as independent entities like telephone lines; they are not a cause of success. They are a product of choices, decisions, desires and the careful handling of other individuals. They have to be constructed and maintained. Clansmen are fiercely competitive; complex business networks and networks of reciprocity exist irrespective of the associations; and leaders of the associations may be selected partly because of their material wealth and their connections. These particular attributes not only convey prestige and demonstrate an ability to handle people and money—they also possess obvious practical advantages. Moreover, dependence upon his connections and financial strength is the leaders' key to the manipulation and control of other members in the association. Constitutional functions are often vague, records may not be kept, and it is expected that the leaders will use their post to their own personal advantage. The rank and file are mostly interested in seeking practical help of one sort or another, or are simply trying to avoid loneliness; a few may be acolytes and status-seekers. For all intents and purposes the leaders are the associations.

Holding an office enables the merchant to advertise his material wealth (and, therefore, his achievements) by actively participating in, or at least contributing to, philanthropic causes, or by pursuing objectives apparently quite selflessly on the behalf of others. This is a manner of exhibitionism which simultaneously allows the merchant to legitimise his pursuit of profit and his material success, and to open up new opportunities for the expansion of his business connections. Indeed, such is the importance of obtaining an office that associations may be riddled with many honorary posts (Freedman, 1957).

The merchant not only advertises and legitimises his success by ostentatiously parading his philanthropy and selflessness—vigorous displays which are competitively ranked (D.E. Willmott, 1960)—but perhaps also plays to those individuals (including other merchants) who place greater esteem on the manipulation of social relations as an end in itself. The selection of leaders is not always entirely nor even largely determined by material wealth. Prestige may also be derived from age, scholarship, natural leadership and proven commitment to community affairs. In some cases leaders with material wealth are drafted purely to serve as executive members: they make exemplary and well-trumpeted contributions to the association's coffers and good causes, but otherwise they take little part in its affairs. Hometown associations occasionally serve as a political base for clan leaders and, as See (1981) notes, some of these organisations were formed precisely for this purpose. Leaders often manage to hold posts in more than one association, thereby forming 'interlocking directorates'. On the one hand, this enables a degree of cohesion among associations which are in any event intensely paternalistic and personalistic and, for this reason, liable to disintegrate or fragment. On the other hand, these interlocking directorates further add to the complexity of divisions and the undercurrents of intrigue.

The association, then, is based upon shifting tides of reciprocity and its soft-framed institutions, and it is this which endows the association with its extraordinary fluidity and multiplicity of aspects. Indeed, the forum of the association may be of little importance, and no more than a gimcrack facade quite incidental to the realisation of the individual's goals. There is often a very poor 'fit' between the formal institution of the association and the strands of reciprocity and reciprocal groupings from which the association is thrown up or is designed to encourage, and upon which it rests. As Omohundro (1981) puts it, Chinese associations may often be little more than 'puffery, camouflage—and very fragile' (p. 109). In much the same way as those familiar roles purported to be 'Chinese' referred to in Chapter 4 (the filial son, the obedient wife, the artist who works all hours to develop his 'cultural heritage' in a highly self-conscious manner, the businessman who sleeps in the factory, the paternalistic leader, the scholar whose dominant concern is to serve the student, and the student whose duty is to obey the scholar) the associations comprise individuals who exhibit little ability, and are not genuinely sincere, in their 'expected' roles which are, in any case, largely for public display. But the pretence is nevertheless a useful ploy for the realisation of certain ends. By helping others, the merchant can advertise his achievements, bolster his prestige, widen his business connections, and, in the eyes of others, reinforce the importance of his business activities. By establishing, joining and, perhaps, taking part in the association's activities, he thereby acts to safeguard and legitimise trade and to defuse the tensions which might otherwise arise either out of envy and resentment of his success or as a consequence of the coexistence of values and moral beliefs that emphasise altruism, giving, and the importance of the ability to handle individuals. Given all that has been said about the association as a nexus of reciprocity, it is clear that the manipulation of sentiments, emotions, values and appearances is often consciously directed at the conduct of trade. But whether the merchant's philanthropy or selfless participation in the association's numerous activities is an expression of self-aggrandisement or is motivated by genuine altruism, or a desire to justify his success (to himself as well as to others), or to expand business connections, his actions, regardless of his motivations, serve to protect and to legitimise trade. The merchant's actions, and the perceptions of his actions, are to some extent conditioned by the feelings of envy and resentment expressed by others, and by the existence of other, alternative and potentially opposing values. In a similar way, so the practice of trade would condition the actions and perceptions of those individuals who place greater importance on contributions to 'the community', altruism and the ability to handle social relations as an end in itself. There is, then, a symbiosis among the different aspects of the association and perhaps even among conflicting wealth values within the individual.

The association as a symbol of elite designation

Almost by definition, the association as a symbol of elite designation provides a clear focus and direction for the energies of individuals who comprise the association and whose activities constitute the association's different aspects. Given the predominance of trade and its underlying wealth values, any practice or value would, in this context, be 'turned' towards the extension and institutionalisation of trade even if that practice or value had no direct bearing upon the conduct of trade.

A symbol of group solidarity may be a 'shared attribute'—most especially the worship of a common ancestor (fictitious or genuine)—but may also extend to many other actions, beliefs and prescribed behaviour including the observance of mourning grades, ceremonies within the kin group, and even the simple gesture of attending a birthday party. A symbol of group solidarity may also include making contributions to a school or hospital or some other cause regarded as popular or worthy. Gifts such as these, which confirm participation in and solidarity with the group, may help the individual in his attempts to gain acceptance into the elite core which effectively constitutes the association, and may also be used to elicit favours in return. In other words, elite symbols are manipulated as one of the soft-framed institutions of reciprocity which may be directed towards the extension and institutionalisation of trade.

Arguably the most important implication of the association as a symbol of elite designation is that the association may act to differentiate and to separate the Chinese more than to unify (Siow, 1983; Emerson, 1937; Freedman, 1960; Weber, 1968). It is true that 'interlocking directorates' make for a wider cohesion; yet, simultaneously, they may also create more complex factions. To this consideration must be added the point that associations are only one expression of reciprocal networks and groupings, and of elite designation. These complex elites may take great care to distinguish themselves by dress, residential location, consumption patterns, life-styles as well as by contributions to religious organisations, temples and associations (Gosling, 1983). The association, then, provides an imprecise, blurred, temporary, and yet clearer presentation of divisions and, to some extent, of economic and social ranking.

By providing a focus for energies and loyalties, and by highlighting the differences among the Chinese groups and between each of these groups and indigenes, the association as a symbol of elite designation may therefore enhance and give clearer direction to competition in trade, or, conceivably, may give a stronger institutional form to divisions which may have stimulated trade in the first instance. Certainly in some instances the divisions among trading activities were sharply defined. In Sarawak, nearly all the members of each of the dialect groups—including the Foochow, Henghua, Chanan, Cantonese, Luichow, Hainan and Hakka—were engaged in a single occupation peculiar to their group (T'ien, 1956), with the exception of the Teochow and Fukienese. Moreover, the dialect associations were particularly well focused: ' . . . the most important function of the dialect associations is in connexion not with dialect, locality or clan matters . . . but with the economic interests of the occupation which is followed by the majority of its members' (p. 17). Similar, though not such clearly defined, divisions existed in Singapore, the Philippines and in Peninsular Malaysia. Competition in the same line of business existed among members of the same association, helping to prevent the creation of damaging monopolies. However, intense and ruthless trading relationships, including the supply of credit, may well have been easier and, given the occupational divisions, only possible between individuals of different dialect groups, particularly if kinship pervaded the associations. In Peninsular Malaysia, for example, each dialect group in the same trade created its own separate guild. Sarawak's urban bazaars, too, contained speakers of many different Chinese dialect groups; and while some of the rural Chinese planters, predominantly Hakka, managed to save enough capital to set up as petty merchants in the rural bazaars, most of the large shops in the rural areas were owned by Fukienese and Teochow from whom the Hakka planters would obtain credit and with whom they would trade their crops (T'ien, 1956). The greater clarity which the associations gave to divisions among the Chinese may also have served to strengthen geographical divisions which, by acting to stress differences and to concentrate energies and loyalties still further, could act as a stimulus for trade. It was argued in Chapter 2 that the markets of Hong Kong were generated in just such circumstances.

Finally, the association as a symbol of elite designation highlights differences between the Chinese groups, their basic wealth values (which, in many instances, are those underlying trade) and, on the other hand, the indigenes. The association thus provides a clearer focus and direction for the energies and activities of the individuals who comprise the association and its different aspects. The association, then, is a focus for a group which makes a virtue out of actions and values which others, most especially the indigenes, may ignore, frown upon or detest. The emphasis on differences, and the attempt to construct an aura of exclusivity and uniqueness which distinguishes them from other Chinese and from the indigenes, create a sense of unity and trust (though very much infused with a realistic attitude to human nature, for this particular aspect of the association does not exist separate from the others) among the participants of the group. It was axiomatic that the symbolic significance of those institutions, values and practices which were conducive to—or at least did not work against—the extension and institutionalisation of trade should be used and thereby emphasise still further the differences between the basic wealth values of each Chinese group and those of the indigenes. The association thereby gave more tangible form and focus to divisions which may have initiated, or given further stimulus to, trade in the first instance, and to values, desires and practices which enabled a group to trade and conduct business in hostile circumstances and over long distances.

A number of interesting questions arise out of this discussion. Why do groups wish to symbolise their distinctions? Why are the symbols of elite designation so chosen? And are these symbols consciously manipulated? The answer to the first question lies with the spontaneity of the human mind and circumstances. The Chinese immigrants lived and operated in territories which were often hostile and which, as in the case of Malaya/Singapore, were governed by colonial powers that encouraged divisions among the Chinese and effected control via the immigrants' own chiefs. Not that much encouragement was necessary: the immigrants originated from different parts of China; they spoke different languages; and rivalry, violence and prejudice among clans and lineages were commonplace. In Southeast Asia mutual physical protection and the safeguard of trade (one of the few means of livelihood open to them), and, therefore, mutual support and loyalty within a notional group, were essential.

The answer to whether or not these symbols of distinction are consciously manipulated is more important to the present discussion. In this regard Cohen's (1981) views on the Creoles* in West Africa are particularly interesting and pertinent. Cohen argues that the interests of the elite are maintained and developed:

---

*The Creoles are said to be descendants of emancipated slaves who originated from other parts of British West Africa, most especially Nigeria.

'. . . by means of an organisation which is complex in structure, being partly associative and partly communal. For example, an elite may include members of different professions, each of which may be organised in a formal association that caters to its own particular interests, but the links across the professional associations may be communal, like those of kinship, friendship, or ritual. On the one hand, even within essentially formal professional associations many communal relationships develop which substantially affect the functioning of the association as a whole; on the other, communal, cross-associational relationships may be partly formally organised as in churches and religious denominations. In all these cases, the associational aspect is clearly visible, and its observation and study pose methodological or analytical problems. It is the communal aspect that poses a challenge to the sociological imagination.

To deal with this latter aspect methodically and empirically, we can study it in its manifestations in symbolic or dramatic performances. For comparative purposes, these can be analysed in terms of symbolic functions, symbolic forms, and symbolic techniques. A symbolic function, such as the achievement of communion between disparate individuals or groups, can be achieved by means of different symbolic forms, such as a church service, the celebration of the memory of an ancestor, or the staging of extensive ceremonials among overlapping groupings within the elite. Similarly, a symbolic form, such as a church service, can employ different techniques of signification, such as poetry, music, dancing, and commensality.

The symbolic process, involving functions, forms and techniques, is geared by elite groups towards achieving solutions to basic organisational problems that are peculiar to the elite. I must hasten to point out . . . that no reductionism of the symbolic in terms of the utilitarian is implied here. On the contrary, the assumption here is that the symbolic process is so basic to the human condition and so powerful in motivating action . . . that it is everywhere manipulated by collectivities of all sorts in their own interests. The result is that symbolic action is always bivocal, being both moral and utilitarian, steeped in the human psyche, yet greatly conditioned by the power order. This is why it is the very essence of symbols that they are ambiguous, and hence manipulable and mystifying. Nor is there an assumption here that the manipulation of the symbolic process is done consciously and rationally. A few leaders may at times do so, but for most of the time and most of the people the process is not consciously manipulated . . . ' (Cohen, 1981, pp. 218–219).

Cohen does not seem to be implying that the Creoles are genetically predisposed to symbolism: merely that actions performed are often motivated by many different considerations, and may vary in their significance, and may result in different effects. In other words, actions possess different aspects, including that of elite designation. Although occasionally they may be manipulated deliberately, the use, or interpretation, of actions as elite designation is not the only, or prime, motivation for the performance of these actions. Indeed, the symbolism of these actions may be quite unintentional. Yet this does not deny a strong, basic desire of individuals to associate themselves with and, at the same time, distinguish themselves from others. It is this desire which may partly account for, or inform, the performance of actions, or for the way in which they are interpreted, even though other considerations are uppermost in the minds of those individuals performing these actions.

This, at least, is the position taken in these pages, although much greater emphasis is placed upon conscious and deliberate manipulation; and the implication that manipulation may be subconscious is, perhaps, best understood as the subsequent reinterpretation of actions performed for other reasons, or as an alteration in the significance of such actions as other circumstances change. Cohen shies away from this 'utilitarian reductionism' and pursues the question back into the human psyche and the spontaneity of the human mind. Certainly, it is accepted in these pages that the goals pursued and the methods employed by individuals are a product of their minds and the choices they make, whatever circumstances and chance events may appear to dictate. And it may often be that the division between conscious and rational action, and the actions and choices which emerge from desires and emotions unformulated in thought and words, are not clear even to the individual concerned. But none of this should be taken as support for the assumption that the human psyche is a template, determined by genetics or some external force which creates a complex but ultimately predictable world of mechanistic beings which form ideal material for the theories of the social scientist. Cohen does not opt for genetic determinism or for chains of cause and effect; but he does leave perhaps too much to the 'subconscious' and thereby room for these forces to operate.

The same is true of the Chinese association. Clearly, on occasions, the association as a symbol of elite designation is manipulated and 'turned' in order to extend and institutionalise trade; it is also true that the *effect* of the association as a symbol of elite designation is to extend and institutionalise trade. The accentuation of divisions creates circumstances more conducive to trade and, when trade and its underlying wealth values predominate, the symbolism of the association intensifies competition and focuses energies on trade.

Arising out of these arguments are three further and more general points which bring the discussion back to some of the ideas presented in Chapter 1. Firstly, institutional patterns, forms of exchange and moral values possess many aspects and therefore create different and changing contexts. These contexts will not determine the actions and decisions of individuals, but they may alter the significance of those individuals' subsequent actions and decisions and therefore alter the aspects of institutions or values especially if, as is likely, those individuals first modify their behaviour to take account of the changing meaning of their planned decisions and actions.

Secondly, it may be suggested that if, for whatever reason, individuals become engaged primarily in trade, then institutions and values may gradually be 'turned' towards the extension and institutionalisation of trade. This may be for the simple reason that the individuals who comprise those institutions or hold those values are also engaged in trade. Or it may be that despite changes in the context (participation in trade and the pursuit of profit) and therefore in the significance of the actions and decisions of those individuals, the new aspects which appear nevertheless serve the extension and institutionalisation of trade. Alternatively, the new aspects which appear subsequent to changing contexts may create tension if individuals are not prepared to alter their behaviour, choices, values and decisions.

Thirdly, institutions and values, regardless of the reason for their appearance, may create or foster conditions which may stimulate trade (or any other form of exchange). But whether institutions become moulded around a specific form of exchange, or stimulate a specific form of exchange, in neither case is there any causal connection. It is the decisions, desires, choices and response of individuals which account for the creation or turning of institutions or for the practice of a particular form of exchange.

Origins: segmentation, Confucianism and the 'three forces'

Ideas concerning the origins of Chinese associations overseas are evident in much of what has already been written. It has been argued in this chapter that the Chinese association is an expression of conscious, pragmatic manœuvring designed to serve numerous and diverse ends; and it is precisely because of the innumerable and changing values, beliefs, desires and goals of individuals that the associations possess many different aspects. However, the view enunciated in these pages is at odds with much of the literature on Chinese associations—institutions which have provided a rich source of material for those who have endeavoured to enlarge that cloak which defines and explains 'the Chinese'. It has been indicated earlier that many of the explanations for the appearance and variety of Chinese associations trace paths far back into what is held to be the associations' cultural, historical and geographical antecedents in traditional China. The cultural determinists, then, maintain that the association can be explained by certain elements of a culture which is uniquely and definitively Chinese. Predictably, the cultural explanation adopts a scientific approach which, with its emphasis on the search for the ultimate origin, for chains of cause and effect, and, implicitly, for the operation of deterministic 'forces', acts to bolster and legitimise a curious blend of romanticism, mysticism and academic elitism. It is important that these cultural explanations should now be examined, partly because of their influence, and partly because they illustrate how both the logic of cultural determinism and the kudos currently attached to the application of the humanities' own brand of science each foster the cause of the other.

Segmentation

Crissman (1967) has constructed a model which, he argues, describes the underlying segmentary organisation of all Overseas Chinese communities. The primary division, or segment, is language which, with the exception of Hakka, coincides with discrete areas in China. Speech segments can be subdivided into counties and into progressively smaller territories down to single villages—the basic units of segmentary organisation. All segments may be regarded as native places with respect to the next largest segment. Thus, a language area may be regarded as a native place within the totality of any particular Overseas Chinese community; and a county of origin may be regarded as a native place within a speech community.

In addition to this hierarchical route, segmentation may also occur on the basis of surname. However, while a common surname (whether genealogically correct or merely fabricated) provides individuals with a reason and justification for association, this segmentary route is different in that surnames cannot form nested hierarchies. And since surnames cannot form nested hierarchies, they cannot, by themselves, comprise part of a segmentary system unless combined with geographical segments. Surname communities, then, must be contained within a geographical boundary to which they are therefore structurally subordinate.

The use of these two 'principles'—geography and surname—is bound by certain rules which Crissman elaborates upon at some length: 'Because each of the two criteria is capable of dividing communities based on the other, if they were applied in all circumstances they would cross-cut the total Chinese community, producing a vast number of interlocking, overlapping segments. Although there is some overlap, the two principles are in fact used intermittently. Only some of the total possible sub-communities are recognised within any one Chinese community, and the pattern of segmentation in each city or locality is unique'

(1967, p. 191). Usually, one or other of the principles is the basis for further segmentation of any particular level in the hierarchy. If segmentation occurs according to surname, the subsequent segmentation will be according to locality. If segmentation occurs according to locality, then further segmentation can follow either surname or locality or both. However, while the subsequent divisions of each segment may be affected by the principles upon which that segment was formed, the route followed from each segment does not affect decisions made in other parallel segments. In other words, the principles upon which any particular segment was formed operates only vertically, not horizontally.

Each segment may find expression in many different types of association. Crissman identifies three categories of association—village community, middle-range community organisations and higher-level community organisations. Village communities are essentially informal associations comprising individuals, each of whom knows the other's face. If the village in question is a lineage village, then its members will be related genealogically. The middle-range associations correspond to small associations, *huikuan*, funeral societies, charitable groups and occupational organisations—institutions which are involved in regulating the business of their members, and which may dispense welfare. Because they are as much political as charitable or commercial organisations, the higher-level organisations, which serve the major communities such as speech groups or the entire Chinese community, are likely to take on various innocuous guises such as hospital committees, school boards and chambers of commerce.

A prerequisite of these organisations is material wealth. Leadership is heavily interlocked and may extend downwards vertically to lower-level organisations. Each leader, then, may hold positions in a number of organisations, including the lower-level organisations which may constitute an important power base.

Further to their responsibilities for the administration and financial affairs of schools, hospitals or chambers of commerce, the leadership of these higher organisations may provide emergency relief to the community locally or to communities in China with which they may establish and maintain 'foreign relations'; formulate policies which affect the whole community; and, most importantly, mediate between the community and host government.

Crissman further argues that the common segmentary structure of the urban Overseas Chinese society elucidated by his model 'provides a basis for inference about the social structure of traditional Chinese cities' (1967, p. 185) in China under the Ch'ing dynasty:

'The urban Chinese abroad are really in the same situation as was the urban population of traditional China. They must govern themselves without having noticeable government institutions and their solution of the dilemma is the same. They use the organisational superstructure of their segmentary social structure as both are presentative political system and a hierarchical administrative system, maintaining a rarely disturbed balance between two aspects of government. The urban Chinese abroad are nearly autonomous and self-governing and their system of government is *peculiarly Chinese*' (p. 200, italics added).

The resemblance between the structure of the Overseas Chinese and the urban Chinese in traditional China as presented by Crissman is indeed striking: emigration overseas was only an extension of long-standing patterns in China; the migrants were men; they were not accompanied by their families; they came from different parts of the Empire but often formed groups from the same areas; each group monopolised specific lines of business or crafts; they were divided into many different ethnic communities, usually on the basis of geography and profession; the small bureaucracies responsible for running the cities meant that the Chinese communities had to be just as autonomous as the Overseas Chinese communities; and the leaders of the urban communities in China acted as mediators between the masses of the community and the bureaucracy.

Crissman also emphasises the links between the rural Chinese and the urban Chinese in traditional China: for example, the rural Chinese had a money economy linked to the cities, and they had a sound knowledge of city life. Crissman thereby reinforces what was in any event a very similar cultural response derived from the segmentary organisation of Chinese social life in rural areas. The similarity of this response may be illustrated by comparing the rural gentry (who mediated between the rural peasants and the bureaucracy) with the *Kapitan China* (who mediated between the Chinese and the Dutch colonial authorities) in the Dutch East Indies, and with the leaders of associations in 'traditional' Chinese cities (who mediated between the masses of the community and the bureaucracy). A further example is the lineage in southeast China— essentially a political organisation which, like the associations in urban centres during the Ch'ing dynasty or under colonial rule, had to retain a low profile lest it should be seen as a challenge to the imperial bureaucracies (Hu, 1944).

Crissman goes on to conclude that:

'The inter-relationship of urban and rural Chinese culture and society provides an explanation of both the similarity between urban social structure in China and overseas, and the surprising structural uniformity of the latter. Although most of the Chinese overseas were from rural areas, they were able to re-create traditional urban forms wherever they went, *out of the common stock of Chinese culture they took with them. The unity underlying the diversities in rural and urban life demanded a parallel, and in a sense inevitably similar, development of Chinese urban society abroad. The basic principles that organise rural life—descent, locality and occupation—are also used to order urban society.* Indeed, it is not just the same principles, but the same facts of stipulated agnation and origin which are used, a use made possible by the putatively temporary nature of migration to cities, whether in China or abroad' (1967, pp. 202–203, italics added).

Crissman's model has a certain attractiveness: a composite of a number of authors' observations of Chinese associations in different parts of Southeast Asia has been neatly arranged within a clear, conceptual framework. And, indeed, even if Crissman is not specifically referred to, many writers have taken up, or have independently developed, the same general theme—that the similarities between Chinese associations overseas and the institutions of rural and urban southeast China suggest the operation of some common underlying cultural principle.

Potter (1970) in his study of land and lineage in traditional China assumes that Chinese cultural values underlay the formation of lineages. Whatever the 'external' conditions which may have facilitated or hindered the establishment of lineages, Potter assumes that 'the Chinese had the desire to form as strong and prestigious a lineage village as they were capable of forming' (p. 131)—an assumption which, Potter argues, he shares with Freedman (1966, p. 8). It may be that Potter has to some extent misrepresented Freedman who appears merely to acknowledge the obvious fact that the Chinese wanted to form lineages. Although Freedman suggests a number of possible reasons, he does not attempt to identify the causal root of the lineage. Why the Chinese should have responded to particular circumstances by establishing lineages, and whether lineages were fostered by or gave rise to the particular wealth values associated with reciprocity are perhaps unanswerable questions. However, Freedman does trace a link between the lineage in south China and the Overseas Chinese clan associations: and, essentially, that link, it is argued, is agnatic kinship. Urban conditions, non-agricultural occupations, the laws and legal institutions of effective governments, and the fact that until the late nineteenth century few migrants were women, prevented or made it unnecessary for the traditional pattern of the lineage to emerge. It was for these same reasons that 'large-scale organisations of Overseas Chinese on the basis of agnatic kinship has typically taken the form of the clan association and not the lineage. But, in turn, the overseas clan association . . . is a development of a form of grouping found in the large towns and cities of China itself; and it follows that, in tracing the fate of agnatic kinship in the Chinese diaspora, we are to a considerable extent watching the general reaction of the lineage to 'unsuitable' conditions, whatever they may be' (1966, pp. 165–166). Omohundro holds a similar view. The association, Omohundro (1981) argues, is a response to political and social conditions in a host country. But although it has undergone 'a process of speciation', the association is, nevertheless, modelled upon the organisation of south China (p. 89). Indeed, Omohundro maintains that the Chinese associations in Iloilo, the Philippines, possess a segmentary structure very close to the ideal proposed by Crissman (1967, p. 106). See (1981) and Wu and Wu (1980) advance less sophisticated arguments, but all three writers appear to imply that the association may be regarded as an expression of 'habits and customs' originating in south China but which now inform the behaviour of the Overseas Chinese. Even L. Lim (1983), who is extremely sceptical of the explanatory powers of 'Chineseness' (see Chapter 4), appears to give nodding acceptance to the determining influence of segmentation. Although she argues that the Overseas Chinese association is immediately derived from the imperatives of communication, immigration and a conscious choice to create networks of information, credit and business contacts, she nevertheless regards the association as a reflection of patrilineal organisation in Mainland China.

Confucianism

Cheng (1985) brings in that ubiquitous strand of 'Chineseness'— Confucianism. Following on from Ho (1966), Cheng argues that locality consciousness among the Overseas Chinese is primarily a consequence of the Confucian code of ethics and law, and the historical imprint left by administrative regulations which governed the appointment of officials on the basis of geographical origin. The associations formed on the basis of common geographical origin by officials in Beijing were later broadened to accommodate traders and candidates for the Imperial Civil Service examination. 'The installation of locality gods and deities at associations in other cities turned some of the associations into temples, while others owing to the trade of its members evolved into trade associations' (Cheng, 1985, p. 36). Although in Singapore the causes and motives for establishing an association during the colonial era may have varied, the principle of locality and kinship remained unchanged: 'It is obvious, then, that the establishment of associations outside China by the Chinese represents the transplanting of a Chinese tradition overseas' (p. 36).

Yao (1984) also gives pre-eminence to Confucianism. Yao argues that the *shetuan* (or voluntary associations) are a product of Confucianist relations within Chinese society. Moreover, Yao (with Freedman in mind) questions the emphasis commonly given to the 'function' of associations in attempts to explain their nature and origins. Focusing on the voluntary associations of Singapore, Yao argues that the fit between their official purpose—charitable works and mutual assistance—and their true activities in the early settlement of Singapore was merely chance. It was partly because of this happy coincidence that 'the idea of social function is an attractive one for many writers when discussing the origins of Chinese voluntary associations in Singapore' (p. 77). Nevertheless, Yao argues, the idea is clearly flawed. The last 30 years or so has been a period of drastic social change and yet Chinese voluntary associations have survived and even increased in numbers. The provision of social welfare by government and the appearance of private and official cultural, occupational and social organisations has made the 'social function of the traditional *shetuan* less than self-evident' (p. 77). Drawing on Carstens (1975), Yao also points out that while charitable works, mutual assistance and social welfare remain the formal objectives of the voluntary associations, the benefits derived from their actions are in practice often insubstantial. A more promising explanation of the origins and nature of Chinese voluntary associations, Yao maintains, is culture:

> 'Chinese voluntary associations are expressions of . . . the key relationships underlying Chinese society. These relationships, based on Confucianism, are structures in the sense that they are constitutive of wider relationships and institutions in day-to-day reality. Indeed what more essentially characterise Chinese *shetuan* are not the charitable activities but rather the type of relationships created. Through social and economic patronage of the leaders, and the dominant ideology of consensus, *shetuan* in effect reproduces the culturally meaningful Confucian relationships. And these relationships are crucial in structuring the differences between leaders and rank and file members in terms of influence, prestige and wealth. In this sense, *shetuan* are important make-ups of Chinese society because they help to produce the underlying social relationships and in turn define the social order' (Yao, 1984, p. 75).

Not surprisingly, then, Yao argues that many features of *shetuan* also create relationships parallel to a Chinese lineage:

> 'Between the members they are supposed to be cooperative, and generous in giving help to each other when the need arises. Ideally, members stand equal to each other under the leaders [with whom they form a paternalistic relationship which is political in nature] and are united in working for the prosperity of the group as a whole' (Yao, 1984, p. 79).

Indeed, Yao argues that *shetuan* are in fact symbolically modelled on the Chinese lineage—the ideal form of 'Big Family'. Forming and joining *shetuan*, then, is 'the essence of being Chinese'. This 'Chineseness' finds expression in the form of the *shetuan* through two related mediums. The first is the 'larger good'—the ethical force of Confucianism.

'This is the concept of *datung* . . . , prosperity of the social collectivity in which people live in accordance with the defined ethical relationships. In such a world, not only individual happiness and familial continuity depend on the wider social order, but also in reverse, that disharmony among the people ultimately affects and even threatens the ideal social order. In this way, conformity to culture implies a sense of social responsibility and even a degree of self-denial. Significantly, such responsibility towards the "larger good" is the best justification for the paternalism of *shetuan* leaders and the subservience of the average members. Thus my contention is that Confucian ideals, crystallised in the metaphor of the family, presents a powerful justification for the Chinese to form *shetuan* in overseas societies' (Yao, 1984, pp. 80–81).

It is worth noting that Yao appears to assign a measure of importance to the conscious manipulation of 'cultural' elements in this explanation of the *shetuan*. However, it becomes apparent that this behaviour is itself predetermined by the second medium through which 'Chineseness' finds expression in the form of the voluntary association. This second medium is the foundation of the first. Citing Ward (1965), Yao argues that while 'being Chinese' may appear a convenient, last-resort answer to difficult and uncomfortable questions about why Chinese behave in a particular way (such as forming and joining associations), it is culturally significant because it may constitute a conscious model of self-perception. Moreover, if the question is pursued, it becomes apparent that there is a diversity of models of 'Chineseness' which, Yao argues, probably stem from one single model—the literati or gentry model. This acts as a yardstick or plan of idealised social aspirations, relationships and practices. This cultural plan is 'not so much deterministic as providing a guiding principle for institutions and behaviour in Chinese society' (1984, p. 81); and it is in accordance with this plan that changes and social processes within Chinese society take place. Individual action, then, may not always correspond with that plan, but the justification of action (by claiming 'Chineseness') does illustrate either unconscious acceptance of Confucian values as a consequence of socialisation; or the mystification of Confucian ideology; or it could convey the message that people do the things they do because they are the only way they know how (p. 81).

This second medium through which 'Chineseness' finds expression in the form of the association is no less than a cultural plan imprinted upon the human mind—a plan which also directs interaction with, and changes in, the institutions and 'social processes' of the wider society:

'Following Giddens then, we consider "structure" and "history" or influence of Confucianism and given socio-political conditions to be both necessary for the production of social process. This is not difficult to understand on a common-sense level. Chinese culture explains the pattern or order of relationships and institutions. Social and political realities, on the other hand, set up conditions which determine such questions as whether these relationships and institution[s] would be formed in the first place, their duration over time, and their connections with other social processes in a society' (Yao, 1984, p. 81).

One problem which is common to both 'segmentary' and 'Confucianist' explanations—and which will be considered again later in this chapter— is that they rely heavily on tautology. The idol of origin—in this instance, culture—is the ultimate cause because it is; and there the argument is left hanging. In Crissman's model, the lineage and the association are organised on the segmentary principle because their basic organisational principle is segmentation (a criticism also made by Sahlins of Evans-Pritchard). In Yao's paper, the origin of the cultural plan is not revealed: the Chinese mind is Chinese because it is Chinese.

The three forces: cultural principles, history and geography

Sangren (1984) incorporates many of the themes running through the 'segmentary' and 'Confucianist ' analyses of the origins of Chinese associations, but Sangren implicitly recognises, and attempts to avoid, the tautologies of these analyses.

Sangren argues that the Chinese establish corporate groups (associations) 'very similar in form and function to lineages on bases other than kinship' (p. 391). Analytical categories such as lineage, kinship and the principle of patrilineage 'constitute impoverished models of a cultural system clearly capable of generating a much greater range of creative organisational responses to changing historical and environmental circumstances' (p. 411). Such categories and their underlying groups of principles—most especially those advocated by writers Sangren describes as British structuralists—assume a relationship between form (such as the lineage), function (such as the transmission of property) and principle (such as rules of descent) which is too close: often these analytical categories simply do not coincide with actual behaviour. Sangren therefore suggests that such categories should be disaggregated in such a way as 'to define more precisely the dimensions of variations of both lineage corporations and other kinds of Chinese groups' (p. 391). Disaggregating a range of specific group forms from function, he argues, allows the analyst to see more clearly the flexibility and range of variations in the Chinese corporation; these disaggregated 'operational features', he argues, provide an insight into a 'deeper culture'.

Although Sangren initially and explicitly makes no claim to explain these organisational features, he does go on to advocate with some vigour just

such an explanation. He rejects the notion of sociological principles—
such as descent, locality and occupation—and in their place substitutes a
collection of equally vague forces. By means of his organisational features—the disaggregated analytical categories—Sangren identifies a
number of key operational norms—such as sequential rotation, decision
by consensus and committee hierarchies—which underlie the variety
and flexibility of the associations and give them their 'Chineseness'.

Sangren is presenting elements of Chinese culture which can be used to
explain the association. These operational norms are the 'deeper' cultural levels 'necessarily capable of generating a variety of adaptive responses, in the present case, in the form of many different kinds of
organisations, from an essentially stable set of normative expectations
about how people can cooperate and ought to behave in groups' (1984,
pp. 410–411).

In addition to this 'deeper' culture, Sangren tentatively introduces two
other forces through which culture finds expression in the association.
The first is history:

> ' . . . attempts to describe what is common in various forms of Chinese organi-
> sations (Crissman, 1967; Jacobs, 1979) go little beyond descriptions of the vari-
> ous particularistic ties (*kuan-hsi*) activated in the formation of such groups.
> But typologies of particularistic ties, like an overemphasis on descent ideolo-
> gy, run the risk of reifying criteria of membership and, most importantly, con-
> founding cultural categories and social groups. In short, the fact that overseas
> Chinese immigrants have been able to "recreate traditional urban forms
> wherever they went, out of the common stock of Chinese culture they took
> with them" (Crissman, 1967: 202) cannot be explained entirely through their
> applications of the "basic sociological principles that organise rural life—
> descent, locality, and occupation" (Crissman, 1967: 203). In themselves, such
> "sociological principles" do not generate Chinese groups. It is more the expe-
> rience of creating and participating in groups than the acknowledgement of
> particularistic ties that accounts for the Chinese talent for organising so adap-
> tively to diverse environments' (Sangren, 1984, pp. 409–410).

Sangren, then, appears to be implying that some force (the force of history) continuously transmits and thereby ensures the constant repetition
of a particular experience, turning the 'deeper' cultural levels he has
identified into phenomena which 'inform all social action but are so
taken for granted that conscious elaboration is unnecessary' (p. 410, footnotes). Sangren's operational norms, in conjunction with the force of
history, produce unthinking and automatic responses which produce
associations.

The second force is that of geography which exercises its influence most
especially through the standard marketing community—the formation

of which depends upon a range of spatial, geometrical and other forces (R.N.W. Hodder, 1993b; Skinner, 1964–65). The standard marketing community is a crucial organisational feature of all associations save one— the county-level surname association whose popularity 'seems to inhere primarily in the opportunity they afford for expanding personal relationships (*kuan-hsi*) beyond the boundaries of the standard marketing community' (Sangren, 1984, p. 405). It would seem, then, that the standard marketing community is a crucial phenomenon through which 'Chineseness' finds expression in the association. Indeed, Sangren goes so far as to suggest that spatial concentration is a concomitant, even determinant, of

> ' . . . other such formal expressions of lineage solidarity as the construction of ancestral halls. An agnatic group is localised at the compound level if most members reside in directly contiguous quarters. Occasionally a compound or group of compounds is large enough to constitute an entire neighbourhood or, less frequently, village, but more common in northern Taiwan are villages composed of two or more agnatic groups. The standard marketing community, as Skinner (1964) suggests, in most cases defines the maximal spatial extension of formal agnatic organisation (Sangren, 1984, p. 398).

Sangren, then, has presented three phenomena which effectively define 'Chineseness' and which account for the appearance and variety of Chinese associations: the 'deeper levels' of culture (or operational norms); history; and, through the standard marketing community, geography.

The trail to China: the idol of origins

If the roots of Chinese associations overseas lie in the forms of group organisations (or as Sangren would have it, their disaggregated operational norms) in China, then it might be expected that these explanations would find support—or at least would not be contradicted by—analyses of the reasons for the appearance, and seeming concentration, of strong lineages in southeast China from where most of the Overseas Chinese immigrants originate.

Potter (1970) integrates a number of explanations into his study of land and lineage in traditional China which begins almost immediately with the assumption that the Chinese would create and maintain lineages provided certain external conditions were present. In other words, the Chinese were culturally predisposed to create lineages but their compulsion to do so was conditioned by a number of variables. The first variable was irrigated rice cultivation, a form of agriculture which was highly productive, required intensive, well-organised and cooperative labour, and may therefore have been the key to common estates (Freedman, 1966). A second variable was the degree of government control: where government control was weak (as in south China) strong lineages were essential for mutual protection and self-help; where government control was strong (as in north China) measures would be taken against lineages whose power exceeded that necessary to maintain effective control over their members (Hu, 1948; Hsiao, 1960). A third variable was the need for defence in the pioneering frontiers of Empire. The Chinese who migrated to southeastern China during the Tang and Sung dynasties were confronted with 'barbarian societies which for the most part they gradually eliminated (to some extent by marrying their women), seizing their land and bringing empty land into cultivation' (Freedman, 1966, p. 163). Here then were the conditions and forms of organisation which appear to bear comparison with Sahlins' (1961) predatory segmentary lineage—a notion which Omohundro (1981) has suggested might be applied to the Chinese associations overseas. As Freedman notes:

> 'Defence was a necessary part of community life, for there were brigands on land and pirates from the sea. It may be that these were the conditions which, acting upon patrilineally organised pioneers (or at least on pioneers bearing a patrilineal ideology), stimulated the growth of relatively independent and tightly settled local lineages which, precisely because they were so organised, built up a system in which interlineage relations were characterised by violence' (1966, p. 163).

The fourth variable was wealth (money), derived either from official careers or from commerce. The former source of funds was possibly more important in north China; the latter source of funds was more important in south China which, from Sung times onwards, was one of the most commercially developed regions in China. 'This can be explained partly by the presence of an agricultural surplus that stimulated trade and partly by the presence of foreign trade. . . . Much of this commercial capital was invested in land . . . ' (Potter, 1970, p. 135).

Clearly, then, there are legitimate reasons based upon circumstantial evidence, to support an argument in favour of a causal link between the lineage—or its underlying organisational principles or disaggregated norms—and the Overseas Chinese association. The appearance and concentration of strong lineages in southeast China (from where most of the Overseas Chinese originate) are surely more than coincidence? Any one of the cultural explanations of the appearance of Overseas Chinese associations could be taken as the cultural determinant left unspecified by Potter, for none of the explanations for the rise of the lineages in southeast China conflict with those for the appearance of Overseas Chinese associations. Indeed, a case could be made that each of the different sets of explanations strengthen the other: the need for defence and Sahlins' notion of predatory lineages could easily be grafted on to Crissman's analysis; Yao's Confucianist relationships and Sangren's operational norms and 'habits' which inform all social action could be integrated into the variables of defence and irrigated rice cultivation.

Three problems

Nevertheless, there are a number of problems with these explanations of the rise and concentration of lineages in southeastern China, with the apparent causal links between these lineages and Chinese associations overseas, and, regardless of any such link, with the notion of an underlying cultural principle responsible for the appearance of those associations.

Cultural principles

The first major problem with cultural explanations is the inability to identify the specific cultural principles for the rise of lineages, or for the appearance of other groups and organisations, including the association, in urban centres in 'traditional' China. Nor is there any clear comprehension of the circumstances that were conducive to the successful operation of any such cultural principle. To begin with, there is some doubt over the distribution of large, strong and differentiated lineages in China, as indeed Potter (1970) notes in some detail. Fairly well-developed lineages did exist in northern China, but Potter is content to assume that such lineages were more prevalent in the southern and central parts of China than in the north although, as Potter again notes, there is 'so little information about lineages in North China that it is simply not possible to say anything conclusive about the distribution of strong lineages there . . . ' (p. 131). This throws some doubt on the apparent coincidence of areas of intensive agriculture and weak government with strong lineages. To this must be added the obvious fact that strong lineages have not been confined to China, nor to areas of intensive agriculture or weak government. These observations do not, in themselves, make it less likely that irrigated rice cultivation or degrees of government control may have influenced or determined the rise and development of the lineage in China, although it is perhaps worth emphasising (as in Chapter 4) that an attempt to argue that similar institutions may be a consequence of different causes appears particularly weak without first having demonstrated any causal link, and can be turned about: similar institutions have many different aspects—that is, they are an expression of different actions, values, beliefs, choices, decisions, desires and goals in the same or very different contexts and circumstances. Evidently, any further argument based upon the assumption of a causal link between irrigated rice cultivation and the formation of lineages is unsound.

The apparently clear-cut relationship between the degree of government control and strong lineages is open to similar doubts. The information is insufficient to allow any argument to proceed upon such an assumption; indeed, it could be argued that an effective and sophisticated government may, because of these very characteristics, provide a framework which encourages the development of strong lineages as a useful instrument for indirect rule.

The relationship which Potter assumes exists between trade, strong lineages and defence, can also be turned about. Potter assumes that trade is a consequence of an agricultural surplus and of foreign trade. Doubts surrounding the notion that agricultural surpluses stimulated trade have been raised in earlier chapters in this book and elsewhere (R.N.W. Hodder, 1993b). It is enough to reiterate two points raised in Chapter 2. Firstly, evidence from other parts of the world strongly suggests that trade and markets are stimulated by external contact either over long distance, between different ecological zones, or among different tribal and sub-tribal groups. Secondly, decisions made by lineages in southeast China may have had crucial significance for the establishment, suppression and survival of markets. External stimulus was not confined to contact across the boundaries of the Chinese Empire. It could be argued, then, that it was the presence of lineages and the tensions and conflicts among them as well as foreign contacts which acted as a stimulus for trade and markets. And after all, the rise of trade and markets in southeast China which, as Potter notes, began during the T'ang and Sung dynasties coincided with the migrations noted by Freedman and, incidentally, with the rise of the ports of Tuen Mun and Hong Kong.

Clearly, the inability to identify the original principle responsible for the rise of lineages—and, therefore, of other groups and associations—in 'traditional' China, and the uncertainty over the circumstances conducive to its successful operation, are both facts which make it impossible, except by an article of faith, to endow with credibility any one of the assumed causal connections between the Overseas Chinese association on the one hand, and the lineage, related groups or their component disaggregated norms or ethical relationships on the other. Again, the debate may again return to the interpretation that, whether in similar or different circumstances, an institution may represent an expression of different values, actions, beliefs, choices, decisions, perceptions, desires and goals, and may possess many different aspects. The Overseas Chinese associations were a response to the needs of immigrants from different parts of China thrown together in alien states to organise their lives and activities and, in particular, to institutionalise, safeguard and legitimise trade. There can be little doubt that they drew on their experiences in China, but only so far as the practice and institutional forms they adapted were relevant and useful. There was, then, no causal connection between the lineage (or its organisational principles or disaggregated norms) in China and the Overseas Chinese associations.

Ethereal forces

A second major problem with the cultural explanation is the reliance on ethereal forces. It has already been noted that Crissman's model leaves the impression that it is based upon a tautology: the lineage and the Chinese association are organised in a segmentary fashion because of the segmentary principle inherent in these organisations; and, therefore, they are organised according to a segmentary principle because they are segmentary organisations. In other words, the force of the segmentary principle determines the nature and organisation of Chinese social structure in rural and urban areas—a structure which, it must be assumed, reinforces its underlying principle.

Yao, too, relies on a variety of curious forces which leaves the impression that his analysis rests upon a tautology. Yao does not specify the causal links between those traits he defines as culture and the Chinese association. A mind which has become ingrained with Confucianism, Yao argues, is by definition Chinese; and it is this fact which, in the absence of external circumstances hostile to Confucianism and the associations, leads inexorably to the formation of the association. There are two questions which must be tackled at this point. How does the mind's infusion with Confucianism occur? And why does a mind infused with Confucianism cause that individual to join or establish an association?

In answer to the first question, Yao suggests three possibilities (for the medium of a Confucianist ethical force is rooted upon these). One suggestion is unconscious 'socialisation'. Another suggestion is the mystification of Confucianism. The third suggestion—and the one preferred by Yao—is the operation of a cultural plan. However, Yao does not reveal the origin of this plan (or 'unconscious model of self-perception'). The plan simply exists and presumably it is the source of Confucianism, for the plan prescribes and guides the way individuals behave (and who, as a consequence, are definitively 'Chinese'), it is the foundation of the Confucianist ethical force, and it is responsible for reproducing 'meaningful' Confucianist relationships in the form of the association—relationships which define 'Chinese'. According to the logic of Yao's argument, then, either it must be accepted that the Chinese mind is Chinese because it is Chinese and so on *ad infinitum*, or it must be assumed that 'the Chinese' psyche is the product of a uniquely Chinese genetic code. The only other alternative is to return to Yao's two former hypotheses. Either the cultural plan simply represents the mystification of Confucianism, or the unconscious acceptance of Confucianist values as a consequence of socialisation.

As for an answer to Yao's first question, the mystification of Confucianism is a phrase which leaves this author no clearer about its meaning or significance. Possibly Yao is suggesting that the framework which Confucianism provides for the conduct of social relations may be endowed with greater potency if it appeals to the non-rational, emotional side of individuals. By providing Confucian values with an emotional coating, these values can enter and operate within the subconscious of the individual more effectively and, given the right circumstances, may thereby stimulate the individual to produce Chinese associations. The notion of socialisation is again somewhat vague, and is therefore, as noted in Chapter 4, a useful term which describes the revisiting of structures and values created by men upon their children, and which possesses the possible advantages of encompassing the guesswork of child psychologists.

However, both the notion of mystification and the notion of socialisation still leave Yao's discussion in a quandary, for both notions imply that the origin and nature of the Chinese association is a consequence of a recursive dialectic between, on the one hand, man, and, on the other, the Confucian values and the associations he created: the two sides of this dialectic being rather like two mirrors reflecting each other to infinity. Again, then, Yao is implicitly presenting the explanation that the Chinese are Chinese because they are Chinese and so on, *ad infinitum*.

As for the reasons why a mind infused with Confucianism causes that individual to join or establish associations, the only link which Yao provides between 'the Chinese' mind and the establishment of the association is the 'ethical force', though Yao does specify that the ethical force is founded upon the cultural plan: the Chinese, it seems, are compelled to form associations because they are Chinese.

Even so, it should be noted that the idea of 'norms' or 'cultural plans' is a popular device. It is utilised in analyses of the Chinese company (see Chapter 4), and it is used by a number of other writers in their explanations of the Chinese association. Wickberg (1988), also following Ward (1965), believes every individual possesses conscious and changing models of 'Chineseness' which are born out of personal experience. Every Chinese individual draws upon these models which that individual derived from, or believes are realised by, other Chinese individuals. 'To be urban and to be Chinese may be to be like people in the market town near one's "home village" . . . in China; or like the county town; or the provincial capital; or like Amoy, Taipei or Hong Kong ' (p. 314). The associations are produced by models of 'Chineseness' and thereby serve as anchors of Chinese identity, enabling the Chinese to integrate into the host community.

In common with Wickberg, Wang Gungwu (1988) argues that Overseas Chinese have multiple identities which must be taken into account if the Chinese are to be understood. These multiple identities, argues Wang, are rooted in various 'norms': 'By norms, I mean the ideal standards which are binding upon members of a group and which serve to guide, control, or regulate their behaviour. Such norms are present in all the concepts of identity [historical, Chinese nationalist, communal, national, culture, ethnic and class] . . . and each of [these] identities is based upon the acceptance of specific sets of norms' (p. 11).

Wang identifies fours sets of norms. The first set are physical norms (such as endogamous marriages which give rise to vague notions of racial purity) which, in practice, are subordinate to the ideological principle of male descent. (No doubt one could also add that, at least until the beginning of the twentieth century, sexual drive among predominantly male immigrants was in itself a sufficient reason to marginalise concerns over racial purity.) These norms, Wang argues, contribute to a keen sense of Chinese ethnic identity. The second set of norms are political. These refer to 'ideas of political loyalty to the state, to the need for commitment to and participation in the tasks of nation-building and, often in the background, to the ideals of democratic rights. In practice, these norms are manifested in different ways and the core principles are focussed on different symbols and institutions' (pp. 11–13). The third set of norms are cultural. Wang identifies two kinds of cultural norms. One is the Chinese cultural norm which includes the written Chinese language, 'the preservation of family ties especially through observing norms about birth, marriage and death, and the support given to clan, district, and other similar organisations which enhance Chinese social solidarity' (p. 14). The other kind of cultural norm is modern. This includes those norms such as educational standards and career patterns outside the Chinese community which the Chinese have found useful to accept.

The fourth set of norms are economic—and it is the elaboration of these norms in particular that the highly deterministic nature of this concept begins to surface. Economic norms, Wang argues,

> 'refer to modern, rational standards of behaviour that influence the conduct of the national economy . . . and those standards regulating each group's pursuit of livelihood and profit. But when so broadly defined, their impact on identity is rather diffuse and not clear. We need to concentrate on those norms which either support the development of a national identity or those which reflect the cultural values of strong economic actors like the Chinese minorities in the region. But perhaps the most distinctive of those norms are those related to and which reflect class interests, interests which cross ethnic and national boundaries. Such norms can directly influence the sense of class identity, but without necessarily undermining an evolving natural identity or diminishing a residual Chinese cultural identity. In the long run, of course, a strong commitment to class identity could weaken the identities of Southeast Asian Chinese as Chinese. But there are other variables involved. For example, economic successes may depend on the persistence of Chinese cultural values and the presence of Chinese ethnic identity and these may override class interests. Also, in most Southeast Asian countries, there are too few working-class Chinese to assert class interests at that level. Without that class, the middle-class economic norms of the Chinese in the region and elsewhere (including China, Taiwan, Hong Kong, North America and Australia) would tend to reinforce appeals to Chineseness. Only strong political pressures from within and powerful multinational organisations which could mobilize Chinese skills from without to serve their interests could keep such a resurgence of Chinese identity in check. No matter which direction the pressures of economic norms might go, I suggest that the normative identity they tend to shape is class identity' (1988, pp. 13–14).

The train of thought in this passage is (for the present writer) difficult to follow. Its meaning is vague, perhaps deliberately so. Wang appears to be suggesting that there are certain undefined economic norms which control Chinese individuals' pursuit of a livelihood and profit. These undefined norms may also support the development of national identity, or reflect the cultural values of strong economic actors 'like the Chinese minority in the region'. But the most distinctive of these undefined norms are those related to and which reflect class interests. These norms may influence class identity without undermining a residual cultural identity. Nevertheless, if economic norms were to become predominant, and if a strong class identity were to persist, then 'Chinese' identity would be weakened. On the other hand, most Chinese in Southeast Asia are middle-class and continued economic success may depend upon 'Chinese' cultural values. There are, then, pressures for a resurgence of Chinese identity—pressures which can only be kept in check by domestic political considerations and the demands of multinational organisations. The meaning of the last sentence quoted above escapes this writer, but Wang may be suggesting that when economic norms predominate over all others, the identity they shape is that of class.

Although vague and obfuscatory, one telling point which appears to rise out of Wang's argument is the interaction between unspecified cultural values which are 'Chinese' (and which presumably coincide with Chinese cultural norms), and the economic norms which govern the pursuit of profit. Wang is suggesting, then, that 'Chineseness', in its different manifestations, is an expression of different combinations of norms which may be affected by external variables: and that economic success may be dependent upon 'Chineseness'. A clear chain of cause and effect is thus established: norms produce 'Chineseness', and 'Chineseness' produces economic success.

Wickberg and Wang, like Yao, Ward and Redding, give the impression that individuals are Chinese because they operate according to norms which are distinctively Chinese and which, presumably, derive from individuals who are Chinese. Either the tautology must be ignored or accepted without question; or it must be assumed that these norms are genetically determined; or that they are endowed by some force; or that they were once generated by individuals who happened to be Chinese but which now revisit the Chinese, again implying that some historical, geographical or cultural force is at work.

Although Sangren attempts to circumvent the problem of tautology, his 'science' requires him to trace a chain of cause and effect even further back to the ultimate idol of origins, leaving his explanation of associations explicitly reliant on the operation of 'forces'. It was noted earlier in this chapter that Sangren presents three phenomena which define 'Chineseness': the 'deeper levels' of culture (or operational norms); history; and, through the standard marketing community, geography. The power of the cultural cloak, strengthened by the twin forces of history and geography, is demonstrated by Sangren's criticisms of Silin (1976) who similarly argues that the link between culture and organisation has not received sufficient attention because of scholarly emphasis on structure. Sangren wholeheartedly endorses Silin's general argument that 'the forms of organisation reflect such primary cultural orientations as assumptions about the nature of man and of social relations, in addition to secondary assumptions regarding the nature of legitimate authority, how decisions in groups ought to be made, and what makes a good leader' (1984, p. 410, footnotes). However, Sangren goes on to argue that Silin 'underestimates *Chinese culture's capacity* to generate cooperative as well as hierarchical/authoritarian organisations. For example, his assertion (p. 128) that "cooperative groups, especially when composed of non-kin, are understood primarily as temporary coalitions rather than permanent alliances" is belied by many traditional corporations. He also argues (p. 61) that Chinese seem to prefer "unipowered" versus "multipowered" organisations, the latter being unstable in Chinese society. Again, many of the organisations described here [by Sangren] appear to qualify as "multipowered" (i.e. there is no single leader), but they are nonetheless quite stable. In short, it seems to me that *culture* rarely limits itself to a single model of how organisations and their internal relations can and ought to operate' (1984, p. 410, footnotes, italics added).

Sangren's use of the word 'culture' to indicate a very real, very active, and tangible force, rather than a passive, abstract concept which describes the outcome of actions, choices, behaviour and events, is not lax expression: it may only be unguarded, but it may even be deliberate for it helps to reinforce Sangren's attempt to dismiss contrary evidence and interpretations which replace a prevalent characteristic of Sangren's operational norms (cooperative behaviour) with the conscious manipulation of groups, and thereby weakens Sangren's deterministic force—the 'deeper' cultural levels. Indeed, Sangren's belief that culture, history and geography are real, tangible forces is more fully developed in later work on Ta-ch'i—the same northern Taiwanese marketing community on which his analysis of associations is based.

According to the criteria now prevalent (especially in the United States), Sangren's (1987) study of *History and Magical Power in a Chinese Community* is scholarly: it is rigorous, holistic, tidy and continues the seemingly endless progression along established lines of thought. It is a 'scientific' work. It presents a world of discrete elements which mix and interact in a fashion predictable to the skilled social scientist, creating a perfect, ordered whole. To understand this world, paths of cause and effect (not always direct and unidirectional) must be traced. Always implicit, and often explicit, in Sangren's argument is the existence of forces: behind each cause is another and then another *ad infinitum*, pushing explanation further and further back into the twilit world of historical 'processes' and programmed minds. The curious world that Sangren manipulates comprises strange external forces which entrance men who know only how to obey the structures, patterns and behaviour identified by Sangren.

Sangren begins his analysis with the assumption (inspired by Skinner, 1964–65) that:

'social systems and human interaction are hierarchically patterned in space and time. The assumption informs my attempt to discern order in the historical development of society and economy in Ta-ch'i. More concretely, I attempt to demonstrate that the appropriate regional context for analyzing the 'external forces' that have shaped developments in Ta-ch'i varies over time.

To locate these forces, the historian/social scientist must be prepared to shift focus in spatial terms from one level to another. Taking the spatially hierarchical structure of economy (and social interaction in general) as an object of empirical study provides a methodological "zoom lens" for this analytical endeavour. By operating on the premise that urban hierarchies and exchange networks define nested hierarchies of economic interaction, vague terms like "local", "regional", and "global" can be given concrete spatial specification. Moreover, these shifts in the relevant arena of economic change mark the most important historical watersheds in development at the local level; hence, they form the basis for my periodization. In other words, attention to the spatially hierarchical pattern of economic organisation underlies my approach to the temporal patterning of economic development' (1987, p. 13).

There is little doubt Sangren believes that the emergence of nested social structures (neighbourhoods, villages, market communities) in Ta-ch'i society is predetermined by a 'culture' that is itself shaped by 'History', although it must be said that Sangren appears reluctant to state his position clearly in straightforward language. On the one hand Sangren argues that 'the emergence of nested hierarchy in social organisation owes as much to culture as to purely economic logic' and he suggests that 'culture' has not been inconsistent with nor wholly unconstrained by economic logic. Sangren then goes on to ask: '. . . what of the longer course of Chinese history? Is it either possible or probable that the economic logic of the system determined the emergent forms of a corresponding cultural logic?' (1987, pp. 231–232).

Sangren further clouds his apparently ambiguous position. First he argues that

'Many of the key cultural distinctions that structure cosmology in Ta-ch'i today have clear antecedents in ancient China. Yin and yang, order and disorder, outsider and insider, and the power embodied in their respective mediations clearly antedate the full-blown emergence of integrated regional economic systems in China' (1987, p. 232).

But then, once again, Sangren immediately backtracks:

'Although it would be incorrect to argue that Chinese cultural logics directly caused the complex patterns of organisation of China's late traditional economy (Skinner's arguments provide a much more plausible and compelling set of explanations), the persistence of the structure of value embodied in these ancient ideas just as clearly suggests that they were also no impediment to the development of those patterns' (1987, p. 232).

Nevertheless, a central theme of Sangren's work is that social institutions (for example, neighbourhoods, villages and market communities and, therefore, associations or corporations) are intimately related to religious ideas and that the power, or *ling*, 'attributed to supernatural entities is better understood as a function of their mediating order and disorder with reference to the entire set of cosmological categories' (1987, p. 230). The pioneers of Ta-Ch'i brought with them from China their 'culture'. Although economic logic to some degree helped shape this 'culture', it was 'culture' that first determined the emergence of nested hierarchies.

Sangren's assumptions concerning the existence and operation of nested hierarchies (a notion borrowed from Skinner, 1964–65) has been questioned elsewhere (R.N.W. Hodder, 1993b) and needs no elaboration here. It is enough to repeat that such a notion is a clear example of the way in which conceptual frameworks distort and limit interpretation and exclude contradictory observations and views. Sangren's 'scientific approach' requires him to trace a chain of cause and effect; yet he is rightly afraid of forwarding a deterministic explanation which is the logical outcome of his attempt to trace cause and effect back to an ultimate origin.

Some may find this criticism harsh. It is true that while Sangren supports a positivistic line of argument he does suggest that cultural anthropology is unlikely to achieve the predictive power of the natural sciences, and that causal paradigms or hypothesis-testing methodologies borrowed from other sciences cannot necessarily be applied mechanically to explain cultural phenomena (1987, p. 234). However, he argues that culture is real, it encompasses the natural sciences, and it should be studied scientifically. He accepts the notion of nested hierarchical settlements and marketing systems—a notion which itself comprises highly questionable beliefs and analytical techniques (R.N.W. Hodder, 1993b). The links Sangren claims exist between ideas (culture), settlements and marketing systems can, then, only be regarded as doubtful. And his defence of positivism is clearly little more than an attempt to justify his own practices. Interpretation is presented (and thought of) as objective fact because it adopts what many hold to be 'good science' and must therefore be erudite.

Sangren's examination of Chinese associations and other organisations is a declaration of faith. Its central pillars are vague forces; impoverished geographical theory; arbitrary 'Chinese' attributes; and unspecified relationships between these forces and those artificial characteristics which define 'the Chinese'. Individual choices and decisions, and the complexity of the humanities, is replaced by a cloak which simplifies, defines and explains 'the Chinese'. Attempts to elucidate and interpret this complexity in clear terms is replaced by obfuscatory intellectual constructs. To explain an imagined, simple, mechanistic and predictable world in complex language is far less taxing than an attempt to interpret a complex, uncertain and unpredictable world in plain and simple language.

In the final paragraph of *History and Magical Power in a Chinese Community*, Sangren is closer to the truth than he perhaps realises:

> 'If the reader is convinced that attention both to history conceived objectively and to what Ta-ch'i's natives make of it provides a fuller understanding of local history and culture, then at the very least this premise [the adoption of a broadly defined positivistic framework] will be shown to possess a certain degree of interpretative, perhaps magical, power' (1987, p. 237).

The question of hierarchies   At this point it is worth remarking on the importance Sangren attaches to the notion of a hierarchy, for it is also a notion which constitutes an important theme in the work of a number of writers who touch upon or are directly concerned with Chinese associations. Siu (1990), in her examination of the chrysanthemum festivals of south China, states that, in view of Skinner's work, little justification is needed to adopt a regional analysis for Chinese history and society (p. 786), and suggests that the literati culture, lineage and community institutions, and popular religion in late Imperial China served to legitimise the state and the people's respective places in relation to it (p. 790). Omohundro (1977) in his study of Chinese trading patterns in the Philippines argues that stable and non-random migration from south China to the Philippines produced ethnic differences (perceptible at least to the Chinese) among the Chinese in different Philippine cities. He accepts the validity of Skinner's intellectual constructs, and argues that marketing networks, even if not complete transpositions of homeland systems, do exist among the Philippine Chinese. The difference among the Chinese in different cities may be due 'to efforts to perpetuate regional subcultures by continued geographic separation in Southeast Asia' (p. 124) One example of this system presented by Omohundro is the patron–client hierarchy of smaller merchants or *ahentes* headed by the *cabacillo*. Crissman's model may owe an implicit debt to Skinner. Crissman wrote his paper shortly after the appearance of Skinner's seminal articles on Chinese marketing systems, and has applied his own version of central-place theory to the study of Chinese society.

Could it be, then, that the very *notion* of a hierarchy has to some extent predetermined the nature of the analyses conducted by Sangren, Crissman and Skinner and, therefore, many of the conclusions reached? Is there something inherently attractive in the notion of a hierarchy which endows it with enduring importance and a wide range of applications? It is certainly a clear, ordered and predictable concept. And, rather like a hexagonal pattern which can be discerned even among randomly distributed points, some sort of hierarchy can be found wherever smaller and larger settlements or groups or organisations exist together in any

particular area, no matter how the boundaries of that area may be delimited. The hierarchy is a flexible tool which can easily be imposed upon
phenomena and then used to explain the relationship of those phenomena to each other and to a greater whole.

Yet, as the discussions presented above and elsewhere (R.N.W. Hodder,
1993b) make clear, there are evident weaknesses inherent in Skinner's
study of Chinese marketing systems and in Sangren's analysis of
Chinese society. The hierarchical structure which Crissman envisages in
his model is also pure assumption, and it is unidimensional, for its sole
purpose is to provide *in fact*, as well as conceptually, a framework
around which Chinese society can form. Although still too rigid, and a
little too limited in its dimensions, Wertheim's (1964) view of Chinese
society raises objections which pre-empt Crissman's assumptions.
Wertheim considers the notion of a 'hierarchy' more as 'stratification'
which, in colonial society, is arranged by 'race'. But Wertheim is at pains
to stress the crucial importance of value systems, for it is from these systems that the stability of a stratified society ultimately derives:

> 'We should not primarily look for the inherent structure of a given society, but
> for the value systems adopted in different layers of society. . . . Instead of
> searching exclusively for integrative expedients, we should with equal intel-
> lectual force try to detect strains and conflict in society, as possible agents in
> future change. Therefore, a description of the society in purely synchronic
> terms seems basically inadequate, as conflicting value systems can only be
> understood in diachronic perspective' (1964, pp. 35–36).

Moreover, Wertheim argues, the social anthropologist should be aware
that any fact may also represent a symbol of these conflicting value systems in society:

> 'A good instance of the insoluble inadequacies into which the one-dimension-
> al approach to society is liable to run, is to be found in Skinner's interesting
> analysis of leadership among the Chinese community in Thailand. The
> author has attempted to analyse that community in terms of an evaluation of
> prestige which is assumed to be generally accepted among the Chinese
> group. But his own analysis presents sufficient proof that the dominating
> values as found by the author are not generally accepted. There are several
> incidents reported in his book which indicate that a large section of the
> Chinese community does not accept the leadership of those assuming that
> role, as for example the mass demonstration for a fee reduction for Chinese,
> or the chain of events during the period of emergency measures after a heavy
> fire in Bangkok' (1964, p. 36).

Crissman's model comprises a composite of observations upon which he
has imposed a conceptual framework with its own internal logic and
mechanisms, inveigling disparate observations into a cohesive, ordered
whole. In the fashion of the true social scientist, Crissman has constructed

an intellectual framework which he has then used to interpret and order numerous fragmented observations and ideas. Once successfully integrated it is then assumed that this accomplishment endows the selected observations with the status of factual evidence which can be used in support of the very conclusions with which Crissman's interpretation began. The result is a limited, mechanistic, unidimensional picture with an internal consistency which acts only to limit thought and to exclude other possible interpretations.

Comparison

A third problem with the cultural explanation is that its proponents do not consider, or pay very little regard to, the nature of similar organisations created by people who are not of Chinese descent. In those instances when comparisons are made, they are used to highlight the distinctiveness of 'the Chinese' association. Once more it must be emphasised that any attempt to defend this charge by claiming that similar phenomena may exist for different reasons, and that, therefore, the appearance of these phenomena in different societies does not represent grounds for rejecting the influence of 'Chineseness', merely presupposes the initial assumptions (including that of a link between 'culture' and the phenomenon) to be legitimate. This defence may also be turned around: the existence of the same or very similar phenomenon (such as an association) in different contexts reinforces the argument that institutions, values and forms of exchange possess many different and changing aspects, and, therefore, that no assumptions can be made about the existence of chain of cause and effect stretching back to an ultimate origin.

The criticism that Crissman's segmentary model is founded upon a tautology is the same as that made by Sahlins (1961) of Evans-Pritchard's work on the Nuer. Another important point raised by Sahlins which is also relevant to Crissman's model is that segmentation (rather like hierarchies) can be seen in all kinds of societies, social groups and organisations. Sahlins gives the example of the University of Michigan which is divided into colleges, semi-official divisions and minimal departmental segments. The same description could be made of the Chinese University of Hong Kong which is divided into colleges that cut across faculties and their minimal departmental segments; and of the University of London which is divided into colleges, some of which correspond to general faculties, and some of which contain their own faculties, and all of which comprise departmental segments. Segmentation, then, is not evidence of a causal connection between segmentary lineages and other forms of organisation.

This last point leads to another even broader and more fundamental weakness in Crissman's segmentary model. Sahlins restricts the term segmentary lineage to predatory organisations in conflict with other tribes. The segmentary lineage, Sahlins maintains, represents a particular stage of social development somewhere between less-developed *bands* and more advanced *chiefdoms*. All these stages are confined to societies at a tribal level of evolution. Doubts concerning the philosophy underlying a fixation with the classification of phenomena and a Darwinian view of society are implicit in many of the arguments presented in these pages. However, Sahlins' views do raise two interesting observations which are relevant to the current discussion. The first, which has already been commented on, is that lineages in southeast China may have been competing predatory organisations. It was precisely because of this characteristic that lineages stimulated the development of trade and markets. There is also the possibility, noted earlier in this chapter, that parallel competing and predatory behaviour (but not causally linked to lineages in China) was exhibited by the Overseas Chinese operating through associations, further stimulating trade. The second observation is that while Sahlins' taxonomy provides a way of marking out one society from another, Crissman focuses upon the 'sociological principles' of descent, locality and occupation rather than on a tightly defined segmentary lineage or organisation. Crissman's preoccupation with sociological principles allows him to establish tenuous causal links between the lineage and the association, but it also wipes out any distinctions between the Chinese lineage, the Chinese association in 'traditional' China or in Southeast Asia, and associations formed by ethnic groups other than Chinese in other parts of the world. Whereas Sahlins provides a method which could be used to compare, Crissman simply ignores comparison, implying that the sociological principles he has identified are well-nigh universal and possess extraordinary explanatory powers. Yet, he claims, the Chinese are peculiarly 'Chinese'.

Indeed, it is interesting and perhaps significant to observe that while most writers on Chinese associations seem content to marginalise or ignore comparisons and, instead, highlight attributes and characteristics which are assumed to be uniquely 'Chinese', studies of associations which have arisen at different times in other parts of the world emphasise practical and logistical considerations rather than deterministic cultural traits.

Citing La Piana (1927), Thompson (1989) notes that in the City of Rome during the first centuries of the Roman Empire, immigrants from the same place of origin 'tended to cluster together for social and psychological security in particular districts and in self-imposed ethnic residential segregation, maintaining their language and their culture for as long as they kept up their links with their native countries . . . ' (p. 116). Clubs and associations of a religious and economic kind supported the immigrants' attachment to their cultural traditions until the newcomers had gathered the means and experience to go out and take their place within the wider society. These associations, which were often organised by co-religionists, also institutionalised the fellowship of slaves and free-commoners and thereby aided the attempts of polytheistic religious cults, and later the Christians, to integrate these groups. The associations were also a prop for self-esteem and thus represented part of a set of ego defence responses which included 'avoidance of potential mockers . . . ; seeking revenge on mockers and potential mockers where possible (by cheating or harming them in some way when it seemed safe to do so); acting out (in protective clowning, for instance) the role or roles suggested by the disparagement; or the more positive defence mechanism of a powerful striving for success in a world whose popular slogans included "A man's purse determines his worth" and "Money makes the ugly handsome", and in which (as Seneca puts it) "none are more ready to trample on others than those whom personal experience of insults has taught to be insulting"' (Thompson, 1989, p. 138). But the main attraction of the associations was the social security they offered, the relief from loneliness, and the guarantee of a decent funeral.

Northrup's (1978) study of the slave trade in sub-Saharan Africa during the eighteenth and early nineteenth centuries provides another interesting parallel to the Overseas Chinese association. One of the most significant trends of this era was the spread of various men's associations among the communities enriched by trade—associations which 'directed the energies of the traders towards the promotion of traditional values and social harmony' (p. 112). These voluntary associations of prominent men were of several types:

> 'In one type, called secret societies because their rites and lore were carefully concealed from non-members, membership was open to all adult men . . . and generally acted to support established authority by restraining the fissiparous tendencies of lineage organisation. In some places these societies replaced village councils of elders as the primary structure of internal governance. Another type of association had a more limited membership, being open only to people who felt or exhibited in some way a special call to a quasi-religious position of influence and power. Men whose lives demonstrated unusual achievement (of which economic success was one type) could obtain special titles recognising their achievement and granting them powerful roles as political leaders, judges and peacemakers. . . . Whatever their earlier functions it is clear that the structures based on the association, like those based upon descent, were very important in explaining the changes in political organisation which occurred as a result of the expansion of trade in the eighteenth century' (Northrup, 1978, p. 108).

Citing Ruel, Northrup goes on to describe how one particular type of secret society—the *Ekpe* or leopard association among the Bayang in the upper reaches of the Cross River—functioned as a way of reconciling the conflicting principle of individual interest with collective action: the wealthy could buy graded titles in the association which endowed them with higher status while at the same time subjecting these wealthy individuals to the discipline of the association's rules which reflected the norms of society at large (1978, pp. 108–109), So, too, in Old Calabar, the *Ekpe:*

> 'helped the Efik society by curbing some of the blind desires of the wealthy and restraining their oppression of the weak and the poor, because the poor had every right to take his case to the *Ekpe* council in order to exact the justice due to him [and even slaves were eligible for membership in the lower grades]. . . . It kept each member of the fraternity striving to ascend the *Ekpe* ladder in order to obtain the privileges due to the next grade so that he could better his social and political status and gain recognition in the community; in this way he had to be as civil and law-abiding as the Fraternity's regulations prescribed' (Aye, 1967, p. 75, cited in Northrup, 1978, p. 109).

The *Ekpe* society of Old Calabar spread to the Arochukwu, the Aro, Igbo and Ibibio and variants of the society spread to the Bende-Ngwa-Ndoki area—other communities which traded with the Efik. Although not exclusively concerned with economic matters, the highest grades could only be afforded by the wealthy, and certainly it appears that membership in this common society greatly facilitated trade among the different communities.

Another type of society which similarly acted to legitimise trade and to defuse the tensions created by trade, was the Title Society among the Igbo:

> 'A title was taken in an elaborate ceremony which symbolised the title holder's transition from ordinary human status to a higher, more spiritual status, which, while giving him many privileges such as great freedom of movement, exemption from assault and bodily labour, and judicial powers, also hemmed him in with strict moral obligations enforced by spiritual sanctions and public opinion. A man could not, for example, hold a title and continue to be a trader. Title-holders were distinguished by special dress and regalia including an ivory "trumpet". Little is known of the early development of this society and of others like it, but by the nineteenth century titles and trading wealth were clearly linked. Wealthy men sought to convert their new money into traditional offices of high social standing, while society sought to regulate and divert "the intoxication of wealth and power" to sober social ends' (Northrup, 1978, p. 111).

Trade, then, had altered the context of existing institutions and values. The associations were an expression of a symbiosis between, on the one hand, the 'turning' of these existing institutions and values towards the extension and institutionalisation of trade and the defusion of tensions which thereby arose; and, on the other hand, the 'turning' of trade towards support of existing institutions and values. There are clear parallels here not only with the Chinese associations but also perhaps with the guilds which arose in England (and many other countries in Europe) during the Middle Ages. Based upon different occupations, the guilds were not only concerned with economic matters; they provided their members with social channels along which the profitable merchant could rise to positions of power within a state that came to recognise trade and the material progress it generated as the basis of national strength.

If viewed as a symbol of elite designation, the associations established more recently in contemporary Africa also suggest parallels with the Chinese association. Cohen's study (1981) of the Creole appears to portray the association as a nexus or expression of reciprocity, and demonstrates how, as the context changed (rising tension between different African groups), the aspects of individual actions and decisions (such as the formation of voluntary associations) changed, and were then directed towards different ends, in this case, stronger and more formal corporate bodies which desired to preserve their elite status which rested upon education, state employment and the ownership of land. This instance, it may be argued, further illustrates the flexibility of institutional patterns and thus the spontaneity of the human mind, for the association was 'turned' not only into a symbol of elite designation, but also towards the institutionalisation of reciprocity and its underlying wealth values. The Creole elite:

'collaborate informally through their exclusive network of amity—a web of primary relationships which are governed by moral rights and obligations, and objectified and kept alive in the course of frequent, elaborate and costly ceremonials.

A man aspiring to identify with the elite may succeed in acquiring such external markers as accent, housing, clothing and other items of conspicuous consumption. But he will not thereby become automatically affiliated within the power elite and partake in their privileges. To do so, he must achieve the much more difficult task of "grafting" himself on to the inner network of primary relationships which link the members of the group together. . . .

The Creole network is . . . complex, consisting of relationships of friendship, of ritual brotherhood and parenthood, of old school links, and so on. But because of their small number, their concentration in one locality, and their tendency to marry within their group, the Creoles have tended to articulate their total network in the idiom of one institution, the cousinhood. . . .

It is indeed no exaggeration to say that any Creole can trace some kind of kinship relationship to every other Creole. No sooner do two Creoles develop a friendship than they "discover" a link that shows they are cousins' (pp. 60–65).

After independence, with the loss of the colonial administration's protection, the Creoles began to play down their distinctiveness, most especially their external markers of identity and, indeed, began to emphasise their kinship ties with other African groups. On the other hand, Africanisation had brought the Creoles into senior positions within the civil service and had also provided an opportunity and reason for other African groups to give vent to their desires for land and jobs. The Creoles responded to these external threats to their enhanced power by strengthening 'family' (reciprocal) ties and intensifying associated ceremonies. And whereas under colonial rule the Creoles had set up numerous voluntary associations—mostly little more than middle-class diversions such as cultural societies, youth clubs, cultural societies, men's clubs, Old School Boys' and Old School Girls' associations—which possessed no fundamental corporate organisational function, the Creoles now established more formal corporate organisations, defining lines of authority and centralising that authority in a few individuals who also held power and influence within the bureaucracy and many other professions:

> 'The most dramatic innovative change among the Creoles during the last twenty-five years has been the massive adoption of Freemasonry by a large proportion of their men. Yet this change, though innovative in a number of respects, and fulfilling new organisational needs, both universalistic and particularistic, is essentially a *new combination* of diverse old symbolic forms and practices.

> Secret societies generally are a tradition in Sierra Leone and in many other parts of West Africa, and many Creole men had been affiliated to traditional secret societies like the Hunters' societies. The Masonic order itself was introduced into the country during the colonial period by the British civil servants and army officers, and a few important Creole men became Masons at that time. The relationship of brotherhood created by the order among its members is yet another version of the more traditional forms of amity relationships, like those of family, cousinhood, godparenthood, and friendship, which already linked many Creoles. Commensality—eating and drinking together—which is an essential part of the activities of almost every lodge, has always been an elaborate tradition in Creole culture. The initiation of new members, and the promotion of members to higher ritual grades within the order, are celebrated in dramas of death and rebirth that are familiar to the Creoles. The exhibition of the Bible and the singing of hymns at every meeting are all standard Christian practices of the sort that have played such a crucial role in the formation of Creoledom. Regular attendance at lodge meetings resembles regular attendance at church, and the existence of two hierarchies in lodge organisation, the one ritual and the other administrative, echoes the organisation of the Christian churches. . . . Moreover, the organisation and administration of the lodges involve skills no different from those already familiar to the Creoles in the organisation of their clubs, societies and other types of voluntary associations. . . .

What is new is that these symbols have been restructured . . . to create a more rigorous and effective organisation, linking some of the most important mature males as the patrons—the decision makers in many walks of life—in a close-knit, confidential brotherhood, in order to serve new organisational functions for both particularistic and universalistic purposes. On the particularistic side, post-Independence Creoledom sought an integrated corporate leadership, supported by a unified authority structure, to take decisions and to ensure the compliance of individuals. The order in effect linked at a higher level a chain of authority running from the grass roots of Creole society to the top: members of a "family" would abide by the authority of their head-patrons. . . .

Simultaneously, these "family" heads occupy important positions in the state bureaucracy and the major public professions. The order solves the problem of the rapid and effective exchange of information necessary to coordinate state institutions and services . . . ' (pp. 153–155).

These last comments and Cohen's observations that the Creoles borrowed Masonry from the British civil servants and army officers raise yet another interesting parallel with Chinese associations. Lupton and Wilson's (1959) study mentioned in Chapter 4 similarly demonstrate that in Britain many of the top decision-makers (Cabinet Ministers, senior civil servants, directors of the Bank of England, directors of other major banks, city firms and insurance companies) were linked together and with other influential individuals by complex networks of close personal and kinship ties. Many of these individuals had attended the same schools and universities and now belonged to the same clubs. In Britain, too, this web of relationships enabled rapid, informal and often necessarily intuitive decisions to be made.

Again it might be argued that the existence of associations similar to those of the Overseas Chinese does not invalidate the argument that Chinese associations are a product of 'Chinese' characteristics, whether or not these characteristics are found elsewhere (and are also held to be responsible for similar institutions). Again, however, such an argument is worth consideration only if it has already been demonstrated that the association is a consequence of 'culture'. It might also be countered that associations elsewhere are not, or are not so clearly, organised according to descent or locality or Confucianism (or any other norms or characteristics taken to be 'Chinese'). Yet this argument, too, merely presumes the existence of an ultimate 'cultural' origin. The fact that associations have been established in different or similar circumstances may simply demonstrate that an institution possesses many different aspects: it is an expression of the human mind's spontaneity, of different values, perceptions, beliefs, ideas, choices and actions and may be directed towards different ends.

Conclusions

L. Lim (1983) notes that personalistic networks and associations such as the Rotary, Lions and Kiwanis Clubs are familiar within the 'Western' corporate world. Yet these associations 'which in their formal and informal functions much resemble the Chinese associations of Southeast Asia . . . are rarely, if ever, claimed to be the main reason for the individual career successes of their members' (pp. 4–5). So, too, the Chinese association cannot help explain sustained material progress. But if taken as an expression of the extension and institutionalisation of trade, the study of Overseas Chinese associations may help in an understanding of that progress.

It is true that Chinese associations are one of the less likely phenomena which could be presented as an explanation for Chinese economic success, although some writers have made just this suggestion (see, for example, Omohundro, 1981, p. 106; Deyo, 1983). However, Chinese associations have proved a rich source of material for the creation of a cultural mantle which can be used to scientifically explain 'the Chinese' and their economic progress.

This symbiotic relationship between the scientific approach and cultural determinism lends itself easily to an explanation of 'the Chinese': 'science' needs respectable 'origins', 'variables', 'elements' and 'forces' to work with; and the delineation and dissection of a notion as diffuse and as vague as 'culture' needs the mock precision and rigour of the humanities' own brand of 'science'. The two together help slake the thirst of some intellectuals (particularly, to use Fisher's phrase, the Californian neophytes, and one might add, those Chinese scholars who enjoy the manufactured exclusivity of 'Chineseness') for a specialised niche in which to foster the kudos and affected mystique attached to a field of research which is alien and of limited access to most others.

The study of 'things Chinese' unfortunately retains its mysterious, romantic and scholarly cachet. By virtue of a professed objective science its practitioners claim to see the world as it is. They produce *en bloc* the phenomenon of 'the Chinese'—a distinct, clear-cut entity which can be more easily manipulated within the intellect in order to uncover an accurate explanation and analysis of these people and their values, beliefs, institutions, problems and solutions. Individuals are pulled, twisted and directed within the intellect by those forces which make up that cultural mantle which defines 'the Chinese'—an expedient notion that nourishes the writer's feelings of exclusivity and the propagation of his ideas. The creation of 'the Chinese', refined and honed by the purported objectivity and precision of science, produces for the American or European writer a golden society which he understands more clearly and intimately than other scholars or laymen. He becomes the channel, as he sees it, between two peoples. It emphasises his uniqueness within his own society and among the Chinese. Science, then, masks a scholarly version of the bare-footed traveller who understands of what he sees only that which he first created within his own mind and then imposed upon another people.

For many Chinese, too, the notion of 'the Chinese' serves as an intellectual justification of a self-conscious uniqueness which must be preserved at all costs. It is a distinctiveness which can be used to appeal to or to demonstrate an individual's sense of righteousness and superiority, or which, like a sun-blind, can be drawn to protect him from external judgements of quality, ability and performance, and from external competition. 'The Chinese', then, is a political device which may be used to 'out-Chinese' opponents; it is an expression of defensive parochialism; and it may serve as a call to nationalism and racial assertiveness.

But perhaps most importantly for a writer, this science allows him to remain dryly intellectual. If he puts aside his science he is at once confronted with a bewildering complex of raw emotions, values, beliefs and practices with many different aspects which the observer may find objectionable, intolerable, unfathomable, inconvenient and contradictory. If he is not already trapped by his education and assumptions, and if he is willing to step outside his own values and beliefs, then that mass of tell-tale characteristics which makes up 'the Chinese' begins to fall apart. Without his science, he must not only be aware, but also consider and try to understand, emotionally as well as intellectually, the raw feelings and motivations of others, the many different and changing aspects of individual behaviour, values and beliefs and, therefore, the different and changing aspects of institutions, forms of exchange and moral imperatives.

The school of anthropology which Sangren loosely refers to as the British structuralists (a term which he uses to embrace Radcliffe-Brown, Evans-Pritchard and Freedman) is clearly aware of the different and changing aspects of phenomena, and of the different interpretations possible of any given observation. This awareness also forms an important part of 'post-processual' archaeology:

'If material culture is a "text", then a multiplicity of readings could have existed in the past. An example is the varied meanings given in British society to the use of safety pins by punks. It seemed to me . . . that individuals would create verbal reasons for such items but that these verbal reasons were not "correct" or "incorrect"—they were all interpretations of a text in different verbal contexts, and in different social contexts. I had the same impression many times in Baringo. Individuals seemed to be making up the verbal meanings of things as I talked to them, contradicting and varying their responses as a social ploy.

As Drummond (1983) has suggested, the interpretation of meaning is not a matter of "getting it right". . . . The cultural reality is a shifting assortment of varied perspectives, so that, when looked at as a whole, there is no one "true" version of events' (I. Hodder, 1986, pp. 149–150).[1]

It is true that Sangren's British structuralists use the language of the modern-day social scientist: they talk of theory and principle and they share the same general aim—'the elucidation of the sufficient causal or functional determinants involved in the observed data of behaviour' (Fortes, 1967, p. 345). In many ways their method and interpretation are too mechanistic: the presumption of structure precedes analysis and within these intellectual constructs are organised relationships, actions, symbols and motivations; and while they achieve something of a balance between an emphasis on individual choice and the constraints placed upon the individual by existing institutions and values, individuals are nevertheless separated from the 'structure' rather than made synonymous with it.

Even so, it is clear that many of these writers are concerned first and foremost with meticulous, painstaking description and interpretation which takes into account different aspects and possibilities; the rather mechanistic impression left by their careful, step-by-step method is reinforced by the enviable ability to present complexity in clear and precise language; yet they leave very little to vague mystical laws or overarching 'structures'. This approach is a very far cry from the application of programmed minds and theory: a technique, not a way of thinking, which provides a ready-made explanation of the phenomena to be analysed—a device to rubber-stamp the identification of causal elements and structures capable of producing specific phenomena. To a large extent Sangren has read a currently fashionable brand of rigorous science into the works of others who attempt to elucidate complexity, to delineate a level of imperfect and cautious generalisation, to understand and to interpret in clear and precise language and often with compassion, humour and integrity.

Much of the literature on the Chinese association represents yet more elements to be woven into that cloak which defines 'the Chinese'. Once again, the scientific approach operates to simplify complexity. An ultimate cause or origin is assumed to underlie the variety of associations and their different aspects, and analysis is pushed further and further back into the gyres of cause and effect which are, in any case, often fixed and delineated in advance by the writers' own predilections for a specific theory and by suspicions about an ultimate origin. In order to help facilitate analysis, individuals and their changing values and desires and, consequently, the different aspects of associations are marginalised or dispensed with altogether. The individual is then replaced by laws and forces which are assumed to exist. Occasionally the assumption of these forces is made explicit. But more often they are either left unspecified or temporarily legitimised by platitudes emphasising the need for further research; or are left unmentioned in the hope that the final explanation and the scientific methodology used to reach that explanation are justification enough; or some form of words is devised explicitly denying the existence of such forces, and yet implicitly allowing their covert existence and operation within theory. Unable to cope with the raw emotions and desires and ploys of the individuals it purports to explain, the scientific approach, reinforced by popular sentiment and moralising, allows the wishes, desires and emotions of individuals and their conscious manipulation of others to melt away—all this is treated as unworthy for analysis or as merely symptomatic of greater things at work.

Note

1. It is with this awareness in mind that Ward's, Carstens's and Yao's observation that behaviour is often justified by the Chinese as 'something which the Chinese do' should be interpreted. To present such an observation as evidence of a cultural plan smacks of preconceptions and convenience.

6

A community in Davao

Introduction

It is evident from the previous chapters that while circumstances are generally taken to be part of any explanation of the economic success of the Overseas Chinese, it is commonly believed that the most important determinant—in combination with certain economic, sociological and psychological forces—is 'culture'. So powerful is this desire to ascribe particular values, institutions, forms of exchange, practices and actions to an underlying culture, that comparisons between the Chinese and non-Chinese are either ignored or marginalised or used very selectively in an attempt to highlight characteristics which are held to be uniquely Chinese and which, it is reasoned, must therefore explain the economic success of Overseas Chinese. It was noted in Chapter 4, for instance, that whereas Redding (1980) ignores similarities between 'Chinese' and 'Western' thought (categories of thought so cumbersome as to be effectively useless except in casual conversation when the conveyance of emotions and feelings may be more important than precise meaning), Mackie (1992b) uses comparisons between Chinese and non-Chinese companies to single out 'Chinese' characteristics. And, in Chapter 5, it was observed that, with few exceptions (L. Lim, 1983), most writers choose to disregard similarities between Chinese associations and those formed by non-Chinese.

An additional example is provided by Omohundro (1983) who devises a curious, Darwinian-cum-Pavlovian, dialectic between 'culture' and Chinese merchants in the Philippines. A selection of Chinese cultural materials exists simply for the merchant to choose from as he acts to secure his livelihood. Slight alterations are made to these materials, and if his choices are successful, these alterations are then retained as part of a specifically 'Chinese' cultural stock: 'All merchants participate in this statistical trend of minute changes over time, even the China-born traditionalists. Individuals are ruined and rewarded and the resulting pool of cultural values and strategies, though based predominantly upon pre-existing cultural material, is altered in emphasis, in proportions' (p. 80).

This culture (some of it 'imported' from China, some of it peculiar to the Overseas Chinese) is taken to be separate from Filipino culture. However, in order to extend social and commercial networks, the Chinese acquire certain fragments of Filipino culture, or adapt elements of Chinese culture in an effort to imitate Filipino practices. Thus Omohundro regards blood-brotherhoods as a 'Chinese' practice 'under-going modification in the Philippine environment' (1983, p. 70), gradual-ly moving closer to the Filipino *barkada* (or clique). According to Omohundro, the 'Chinese-style sworn brotherhood was usually formed between young single men who became close through co-residence, school, work, wartime or immigration experiences' (p. 70). The Filipino *barkada* 'system' (as Omohundro describes it) is less formal and multi-stranded. The *compadre* (godparent) 'system', on the other hand, is a Filipino custom being reworked by immigrant Chinese. Arguably, how-ever, Omohundro—with one eye on cultural determinants, and the other on methodological convenience—has allowed thought to be limit-ed by artificial analytical categories of relationships and reciprocal net-works. By first creating mechanistic categories, the focus is shifted from the substance of actions and relationships towards superficial and, more often than not, contrived and selected differences between these cate-gories—differences which are taken, by definition, to be 'cultural'. Thus *compadre* and *barkada* are presented as rather awkward, involved and scripted performances rather than as imperfect descriptions of soft-framed institutions. Arguably such relationships are better understood as friendships and reciprocal networks. Indeed, if comparison is made with Freedman's (1957) observations on the importance of friendship (see, for example, pp. 87–88), which, in Singaporean Chinese society, implies networks of obligations comprising the reciprocal exchange of favours, then it is difficult to discern the purported 'cultural' peculiarities identified by Omohundro, even if one first accepts his analytical cate-gories of relationships among Filipinos. The soft-framed institutions of reciprocity may differ in detail but not in substance.

Gosling (1983), on the other hand, puts 'Chineseness' into a more bal-anced comparative perspective and yet nevertheless places considerable emphasis on the fact that institutions exist and that they are efficient. In his study of Chinese crop dealers in Malaysia and Thailand, Gosling writes:

'The orientation of the Chinese, their values, prior experience and goals compared with those of the Malays and Thai inclined the former into commerce and inhibited the latter. The economic situation in Malaysia and Thailand in the 19th and 20th centuries prior to independence facilitated Chinese dominance of commerce. Chinese socio-economic organisation also contributed to their success, as did their situation as an ethnic minority. The existence of other ethnic groups as middlemen indicates that there is nothing inherently unique about the Chinese which explains their dominance; they were simply the right people in the right place at the right time; once established, their organisation and efficiency made them formidable competitors.

The orientation of Chinese migrants to Southeast Asia was a product of culture, experience and aspirations rooted in the land from which they came. Extreme population pressure in Southern China's limited land fostered the values of acquisition and accumulation. Lineage adherence was vital for the provision of land from direct ancestors and involved the obligation to increase and pass on wealth to maintain future generations, saving for the survival of the lineage, in a situation of increasing pressure on resources. Lineage groups and other associations and organisations were vital for ensuring current survival by sharing available resources. Upward mobility could not be achieved from farming where land was not available and escape from rural poverty was best achieved through accumulation of capital and contacts to be used in non-agricultural occupations.

South China was long the centre of China's foreign trade and its urban population included merchants, artisans and labourers to support a vast network of local and long distance exchange. Merchants were able to achieve status through wealth, used to educate their children and to compete for positions in the bureaucracy or to purchase privileged positions. The rural and urban poor, acquisitive, hardworking and frugal by necessity, found in the merchants of South China both the model and means for the better life. Trade was the way to achieve higher status and South China's trade with Southeast Asia provided the opportunity' (p. 150–151).

There is much to agree with in this passage. One important point which arises out of Gosling's discussion is the notion that minorities are likely to engage in trade because they are free from the restraints which normally exist within their own communities and those of the indigenes. The same idea is also inherent in the interpretation of markets as exogenous phenomena (Chapter 2). If this idea were taken a little further, then it could be suggested that it is not so much freedom from prevailing institutions and values and forms of exchange as the freedom and desire to 'turn' existing institutions, values and other forms of exchange towards the extension and institutionalisation of trade which helps provide an understanding of economic success.

Nevertheless, Gosling is clearly suggesting that circumstance and Chinese socio-economic organisation and values led to the formation of institutions which made the Chinese competitive and economically successful. The impression left is that of deterministic structures rooted in a culture which is specifically Chinese and in a 'force' which appears to be that of socialisation: efficient institutions, it would seem, arose out of, but now revisit, the Chinese, and explain their economic success.

Although it may seem a little unusual to concentrate upon a Filipino community and its Filipino and Chinese merchants, an understanding of a specific indigene community is pertinent and important. The analysis helps to illustrate that apparent differences between 'The Chinese' and 'The Filipinos' is a consequence of a unidimensional interpretation of values, institutions, actions, beliefs and ideas. If attention is focused upon wealth values, forms of exchange and their associated institutions and moral imperatives, and if these phenomena are 'turned' in order that their many different aspects may be viewed, then apparently profound differences begin to dissolve, and striking similarities among groups emerge. It is the manner and ends to which multidimensional institutions, values and forms of exchange are directed, rather than any inherent difference in such phenomena, which help towards an understanding of material progress. Ethnicity and 'culture' come to be seen for what they are: unidimensional presentations, interpretations and perceptions of institutions, values, beliefs, actions, ideas and behaviour which possess many different aspects. Moreover, these unidimensional presentations, interpretations and perceptions may themselves come to form part of the institutional patterns and moral imperatives associated with a particular form of exchange.

The pre-eminence which Chinese and Filipino merchants give to trust (only one of reciprocity's soft-framed institutions), introduces a stability and predictability into relationships which enables complex networks of reciprocity to be directed towards the prosecution of trade. So, too, within the company, whether 'Chinese' or 'Filipino', the emphasis given to trust and loyalty—combined with certain organisational methods of control (which help monitor the performance and actions of individuals), and with the institutionalisation of certain patterns and standards of behaviour—enables reciprocal exchange and its soft-framed institutions to be distanced from, and focused upon, the prosecution of trade.

There are, then, striking similarities both in the way in which greater predictability and stability are brought to relationships, and in the values and the broader institutions used to safeguard, legitimise and facilitate trade: 'Filipino' and 'Chinese' are not, and should not be treated as, analytical categories; rather, they are concepts which constitute part of reciprocity's soft-framed institutions. It is not surprising, then, that there are extensive networks among merchants, irrespective of artificial analytical categories of 'culture' or 'ethnicity'. These networks find expression, not only in business transactions, but also in the ownership and organisation of companies, in political machinations and in the associations.

Davao City

The city of Davao lies on the southern coast of the island of Mindanao in the south of the Philippine archipelago. The city is one of seven provincial-level units which make up Region XI (also referred to as Southern Mindanao)—one of 12 regional administrative units into which the Philippines is divided.

Large rural areas are encompassed within the city's administrative jurisdiction which covers more than 2400 km². About 40% of this area is classified as agricultural. Most of the urban and urbanising parts of Davao are on a coastal plain which forms a narrow band averaging 5 km in width and extending 36 km from Lasang to Binugao. The city proper, which occupies a thin 25 km slice of this coastal strip from Toril to Panacan, comprises a hotchpotch of agricultural land, residential estates and denser agglomerations of commercial and residential buildings in and around the city, Poblacion, Toril and Panacan. These beaded settlements are strung together by the main highways which are, in places, lined with coconut-wood houses, shops, *sari-sari* stores and the occasional light-industrial plant—most often food processing, bottling or ice manufacturing. Although the city proper covers only a small percentage of the total area of Davao City, it contains about 85% of the total population living in the urban and urbanising areas, or almost 60% of the city's total population (estimated at over 923 000 in 1993).

Agriculture still forms the mainstay of Davao's economy: more than 40% of the city's GDP is derived from agriculture, 32% from services and 24% from industry. While the number of people employed in manufacturing fell by 7000 between 1989 and 1992 as per capita productivity increased, the number employed in the agricultural sector (which is taken to include related businesses as well as fisheries and farming) increased by 8000 each year. Of the total labour force, over a third is now employed in the agricultural sector, 10% in manufacturing activities unrelated to agriculture, 22% in mining and 20% in retailing and wholesaling; and a little over 12% in the service sector.

Moreover, agricultural goods are the most important, if not the only, source of foreign exchange. The share of Southern Mindanao's exports handled by the port of Davao fell markedly from 70% in 1987 to 57% in

1993. The value of goods produced in Davao and then exported rose from US$327.4 million (f.o.b.) to US$351 million (f.o.b.) in 1988; the value of these exports fell to US$271 million in 1992, and then recovered only slightly during 1993. The vast majority of Davao's exports comprise bananas (fresh and dried), coconut oil (crude and refined) and coconuts (fresh and semi-processed). Goods imported into Davao and used in the city mainly comprise chemicals. The value of these imports rose from US$53 million (f.o.b.) in 1987, to more than US$75 million in 1993.

It is also in the agricultural sector that investments are concentrated. Within Region XI around 80% of investments registered with the Board of Investments are in this sector, and half of all investments are concentrated in Davao. Only around one-quarter of these projects are joint ventures, and a mere 4% are undertaken by foreign interests. It is true that in 1993 (after six years of stagnation at under 1 billion pesos per annum) the value of investments leapt to over 4 billion pesos with the start of a single joint venture between Davao Union Cement and the Sumimoto Cement Corporation (Japan). However, it remains to be seen if this level of investment can be sustained, or increased, and the range of activities broadened.

Local government economic strategies—with their references to hierarchies of growth centres, to integrated development areas, to economic zones and to 'balanced and equitable development'—read more like undergraduate essays than a serious attempt to realise sustained and rapid material progress. On the other hand, despite the presence of these grandiose and now-fading volumes on the bookshelves in some government offices, there has, in practice, been a clear shift in philosophy. Government, central and local, now sees itself more as an instrument to facilitate the private sector rather than as a framework of institutions and ideology into which business must fit (R.N.W. Hodder, 1991). Complex planning institutions and procedures are no longer held in much favour. The emphasis is now upon: direct consultation and clear communication between government and business; attempts to entice foreign investment; the promotion of exports; and the construction of a trade polygon which embraces Davao, Sabah, Labuan, Brunei, Sarawak, Kalimantan, Sulawesi and Maluku. Industrial parks at Ilang and Panacan in Davao are planned, and a range of fiscal incentives (including income tax holidays and exemptions from a stream of other taxes, duties and fees) are now available, and restrictions on the employment of foreign nationals have been loosened. There is also a stronger commitment from local government to embark on those infrastructural projects which are of real value to business.

All sectors of the city's economy are dominated by privately owned enterprises. For most of the period from 1985 to 1989 the free-enterprise economy of Davao led the region's economic growth which averaged 4% per annum, while over the same period the Philippines' GNP rose from a negative figure of –6% (1984) to a little over 1% in 1986, and then rose still further to 6.7% in 1988 (*Inquirer*, 2 January, 1989). From 1988 to 1992, the city's GDP showed a steady, if slow, improvement of no more than 2% per annum.

Market Davao

Davao's growth since its establishment in 1848 has in many ways been similar to that of Hong Kong. As in Hong Kong, docks and markets, which formed the primary foci of Davao City's development, later gathered around the more sophisticated expressions of trade's institutionalisation which had formed the secondary or tertiary points of growth. In Davao, however, the city's physical development has been much slower and less sure, for the city did not lie on any international trade routes, and, while the Muslim Maguindanaons (or Moros) had arrived as merchants, the indigenes, whose societies were dominated by reciprocity, had little experience of trade or handling money.

Under Spanish rule the small outpost was connected to the outside world only by a long and difficult journey to Manila. The settlement was poorly administered and the energies of its small population, which numbered no more than around 4000 missionaries and soldiers together with their families, were directed primarily at converting the indigenes to Catholicism. Nevertheless, it was the Spanish who gave the first stimulus, limited though it was, to Davao's growth. The settlement lay a little to the west of Landing and 'Trading' (an area which still bears the same name today) near the mouth of the Davao River. By the late 1850s the settlement comprised three streets running parallel from north to south. The northern end of these streets was marked by a church; the southern end was crossed by two streets running east to west.

When the Americans took over from the Spanish in December 1899, they encouraged discharged American soldiers to turn their hand to the cultivation of abaca and coconuts. They, in turn, encouraged unemployed Japanese manual labourers (who had originally emigrated to the Philippines from Kyushu, Okinawa and western Honshu to help with the construction of roads) to work on the new plantations. American rule also marked the beginning of more intensive waves of immigrants—including professionals, homesteaders and small businessmen—

from central Bisayas and Luzon. Many of these Filipinos and mestizos (predominantly Spanish–Filipino and Chinese–Filipino) bought large tracts of land from the government or from the indigenes, or acquired the land through marriage. Initially, the migrants organised themselves into fairly distinct groups based on language and place of origin. These groups included, among others, the Cebuanos, the Boholanos, the Capizeno and Antiquenos (two mutually intelligible variants of Ilongos in Negros), the Ilocanos, the Tagalogs, the Kapampangans and the original pioneer families who had married with the indigenes, spoke their own distinct language, and regarded themselves as the true Dabawenyos. It was at this time that the Chinese began to arrive. But it was the Japanese who came to dominate the economy: they came to control many of the industries such as hemp, fishing and lumber; it was they who proved more successful at managing plantations; they dominated in retailing; and it was they who acquired large tracts of land, despite legislation introduced in 1919 which attempted to limit their economic success.

The city's growth from the beginning of American rule until the outbreak of the Second World War took place along the grid comprising: the three streets—Claveria (also known as Recto), Bolton and Anda—running north–south; and four streets running east–west—Magallanes, San Pedro, Rizal and Escario. Within this grid, development was focused on Trading (which lies just to the east of the junction of Claveria (Recto) with Magallanes and San Pedro), and on a market at the junction of Bolton and Magallanes. From these two foci development spread westwards along Magallanes and San Pedro, and, to a lesser extent, northwards along Recto (Claveria), Bolton and Anda. To all intents and purposes this—the commercial heart of Davao—was Japanese. The main commercial buildings housed Japanese enterprises, and many of the smaller businesses run by Filipino or Chinese dummies were owned by Japanese.

Another small settlement, which later merged with the main settlement, was Santa Ana—a small wharf constructed by Japanese and American plantation owners nearly two kilometres to the northeast of Trading. The two settlements were separated by a mixture of agricultural land and swamp. By 1928 a public market had been established at the north end of the main settlement at the junction of Recto and Uyanguren (now Magsaysay Avenue)—the main roads which connected Santa Ana with the city proper (as it then was)—in an area which was largely agricultural except for a few houses scattered along the two main thoroughfares. By 1939, another market had been established at Bangkerohan at the western end of Magallanes Street. This completed a triangle of markets which marked the southern, western and northern points of the main settlement, and from which development crept along the street grid. A second adjacent triangle was marked by Trading and the market at Bolton, the market at the junction of Recto and Uyanguren, and Santa Ana (Figure 6.1). Within this adjacent triangle, however, growth remained even more limited until after the Second World War.

Seven or eight kilometres to the southwest along the coast from Trading is Toril. Before the road from Toril to Bangkerohan was constructed during the early twentieth century, it was possible to reach Davao from Daliao (a small coastal settlement just to the east of Toril) by walking along the beach at low tide. Toril was a market town. The market was originally established on Sa'avedra Street. It was a continuous market which swelled on Sundays when farmers from the rolling hills and valleys to the north and west, and from the mountains three kilometres further to the south and southwest, came down to sell and buy goods. Even in the mid 1950s, the settlement, such as it was, comprised little more than a market, a church and a few nippa huts scattered along the streets which radiate from the market in all directions (Figure 6.2). But it was around the market and along these streets that shops and other commercial buildings and also residences gradually appeared, forming what is now the commercial heart of Toril. In the mid 1970s the original market was moved to the bulking-up point for goods destined for Bangkerohan by truck—a larger site a little way to the north just off Agdon Street at the edge of town. The original market-place is now a small, empty square overgrown with grass and a few palm trees, surrounded by nippa huts and, along the side which faces the road, a handful of wooden stalls selling meat, fruit and vegetables.

During the 1950s and 1960s Davao was stimulated by new waves of migrants, including a number of future tycoons such as Alcantara, Dizon, Floirendo and Pamintuan, who were attracted by the opportunities for acquiring logging concessions and huge tracts of land for lucrative plantation crops. Until the mid 1950s the city's growth remained concentrated within the primary settlement triangle, though beads of growth took place along Claudio which runs northeast from Bangkerohan to Recto (Claveria) and Uyanguren, forming the western edge of the settlement. Within the secondary triangle marked by the junction of Recto and Uyanguren, Trading and Santa Ana, scattered growth took place slowly alongside the main thoroughfares running from the docks westwards to Recto, southwest along the coast to Trading, and from mid-point along that coastal route to the junction of Recto and Uyanguren.

In 1954 another market was established in an unpromising area of

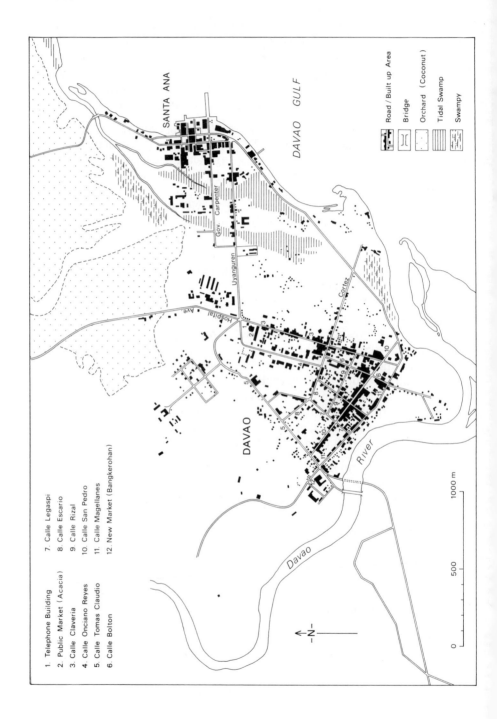

1. Telephone Building
2. Public Market (Acacia)
3. Calle Claveria
4. Calle Onciano Reyes
5. Calle Tomas Claudio
6. Calle Bolton
7. Calle Legaspi
8. Calle Escario
9. Calle Rizal
10. Calle San Pedro
11. Calle Magellanes
12. New Market (Bangkerohan)

SANTA ANA

DAVAO GULF

DAVAO

River

Davao

Gov. Carpenter

Uyanguren

Hospital Ave

Cortez

Road / Built up Area

Bridge

Orchard (Coconut)

Tidal Swamp

Swampy

0   500   1000 m

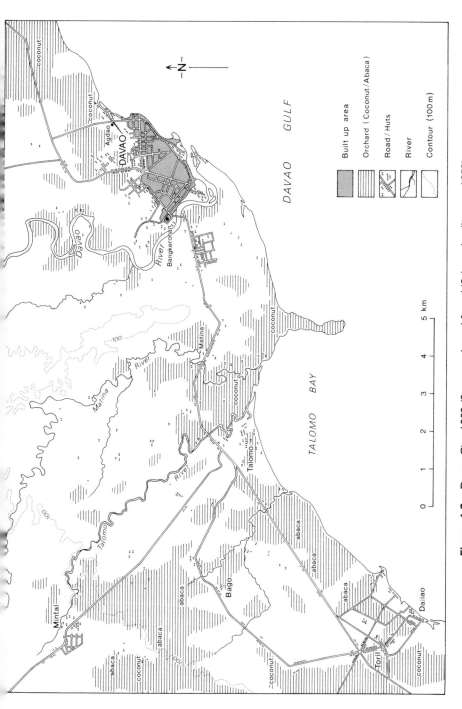

Figure 6.2 Davao City, 1953 (Source: adapted from US Army Intelligence, 1953)

swamp and agricultural land at Agdao, just to the north of Santa Ana. Since that time, growth has spread slowly northwards from Agdao along J.P. Cabaguio Avenue to meet J.P. Laurel—a central thoroughfare which runs northwards from the junction of Recto and Magsaysay (Uyanguren) before turning eastwards to meet J.P. Cabaguio at Buhangin, and along which the commercial heart of Davao has recently begun to overflow. Commercial buildings have also crept very gradually from Agdao northwards along Dacudao, and west along Lapu-Lapu to meet Santa Ana, Monteverde and Magsaysay Avenues which run north-west from Santa Ana. The warehouses and shops which line these three avenues begin to thin out as Recto and J.P. Laurel are reached.

The main foci of growth of Davao City, then, were: Trading (the original landing place of the Spanish) and its attached market at the junction of Bolton and Magallanes; the market sited at Recto and Uyanguren—the junction between the commercial heart of Davao and the wharf of Santa Ana; the market at Agdao where marine products were landed and where agricultural goods from the north and from the south (via Bangkerohan) were brought for sale; the market at Bangkerohan, the main point of entry into the city for agricultural goods and other food-stuffs and specialties from the hills, valleys and mountains to the south-west and west; and Toril—a junction between the coastal plain and the highlands to the west.

These markets were an expression of the institutionalisation of trade and had clearly been stimulated by external contact. But their appearance at these strategic locations was also an expression of deliberate and con-scious decisions. The markets established at the junction of Recto and Uyanguren, and at Agdao and Bangkerohan were part of attempts to encourage the growth of the settlement, for both officials (such as Simplicio Montaño) and merchant landowners had noted the relation-ship between markets, land values and urban growth. In the case of Agdao, a market on this site would facilitate the dissemination of goods to the northern part of the settlement, and provide stimulus for further urban growth. The land for the market was donated to the government by the Iñigo family whose motives were not entirely altruistic, for it was expected that the value of land surrounding a successful market would soon increase. Similar considerations influenced the establishment and siting of markets at the junction of Recto and Uyanguren, and at Bangkerohan (where land for the market was donated by the Mafori family).[1] In all instances it would seem that the strategy was successful. Certainly, Bangkerohan and Agdao are classic examples of the relation-ship between markets and urban growth. On the original lot (just over 1 ha) at Bangkerohan are the concrete buildings of the public market into which 1000 stalls are packed, each one rented out at more than 110 pesos per day. Surrounding the public market is a maze of streets and alleys which barely manage to hold the innumerable stalls and shops (including the prosperous Fortuna Supermarket) which comprise an even larger private sector. Yet another private market has spilt out westwards into the original unloading area for trucks bringing goods to Bangkerohan, and from which a tangle of small alleys lined with wooden huts, stalls and *sari-sari* stalls has spread. The whole complex merges with modern commercial streets to the south and east—the more sophisticated expressions of trade's extension and institutionalisation which grew out from, or became attached to, the market. A public market—which has room for around 700 permanent traders, and another 500 'transient' vendors—also forms the core of Agdao. The public market is surrounded on all four sides by modern shops which flow into the main thoroughfares to the north, west and south. To the north runs J.P Cabaguio and Dacudao. To the west and the south of Agdao the intense commercial landscape begins to decay until the northern point of the original primary settlement triangle is reached and, to the south, the carpentry and sawmill workshops which coat Leon Garcia Street give way to the shipping lines and warehouses of Santa Ana.

The deliberate and successful use of markets to stimulate urban growth in the past is highlighted by a more recent project—the Matina Public Market—which has all but ended in failure. The market is sited on the south bank of the Davao River, a little to the east of the highway which leads into Bangkerohan. The construction of the market building—a large, two-storeyed, concrete shell with enough ready-made stalls for 300 or more traders—was paid for by the World Bank (which contributed 70% of the funds), and local government. However, the market had managed to attract only 80 traders (by the summer of 1994), most of whom have been unable, or have found it very difficult, to pay the rent for the stalls, and have consequently gone out of business or moved elsewhere: most goods bypass the market and are taken directly to Bangkerohan; and it is from Bangkerohan that many traders in other markets in Davao buy some of their stock. The response to one poor decision is an attempt to persuade local government to introduce legislation compelling individuals destined for Bangkerohan to drop off a proportion of their goods at Matina, and to reroute traffic from the main highway past the market.

As in Hong Kong, markets in Davao have also gathered around the more sophisticated expression of trade's institutionalisation which had formed the initial or secondary foci; in some instances, then, markets only subsequently became additional foci of growth. There are also instances of markets now made redundant by the growth they initially stimulated, but on whose former sites new markets have now reappeared. Of the markets originally established at the junction of Recto (Claveria) and Uyanguren (now Magsaysay), and at Bolton, Agdao and Bangkerohan, only the latter two markets remain. The site at Bolton was turned over to other uses as the commercial heart of the city moved west and north, and as it was superseded by markets at the junction of Recto (Claveria) and Uyanguren, and at Bangkerohan. Part of the former market-place at Bolton is now a school; part is an overgrown field; and, to the south of the former market-place, Bolton itself has become a sleepy lane dotted with a few coconut-wood houses and sari-sari stalls. In 1985, however, a few traders selling goods from mats and flat wicker baskets (or nigo) on the ground occasionally gathered at a site in Trading a few hundred metres to the east of Bolton. The mats and nigo were later replaced by stalls, and this ad hoc group of traders (known as a talipapa), following the usual practice in Davao, organised itself into a market association. Dues are paid by each trader for lighting and water, for the upkeep and improvement of stalls, for rent and for the eventual purchase of the land on which the market is sited. On behalf of all its members, the association negotiates with the authorities over such issues as location, health and safety, tax and the purchase of the land. Once it reaches a certain size, and it becomes an institution of some importance, the talipapa may be taken over or bought out by local government (for some markets are registered as a corporation with the Securities and Exchange Commission in the names of the associations' members) and thereby transformed into a public market.

In the early 1970s another market was established between the junction of Jacinto and Recto (Claveria), and the junction of Jacinto and Bangoy just to the north of the Philippine Long-distance Telephone Company— the site of the original market at Recto (Claveria) and Uyanguren. The area in which the present market is sited is known as Acacia (for at the time of the market's establishment the area was covered in acacia trees). There are now three markets which have begun physically to merge with each other. As at Trading, the markets in Acacia, which first appeared in the late 1960s, began as small, spontaneous collections of traders illegally selling goods from the ground. At this time there was a Caltex service station, a few nightclubs and, along Claveria (Recto), residential houses and the occasional hardware store. The markets now comprise more than 100 stalls, and the streets to the south and east are now peppered with shops.

Indeed, markets have sprung up all over Davao. In common with those echoes of the past at Trading and Acacia, these other markets are often associated with new and expanding residential areas and squatter settlements, or with some institution (such as a hospital or school or new shopping centre), or are located at a strategic point (such as a transport node). The market vendors generally obtain their goods from Bangkerohan or Agdao or directly from manufacturers or farmers, or from those supermarkets which also sell goods wholesale. In Davao, as in Hong Kong, it may be that once markets, permanent stalls, shops and groups of residential houses are established, the appearance and evolution of markets are accelerated because demand increases; trade over long distance becomes faster and easier; and a greater variety, and a wider source, of goods becomes available. After all, markets are not the product of local surplus and local exchange; rather, they are, by definition, exogenous phenomena.

Along Quezon Boulevard—the coastal route which links Trading with Santa Ana—two markets have appeared at Mabini (the junction of Quezon and Mabini) and Piapi (the junction of Quezon and Jacinto) (Figure 6.3). Piapi, which was established in 1969, now has an association comprising more than 120 members. Mabini was established in 1972 and now comprises more than 100 members. Attached to both markets are a number of traders who are not association members. When the markets were first established, the road was lined with wooden houses, the odd sari-sari stall and workshop making furniture. Shops of all sorts are now scattered along Quezon Boulevard, but are particularly concentrated opposite to, and on either side of, the two markets behind which squatter settlements have grown. Two additional markets have also been established near Santa Ana—the New Salmonan Fish Vendors Association and Magsaysay Sidewalk.

Another example is to be found at Buhangin—a mixture of squatters and private residential estates which sprawls northwest from the junction of J.P. Laurel and J.P. Cabaguio avenues. A market was first established at a point which is now the junction of the Cabantian, Buhangin and Panancan roads. It comprised a few traders who positioned themselves near a sari-sari stall which was first set up in 1963. As the market grew, and as new roads were constructed, the market had to move on four occasions. The market now comprises more than 130 stalls and is sited on rented land along the Cabantian Road, a few hundred metres to the north of its original location. The market faces a row of some 30–40 stalls which are strung out along the opposite side of the road. These stalls comprise a second market with its own separate association. Both markets are continuous but swell markedly on Sundays. The original sari-sari stall evolved into two large concrete buildings: one is a well-stocked supermarket; the other is a grain shop. The road between the market and these two shops is lined with shophouses built of cement.

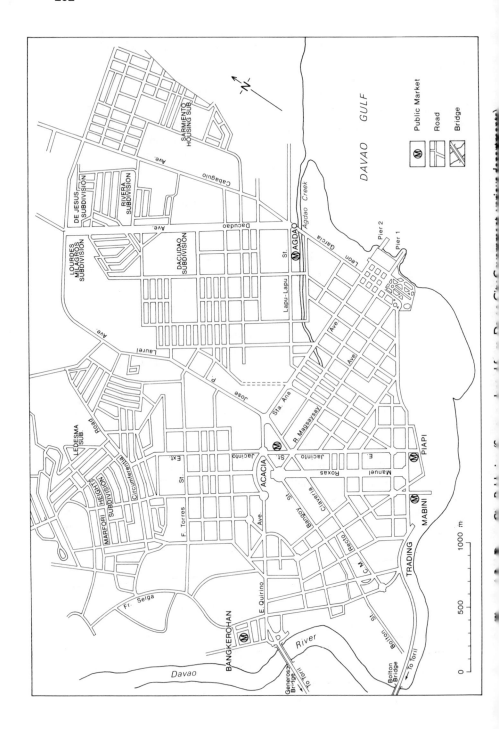

Immediately opposite the Davao Regional Hospital, which lies next to the junction of J.P Laurel and J.P. Cabaguio to the south of Buhangin, around 60 or more wooden shops are organised into a clearly defined square, though one side of this square still comprises the occasional itinerant traders selling their goods from old tarpaulins spread on the ground or from makeshift stalls. The market, which is attached to the tail-end of a former squatter settlement, began in the late 1970s as an occasional gathering of traders selling fruit and *saging* (bananas coated in brown sugar and then toasted over a fire). *Nigo* were replaced by stalls, and, in the last four years, the stalls have been replaced by the large wooden shops.

In Matina, another, but more successful and very lively, market (which is also known as the Matina Public Market) is sited near a private residential subdivision along the MacArthur Highway. The market was first established in 1980 just outside the GSIS (Government Social Insurance Scheme) subdivision, but was later moved a little further away from the residences to its present site after complaints about the smell emanating from the market. The market, which began as a handful of traders, now comprises more than 100 stalls; to this may be added another 20 or 30 permanent stalls run by traders who are not members of the association; and, as at Piapi and Mabini, shops (many of which began as *sari-sari* stalls) have appeared opposite to, and on either side of, the market over the period since 1980.

Ethnicity and statistics: the strength of Chinese merchants in Davao

Most companies in Davao, Filipino or Chinese, are family businesses. Of the 1277 top companies in Davao, a little more than 700 are either corporations or partnerships, and the remaining 570 are owned by single individuals. Of the 700 or so corporations, 240 are registered in Davao. Around two-thirds of these corporations registered in Davao are included in the top 325 corporations. From these records, and from lists of all enterprises (which number around 17 000, companies, partnerships and single proprietorships), it is possible to obtain some indication of the relative strength of 'Chinese' and 'Filipino' businesses in Davao, provided a rather limited perspective of these definitions and figures is adopted.

Just under 50% of all sole proprietorships within the top 1277 companies are Filipino, and, correspondingly, slightly more than 50% are Chinese. Around 32% of the top 325 corporations are Chinese; 40% are Filipino; and 28% are owned jointly by Filipinos and Chinese. Based on an estimate of sales and profits (see note 2), all of the top 251 companies—

corporations and partnerships registered in Davao, and single proprietorships—may be divided into four echelons.[3] In the lowest echelon
there are 167 companies, of which 45% are Chinese, 40% are Filipino and
14% are owned jointly by Chinese and Filipinos. In the third echelon
there are 66 companies. Around 48% of these companies are Chinese,
33% are Filipino and 18% are owned jointly by Chinese and Filipinos. In
the second echelon, which comprises only 12 companies, 7 are Chinese,
4 are Filipino and 1 is owned jointly by Chinese and Filipinos. Of the top
6 companies in Davao, 3 are Filipino, 2 are Chinese and 1 is owned jointly by Filipinos and Chinese. Of the top 251 companies within all four
echelons, 38% are Filipino, 46% are Chinese and 15% are owned jointly
by Chinese and Filipinos. It is true that the good showing of Filipino
companies within the top echelon is also repeated nationally. According
to one recent survey, San Miguel, Ayala Corporation, Bank of the
Philippine Islands, Purefoods, Philippine National Bank and A. Soriano
Corporation were among the top 10 companies in the Philippines. Two
Chinese companies—Shoemart and Jollibee—were ranked fourth and
fifth in 1993, and fourth and second in 1994 (*Far Eastern Economic Review*,
6 January 1994, cited in *Ayala Executive Bulletin*, 4 January 1994; Tiglao,
1994b; *Far Eastern Economic Review*, 5 January 1995). But, taken as a
whole, it would appear from this information that Chinese companies in
Davao have a slight edge over Filipino companies. If the number of ethnic Chinese as a proportion of the total population in the Philippines is
put at 2% (a proportion which is generally accepted in the literature),
then the Chinese lead becomes particularly marked.

Presented in such a manner, information of this kind could easily be
seized upon to confirm, or give weight to, the academic and lay belief
that Chinese either dominate the economy or, at the very least, are much
stronger economically than their numbers suggest, and that their
strength must therefore derive from the fact that they are Chinese.
However, quite apart from the problems with the collection and interpretation of such data (see note 2) there are a number of even more fundamental weaknesses with evidence of this kind. In particular, how may
one identify 'Chinese' and 'Filipinos'? The use of names is extremely
unsatisfactory: many Chinese adopt Filipino names, while many individuals with Chinese names may neither speak, nor read, nor write
Chinese, or may do so only haltingly. Yet this method of identification is
no less unsatisfactory than any other. Any attempt to pin-point hallmarks of 'Chineseness' is plagued by similar weaknesses. Somers-
Heidhues (1992), for example, puts the number of 'Chinese' in their
broadest sense (that is, Filipino citizens with some Chinese ancestry) at
between 2 and 4 million. Somewhere between 600 000 and 900 000 indi-

viduals are thought to be 'ethnic' Chinese—a category which is said to describe individuals who speak Chinese, belong to Chinese associations, socialise with other Chinese, and have been admitted to Filipino citizenship only recently (cf. Weightman, 1985). Given the difficulty in gathering the data, figures such as these can only be regarded as guesswork. More importantly, on what grounds are the criteria of 'Chineseness' founded? Could the term 'ethnic Chinese' be applied to an individual who was born of Chinese parents, who does not use or possess a Chinese name, who speaks Chinese, but does not read or write Chinese, who is Catholic (or even Protestant), who belongs to a Chinese association (which, it was argued in Chapter 5, is not in any case a unique product of individuals who are Chinese), and who socialises with Filipinos as much or, perhaps just as likely, more than he does with 'Chinese'? And what of an individual who is born of Chinese parents, who possesses and uses a Chinese name, who does not speak, read or write Chinese, who socialises equally with Chinese and Filipinos, who belongs to associations with Chinese and Filipino members, and who runs his company in a manner that cultural determinists would regard as being peculiarly 'Chinese'?

Undoubtedly the delightful pedantry of such questions will keep cultural determinists harmlessly occupied. No doubt, too, it will be suggested that the Chinese in Davao, and perhaps even in the rest of the Philippines, have been assimilated and integrated to an extent greater than elsewhere in Southeast Asia, and that the Chinese in Davao are of a particular 'type' or category (see Chapters 4 and 7). Yet this is an argument which smacks of convenience, and, unless one is prepared to conjure up an imagined past world against which a false comparison with the present can be made, then the acceptance of this argument first requires validation of its initial assumptions—that there are profound cultural differences, and that these differences affect and determine the organisation and practice of trade. Moreover, if Davao is a special case (as are most instances which do not fit prescribed theory and ideology), then this merely demonstrates that change, circumstances and individuals will always throw doubt upon any notion of 'the Chinese' community. If the advocates of a theory of 'Chinese' society have stopped short of claiming that the only true 'Chinese' society is the one which meets their own expectations and criteria, then it must be concluded that 'science' has produced a caricature. The use of any professed hallmark merely predetermines the conclusions in the definition: those individuals who behave in a manner the observer deems 'Chinese' cannot be anything else but 'Chinese'; and those individuals who are 'Chinese' cannot behave in any other manner except that which the observer has decided is specifically 'Chinese'.

The problem of identification, and the confusion it produces, lies not with the criteria by which 'Chinese' is identified, but with the very assumption that such a distinction must be made. 'Chinese' are not 'Filipinised', nor are Filipinos 'Sinified', for neither 'Chinese' nor 'Filipino' are analytical categories, or at least should not be treated as such. Even the cautious statement that an emphasis on trade crudely follows lines of ethnicity can only strengthen myth and prejudice. The comparative economic success of the Chinese began with their desire, or need, to pursue trade and, initially at least, in their circumstances as immigrants—a status which freed them from the pressure and desire to conduct reciprocity as an end in itself, and a condition which provided them with external contacts throughout Southeast Asia and beyond. But not all Chinese chose to trade. And, as the remainder of this chapter shows, there are Filipino merchants who have achieved comparative economic success and who have constructed institutions and adopted values which are often thought to be specifically 'Chinese'. Many of the first pioneers, and many latter-day migrants to Davao, originated from other parts of the Philippines where different languages are spoken. They, too, created or brought with them associations based on language and place (including town of origin), and some of these associations reflected the interests of a few dominant families. They, too, chose to give pre-eminence to trade and to redirect reciprocity to that end; and they, too, redirected the Filipino family towards the prosecution of trade (indeed, as noted above, most companies in Davao—Filipino or Chinese—are family businesses), creating an institution very similar to that which is often said to be a specifically Chinese company and which has been described earlier in this book as a nexus of reciprocity.

This is not to say that obvious differences between groups do not exist. But such differences do not reflect or constitute a deterministic culture. Certainly, profound differences in institutions, values and behaviour may appear to exist if the perception chosen is unidimensional and if the observer chooses to concentrate upon unique cosmetic details which are an inevitable consequence of the human mind's spontaneity and its variety of responses to circumstances. The apparent profundity of these differences is deepened further by the individuals who create, adapt, alter or adopt institutions, values and behaviour which they view or present as unidimensional phenomena—as symbols of their uniqueness. But if the observer, or participator, is aware of the many different and shifting aspects of institutions, values and behaviour, and if these phenomena are 'turned' so that their other aspects may be viewed, then the significance of cosmetic details begins to recede, and similarities among groups begin to emerge. As the perspective turns and other aspects begin to slide into view, it becomes clear that differences between 'cultural' or 'ethnic' groups reflect the manner in which institutions, values and exchange are used, and the purpose to which they are directed. The redirection of reciprocity from its conduct as an end in itself towards its use to construct relationships which facilitate or safeguard trade requires only a subtle alteration in values and attitudes. The family whose status and reputation depend upon the conduct of reciprocity for its own sake may appear to be very different from the family company, yet each requires only a shift in aspect to become the other. Even the very concepts of 'ethnicity' and 'culture' are drawn in to become part of the institutional patterns and moral imperatives associated with a form of exchange: both 'Chinese' and 'Filipino' are often deliberately presented as unidimensional phenomena in order that they may be used as positive kernels to help establish reciprocal relationships and facilitate trade, or as a negative opposite against which the individual may create the 'cosmopolitan' and thereby expand opportunities for trade.

'Filipino' and 'Chinese', then, are no more than imperfect descriptions. They describe mixes of wealth values, forms of exchange, moral values, institutions, ideas and beliefs which may possess many different aspects, and which differ from one group or society to another, not in any fundamental sense, but in the ends to which they are directed and thus in the aspects which they present to the observer. The question for analysis, then, is not ethnicity (or whatever other name some imagined cultural determinant may adopt), but sets of wealth values and their associated forms of exchange, institutional patterns and moral imperatives. With these ideas in mind, the confusion, contradiction, inconsistency and pedantry that surround attempts to define and explain 'Chinese' become easier to understand.

The nature of reciprocity

It is important to stress that the observer should be aware of the many different aspects of individuals and of reciprocity and its soft-framed institutions. To argue that reciprocity and its associated wealth values are dominant in a particular society or group is not to imply that the wealth values associated with trade, and the practice of trade, are absent. There is often an uneasy alliance which exists between these two sets of wealth values and forms of exchange.

With this qualification in mind it is no exaggeration to say that, despite the obvious significance of trade in Davao, the dominant wealth value by which prestige is measured among Filipinos is that associated with reciprocity. The orientation of the individual is thus strongly towards the establishment and maintenance of networks of reciprocity as an end in itself. The realisation of specific goals, such as gaining professional qualifications, is a secondary consideration, even though these may perhaps be necessary for physical survival, for satiating intellectual and creative abilities, or for enhancing prestige by demonstrating character, perseverance and—according to accepted, if limited, criteria—intelligence. Indeed, the advancement of an individual within a profession often depends upon the careful management of reciprocal relationships. Moreover, the profession itself may also provide the opportunity for the establishment of broader and more influential connections, and, in some instances may thereby serve as a springboard for the development of another career, perhaps in business or politics. In addition to the construction of reciprocal networks and the attainment of professional status, the expansion of reputation and the development of prestige may derive, in part, from the accumulation of land. Unlike profit realised through the conduct of trade, land is regarded as a reliable form of insurance, as a safe material basis for the advancement of the individual and of his family, and as a sound material instrument for the extension of favours and the creation of dependency in others upon oneself.

For many, then, power and prestige derive from networks of reciprocity. Land, professions and even desultory business concerns and political office are merely symbols of, as well as the means of funding and bolstering, that power and prestige. It is to this end that the family is also directed. The family provides an emotional foundation and a refuge for help and protection. But for each of its members it is also a symbol of, and a means to achieve and claim, the prestige derived from extensive networks of reciprocity built up separately by its members and through which each member can further expand their own networks; from the professional and educational success of its members; from the accumulation of land; from procreation; and from the public knowledge that the children are obedient and the parents are conscientious.

Frequently, networks of reciprocity are built around some convenient or natural point of common interest, activity or circumstance. These kernels include, among others: place or origin; language; workplace; neighbourhood; attendance at the same educational or professional institution; a shared diversion, such as attendance at the same music or dance school; or kinship, however distant and however tenuous. Or it may be that relationships are built around a mutual connection already established by an individual or by other members of his family. The complexity of these networks is illustrated in Figure 6.4 which summarises the relationships emanating from one family (Y) who live in a squatter settlement (which has now been bought up by the National Housing Authority) referred to here as 'The Community'.

Within The Community the oldest relationships of the Y family are with the families DG, CZ and SD. The fathers of DG, CZ and Y families have all worked for the Bureau of Internal Revenue. The father of the family SD had worked as a payroll courier for the Alcantara—a prominent business family. The relationships between the parents of these four families were passed down to the second generation, for the children in each family became part of one another's *barkada*. Another long-standing relationship of FY (the father of the Y family) is with the CBS family who live outside The Community. The father of CBS originates from the same town in Ilocos, speaks the same tongue, and has also worked for the Bureau of Internal Revenue. The wife of CBS was born into the C family which had, at one time, owned large tracts of land and enjoyed considerable influence in Davao. Her brother's daughter, BC, had attended Assumption College—a noted high school—with SY (the daughter of FY). Other contemporaries of SY and BC at Assumption were RZ and JR. The three girls, BC, RZ and JR, constituted another of SY's *barkada*. RZ is a member of a successful landowning family which is related by marriage or blood to a string of other influential landed families, past and present: RZ's grandmother was a Palma-Gil (PG); the sister of her grandmother was Hizon's wife; the wife of Pichon and Rota were Bangoys (BG) (who were related by marriage to the Palma-Gil; and an aunt of RZ was a Villa Abrille [VA]). SY, BC and JR had also attended the same ballet school; they had attended parties together; they had known each other's boyfriends and, at a later date, each others' husbands. JR married the youngest son of P, a Filipino merchant family with interests in real estate and rubber. The core of their business interests is the AV Hotel where both JR and her husband set up their own separate businesses, and where the sister of BC later came to work. BC's sister (MC) had obtained her position at the hotel through the help of Congressman MG (whose mother is the sister of CBS's wife and BC's aunt) who is also a good friend of JR's brother-in-law. JR's sister had been courted by DZ, a former Vice-Mayor of Davao who subsequently came to work as a legal and political advisor to the Gaisanos (GAS)—a large Chinese merchant family based in Cebu. BC married WM who came to work at the same branch of the Bank of the Philippine Islands as RZ.

SY's brother, RY, developed a *barkada* around his involvement in a mahjong parlour. His mother, MY, had formed a close relationship with the owner, MW—a Filipina whose Chinese husband had died many years previously. MY had originally been introduced to MW by D who had once lived in The Community and whose grandfather was the brother of MW's grandmother. Regular players at the mah-jong parlour, and who came to form RY's *barkada*, included JL, a neighbour in The Community; MT, a member of a powerful landowning family in Davao with considerable political influence in Manila; and RO, whose brother was Director of Immigration for the Republic of the Philippines. Both RO and his brother, IG, together with RCAS (the brother-in-law of SY's eldest brother) and a member of the Gaisano family, formed an *ad hoc* partnership for a real estate deal in Davao. Another member of RY's *barkada* is YP who ran a medical practice at a nearby hospital—the Ricardo Limso Medical Centre (a private hospital named in honour of a Chinese merchant's youngest son who had been murdered by kidnappers)—where SY had practised dentistry in a clinic rented by RCAS's brother, RUCAS. The sister of RUCAS's secretary worked for RCAS's wife, LCAS, whose sisters were married into two Chinese merchant families—the Uy and Co.

DY—SY's eldest brother—had read medicine at DMS (the Davao Medical School) where RUCAS also taught dentistry and where SY had secured her first post as Lecturer after graduating from a medical school in Manila. It was at DMS that DY met his wife, TY (the sister of RCAS and RUCAS). Both DY and TY went on to complete their residency training at Davao Doctors (the most reputable hospital in Davao) where one of TY's nieces works as an accountant, another works as midwife, and where one of her nephews works as a medical technician. The *ninong* at DY and TY's wedding was Congressman GC who is now, by virtue of his sponsorship, the *compadre* of MY and of TY's mother (DCAS)—a vivacious *barangay* captain and successful business woman. At Davao Doctors, DY worked under BEL (a member of a prominent Filipino merchant family for whom MY had once worked as a land sales agent) and APT, whose brother was Chief Mason—an organisation that DY and RCAS later joined. DY also formed a close and influential relationship with the Chief of Police whose life he had saved following a shooting incident at the AV Hotel.

Another member of SY's *barkada* at Assumption College was YN whose family owns a chain of pawnshops in Davao (and who make regular visits to Hong Kong to buy additional merchandise). The YN are the *compadres* of the MD family who originated from Cavite—the YN family's hometown in Luzon. MY had been introduced to the MD family via the C family. The brother-in-law of MDD (the eldest daughter of the MD family) worked in the Housing and Land Use Regulatory Board—a national line agency in Davao—and her husband worked for the Alcantara family (AL).

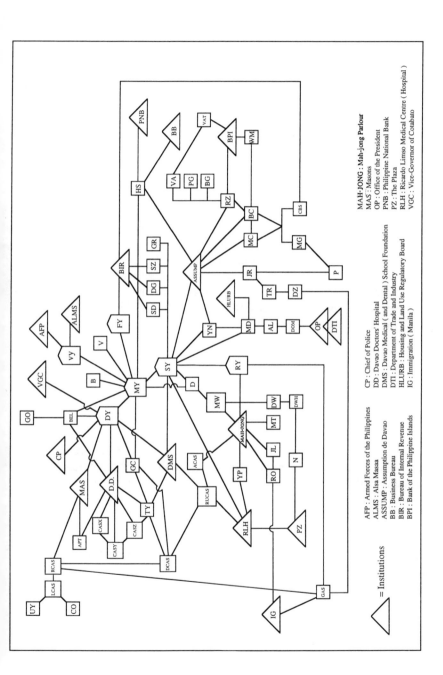

Figure 6.4 Networks of Kin and Reciprocity Focused on 'The Community', Davao City

△ = Institutions

AFP : Armed Forces of the Philippines
ALMS : Alsa Masaa
ASSUMP : Assumption de Davao
BB : Business Bureau
BPI : Bank of the Philippine Islands

CP : Chief of Police
DD : Davao Doctors' Hospital
DMS : Davao Medical (and Dental) School Foundation
DTI : Department of Trade and Industry
HLURB : Housing and Land Use Regulatory Board
IG : Immigration (Manila)

MAH-JONG : Mah-jong Parlour
MAS : Masons
OP : Office of the President
PNB : Philippine National Bank
PZ : The Plaza
RLH : Ricardo Limso Medical Centre (Hospital)
VGC : Vice-Governor of Cotabato

A key figure in these networks of relationships emanating from the Y family is MY. Her pivotal role is not apparent in Figure 6.4 since her influence is wielded primarily through her children. It was MY who encouraged the development of relationships with the children of members of her and FY's *barkada*, who carefully selected the right schools, who encouraged her children's extra-curricular activities, who strongly supported (and probably determined) FY's decisions on the profession which each child should pursue, who followed up the relationships established by her children with the formation of relationships between herself and the families of her children's contacts, who encouraged the maintenance and expansion of these networks (by throwing parties, for example, and by permitting her children to rotate as house guests among those contacts of which she approved), and who provided introductions for her children through her own networks and those of her husband. Moreover, it must be remembered that networks of reciprocity built up in Davao were only part of networks which she had helped foster in Cebu and Manila where she and FY owned property and where a number of her children later studied for their professional qualifications. Her own extensive network of contacts included the merchant family Aboitiz for whom both she and her father had worked (her father as engineer and herself as maid) in Cebu, and HS—a landed family with additional interests in a number of companies. One son of the HS family is a high-ranking official in the Philippine National Bank in Manila, and at the same time handles a number of the family's trading interests; one daughter is Chief of the Business Bureau at City Hall (Davao); her sister had taught SY at Assumption College and later became the College Principal. MY had worked voluntarily as chief cook for SEARCH—a civic organisation which organised a programme of seminars for graduating schoolchildren at a Catholic school in Davao. Members of the HS family had spoken at these seminars, and it was here that MY developed her relationship with the family. When MY later visited the United States, she stayed as house guest with other members of the HS family.

A sketch of this kind may at least begin to indicate the complexity of the relationships which individuals develop. However, it is obviously restricted, for it does not trace all the relationships in which each individual is involved, nor the ramifications and potential conflicts which can occur, nor the constant change. More importantly, it does not do justice either to these relationships' undulating shades of intensity, or to the many different aspects which they exhibit.

Place of origin, language, workplace, profession, kinship, mutual contacts, attendance at the same educational or professional institution are among the many and unformulated kernels frequently used for the establishment and development of a relationship, or for the continuation or legitimisation of a relationship already established. The relationship itself, however, comprises the give and take of material items, money, favours and skills around which soft-framed institutions—such as affection, trust, humour, sarcasm, respect, pride, jealousy, envy, guilt, loyalty—collect and are manipulated in order to build upon, strengthen, support, legitimise, cement and maintain initial reciprocal ties. Included among reciprocity's soft-framed institutions are *barkada*, *compadre*, *ninang*, *ninong* and *pakikisama*—notions which have been mentioned earlier in this book, and which will be referred to again later in this chapter. It should be noted, however, that while it may be tempting to associate these notions with something peculiarly 'Filipino', partly because they are described with exotic terms and partly because they imply, and can certainly be used as an excuse to create, a sense of obligation, they are no more than expressions of, and vary in their intensity to the same extent as, existing relationships which have no special designation and which may easily be understood by more familiar notions such as 'friendship', 'godparent', 'family friend', 'gang' and 'group loyalty'.

It would be cynical to cast these networks of reciprocity in purely utilitarian terms. The relationships outlined above may indeed constitute no more than temporary expedients formed upon vague claims of distant kinship or some conveniently loose connection (such as attendance at the same school or a dimly remembered mutual acquaintance). On the other hand, they may comprise the give and take of favours (such as free consultations or an introduction to a useful contact) cemented by the occasional social gathering and a kernel such as membership of a brother's or sister's *barkada*. Or it may be that reciprocity's soft-framed institutions may comprise strong affection (free of ulterior motives), creating lasting relationships between individuals who may rely on each other in moments of extreme difficulty, regardless of the inconvenience. These networks of reciprocity may also fulfil the desire for conversation, for the cut and thrust of social contact, and for the expression of affection. They may also fulfil a simple need for company: conversations may often degenerate into boring and desultory exchanges in which little is said, for it is the time that the individual is willing to give, the conveyance of emotions and sympathy, and the confirmation of the other's feelings and ideas which are of importance.

Nevertheless, these relationships are also supremely pragmatic: they are the measure as well as the means of wealth and wealth creation and thus of prestige and power; they strengthen the braggart's claim to status, and are the means to shore up his confidence; they provide access to institutions, services and materials which might otherwise be impossible to obtain; they enable individuals to operate more effectively within and among institutions and to further their own career; they are essential for the prosecution of trade either as an end in itself, or merely in order to secure a solid financial base for the conduct of reciprocity and for the accumulation of material items which are taken as a symbol of an individual's worth according to reciprocity's associated wealth values. The utilitarian aspect then not only exists—it is a crucial aspect of these relationships.

Equally, however, it would be misleading to cast the manipulation of reciprocity and its soft-framed institutions as predominantly cold and calculating. Certainly, individuals show a remarkable ability to memorise and take into consideration many different relationships, characters, nuances of behaviour and possible reactions. But while broad objectives (such as the preservation of a family's reputation and unity, the accumulation of land, the realisation of professional status for oneself or one's children, or the establishment of a particular relationship) may be clear, the strategies to realise these goals are often intuitive and 'felt' rather than carefully formulated and reasoned. It is a world of feelings and perceptions that requires sensitivity to changes of rhythm in the conduct of relationships: alterations in the emotions and moods of individuals, and an understanding and awareness of the different sides of their characters, require modification of one's own behaviour. Moreover, in order to realise the clear objectives which may punctuate the crafted haze of emotions, there is much chopping and changing. Circumstances, chance events and, most especially, the very nature of reciprocal networks, the formation and maintenance of which are given priority, create all sorts of distractions and tangents. Individuals may find it difficult or impossible to meet current obligations or to fulfil their own desires and objectives immediately or as planned, for there is always the very strong likelihood that events, other individuals, new or prior obligations, and conflicting objectives will come into play. For these reasons, too, cold and calculating manipulation is more usually replaced by hope and opportunism, and by subservience to the wheels of circumstance which, turning slowly and unnoticed, may bring together extraordinary combinations of relationships and events. The ability to handle such networks, to be aware of (and to take into consideration) their complex ramifications, implications and cross-currents, and yet, despite all this uncertainty, to put together a strategy, to create opportunities and to justify actions, is much to be admired.

Although this world requires the skilful juggling of individuals and their emotions, it is, for the most part, intellectually barren. In contrast, the chants, anthems and sung prayers of the Mass, which drift through the cloisters and over the well-kept lawns and bougainvillaea of the church, are extremely powerful and moving. The choir, packed into the chancel with a guitar and cheap organ, is untrained save for the occasional weekly practice session, and yet spontaneously produces elaborate melodies, complex parts and harmonies from a simple chord progression or one-line melody. The Church, too, like the ballet school or piano class, the aerobics club and gym, the plazas, the roller-skating rinks and bowling alleys, is, in part, no more than a diversion. It provides an excuse and opportunity to gather and to talk, to continue the endless cementation of relationships, to enjoy the camaraderie, to gossip, and to avoid the appearance of retreat, introspection and selfishness. For many attendance is also an expression of their superstitions. Concerns lie not with the nature of their faith, nor with the development of a philosophy for life, nor with questions of absolute right and wrong: the poor come to ask for something better; the well-to-do hope to delay that time when they believe their good fortune will have to be paid for. But attendance is also driven by religious conviction, though often the individual no longer doubts whether his faith is genuine or born out of justification and rationalisation which have now become second nature. Religious faith may thus serve to mask the reality of the individual's manipulation of himself and of others, and therefore takes on the greater power. Once a course of action is chosen, sentiments, girded with religious dictums, may be used to justify and legitimate that action: individuals who give and sacrifice may, when the time is well-judged, quickly and sharply draw attention to their own selflessness and generosity which must be respected and repaid.

Although compassion, genuine altruism and affection are not marginalised (for they are often important aspects of reciprocity's soft-framed institutions), this highly manipulative society, which places so much emphasis upon giving, is no less selfish than one in which sustained material progress is the greater creed. Prestige and power are derived from the establishment of reciprocal connections with a large number of individuals. Real or apparent influence, in the eyes of others, is achieved by a balanced give and take with those of a similar or higher status, or by placing another in one's debt. But the strongest and most assured influence is achieved by creating in others dependence upon oneself. Their dependence gives the individual a real ability to affect events, and thereby to raise his status. And by achieving a reputation for kindness, generosity and compassion, the individual's position becomes unassailable: motives cannot be questioned, and trust must be given. Indeed, such is the attention paid to reciprocity and to the construction and maintenance of relationships, that others, including close friends and kin, may be handled as if no more than convenient assets. The family may be the cornerstone of trust, but if the status of the individual and family is to be raised, then external obligations may come before duties to children and kin who become pawns for the realisation of that greater end while those who are not embraced by these reciprocal networks are ignored or harshly dismissed.

That the utilitarian and altruistic, manipulative and compassionate may coexist without contradiction is illustrated by MY's relationship with the SZ family. Since the mother of the SZ family eloped with a driver some 25 years previously, MY, the *ninang*, has provided emotional support, advice and financial help; her son, a physician, could be called upon free of charge; and while the daughter of the SZ family studied in Cebu for her exams (which she eventually passed to become a civil engineer), she lived in MY's house there free of charge. MY made no call upon the obligations or debts incurred. The fact that MY was the children's *ninang* did not imply any moral compulsion to help: MY had been made their *ninang* because of the relationship which had already existed. But members of the SZ family were happy to repay MY with a range of services and favours. When MY's second daughter failed the government board examination for a professional qualification, neither MY nor her daughter had enough money to pay for a second attempt. MY asked the SZ family to pay, which they did. MY paid back the money, when she was able to, several months later. When it became too inconvenient to use the telephone at the house of the ageing SD family, SZ made their telephone available to MY's family at any time. And when necessary, their car was put at her disposal. The SZ family, then, were part of MY's very complex and skilfully handled web of social networks. It was this network which gave MY her reputation and status, and it was her status and reputation which enabled her to forge links with more powerful families, enhancing her reputation and her ability to help and influence still further. Yet none of this denies feelings and expressions of fondness and genuine compassion.

The different aspects—and, therefore, the ambiguity—of reciprocity and its soft-framed institutions is particularly evident in MY's relationship with her immediate neighbours. The PL family, comprising three brothers, their wives and children, had squatted on MY's land (on which she herself squatted) just 25 metres from her home. MY had taken on one of the brother's eldest children—a nine-year-old girl, L—to help around the house. L was but one of a string of girls aged between 9 and 18 who had at one time or other worked for MY. L attended school during the afternoons, but spent her mornings (from about 5.00 a.m. until midday) sweeping, cleaning, buying drinks or rice from the local *sari-sari* stall, and carrying messages between MY and others in The Community. MY paid for L's school activities, for equipment and clothes, and she fed her. Occasionally L would receive a few pesos, or a bowl of ice-cream, or a slice of chocolate cake as a treat. Both MY and SY (and the rest of the Y family) were reluctant to give L money because she would almost immediately hand it over to her parents. This arrangement, and the unwillingness to channel money through L to her parents may seem harsh at first sight, but this relationship was part of a much more complex and, at times, ruthless play of individuals.

There was no sentimentality in the relationship between MY and L. L was not comforted when she cried; she was not lavished with presents, clothes or shoes; there was no demonstration of affection. But L knew order: she was told when to eat; she was told to do her homework; she was given spot tests on her spelling and mathematics; MY visited her school to ask of her progress; L's tasks, though numerous and repetitive, were light; she was taught the difference between acceptable and unacceptable behaviour; and in the evenings, over a meal, she could share in conversation dominated by the lull and sway of current and past relationships. She was, in effect, being presented with a set of goals to which to direct herself and from which to extract self-esteem. Marriage and children would signify her entry into an adult world: the further development and expansion of social networks, a publicly successful marriage, children who proved capable at school, and perhaps even a profession for herself, would all add to her worth, status and self-confidence. The 'sisterhood' into which L was being brought, the order that she knew, and the domestic and social skills she was being taught by a prominent family amounted to a compassion she had not known in the chaos of her own home. Moreover, should she prove honest, MY would help her achieve professional status as a nurse. She would then be in a far stronger position to assist her parents and younger brothers and sisters.

L's father had been unemployed until MY secured him a job as a traffic policeman through her son VY and a former policeman who was a member of VY's extensive *barkada* within The Community. L's father earned no more than P2000 a month, but this could at least provide the bare minimum of food for his five children. MY would also supply coffee and rice if the family was short, and, if they requested medical advice, DY would provide free consultations and drugs.

Nevertheless, tensions stemming from a number of incidents and from the circumstances in which MY and the PL family found themselves, had developed between MY and the PL family. DY had administered treatment to L's grandfather who suffered from tuberculosis. The old man, however, had sought help only when it was too late. When he died, MY had made available her jeep as a hearse which, although too short (the foot of the coffin stuck out of the back of the jeep), it served its purpose. During the vigil over the body, MY also provided a lamp and an extension cable plugged into her house; and VY supplied the bananas and coffee. After one week of vigil, and despite the heat, the body had not yet been buried. L's parents were still holding 24-hour mah-jong sessions (for every winner paid a five pesos *tong* which goes to the bereaved family); and since they had been unable to secure the *barangay* captain's chop (for he was away at the time), they had been making good use of MY's signature on a sheet of paper which they had been presenting to all the families in The Community to collect donations. There was a humorous side to these events which most people, including MY, had appreciated. Yet L's parents had begun to offend MY's sense of propriety. Her objections were partly that they had used her reputation to collect donations in a rather cynical manner, and partly that the funeral was not the demonstration of respect that it should have been, for an important aspect of the ritual is to mark the status of those who live and of those who are dead. A cheap coffin will quickly become the talk of The Community: 'Oh, he died like a rat'—is an English phrase (yet understood by all) which is pronounced with a staccato roll of the 'r' and a spitting 't', and likely to burn itself into the memory of the family. The money spent on the coffin, the numbers of people who attend, and the right atmosphere created by the public wailing into which genuine feelings of despair are channelled, and by the more relaxed, cheerful wake given in thanks for life: all this is part of the performance. The material symbols and the fulfilment of the expected roles serve to confirm the achievement and purpose of being.

Another incident which created tension between MY and the PL family had involved the PL's children. They had five children including L who was the eldest. Three other children had died from dysentery, measles, bronchitis, poverty and ignorance, for their lives might have been saved had the parents sought help from MY or from a nearby government hospital (outside of which their grandmother had sold fruit) when the children had first become ill. So it was all the more aggravating when it was discovered that L's brother (who was four years old) and sister (a two-year-old) were afflicted with sores. The girl's mosquito bites on her feet and legs had become infected; and an untreated cut between the toes had caused the boy's left foot to become swollen and extremely painful. Their parents would not tend the sores. MY's daughter, SY, took the initiative to clean the wounds, hoping to shame the children's mother into doing something to alleviate their discomfort and perhaps avoid another death. As she cleaned the girl's bites and the boy's foot, now covered with flies, both children cried pitifully. Their mother's reaction was an expression of disgust at the sight of ugly sores she would not touch. When SY had finished, the boy's six-year-old brother carried him back to their hut. Within two hours the boy, again barefooted, limped out on to the poorly drained soil of the compound. MY would not arrange for the children to be vaccinated against tetanus, nor would she herself, nor through her son, dispense any medicines, nor would she attempt to persuade the parents to consult a doctor, partly because she was not prepared to be blamed should anything subsequently happen to the children, but mainly because she suspected that the parents were deliberately neglecting their children in order to play on her emotions and charity.

When L's father returned home, MY was angry and indignant. The father blamed his wife, and then asked MY for P20 to buy rice. MY gave him the money—an act which raised much laughter in her household at the irony and futility of the situation. L's mother then appeared and began sweeping the compound—a ritual she went through whenever she wanted a few pesos, ostensibly for food, but always spent on cigarettes. A short time later that same afternoon, the father, with a cigarette hanging from his lips, ostentatiously paraded his concern for his children by sitting his two-year-old daughter upon a wooden stool in the middle of the compound in plain view of all, and began to shave her head, for she had made a bald patch by using a razor she had found in the midden left by her mother.

A third incident, or rather series of events, led to a demonstration of MY's power, for now she intervened directly in the affairs of the family. It was learnt that the father had been suspected of taking bribes. Since MY had secured his job for him, his behaviour might reflect upon her. VY asked MY to contact Congressman GC (her *compadre*) to intervene, but MY refused: if the father offended again her reputation and her ability to judge character might be sullied. It was then brought to MY's attention that L had missed school for two consecutive days. It turned out that L had been ordered by her mother to stay at home and wash her father's clothes. Her mother would no longer do his laundry for she and her children were being forced to rely upon hand-outs from MY and upon L's pocket-money: most of her husband's salary was being spent on palm-wine and another woman. MY was furious and threatened to impose various sanctions, including the withdrawal of all help. To show his contrition the father and his daughter, L, later walked slowly across the compound hand in hand on their way to church. The following day, however, matters finally came to a head. The father would not sleep at home out of fear that he would be killed by his wife. MY said nothing to his wife except that she wished L to attend school. That same night, L ran crying to MY's house. Her mother had whipped her out of jealousy over the special treatment MY had shown her. MY then told L's parents explicitly that they had been abusing their relationship with the Y family. MY made it clear that she did not need L, and that if they continued to take advantage of the generosity she had shown L and the rest of the PL family, she would release L and break off all contact with them. In addition, the mother was chastised for her inadequacies, the father for his deceit and infidelity.

Clearly, a range of complex emotions and motivations were at work. MY's relationship with the PL family added to her reputation, but that relationship was only a small part of her extensive and intricate networks of reciprocity. The dependency she had fostered had brought with it frustration at that dependency, and jealousy of MY and L. It should not be thought that this example is a special case. It is merely one example of the extent of involvement by one individual in the lives of others, for these relationships are not merely diversions; they are wealth creation, and the exercise of power is one demonstration of the status which the individual has achieved.

The construction of reciprocal networks and the breeding of dependency are particularly intense within the family; it is also within the family that relationships are, in many respects, even more ambiguous. Tension among family members often finds expression as disputes over money (for transactions within the family are commonplace) and over possible changes to inheritance as the parents become older; or as jealousy over the apparent favouritism shown by parents to one child over another (for this may be taken as an indication of the likely apportionment of assets), particularly when illustrated in greater trust or in gifts; or as disagreements over the behaviour of one member which the others consider damaging to the family's reputation and standing. Indeed, it is fair to say that, in addition to the construction, maintenance or expansion of relationships, such transactions, tensions and disputes absorb the energies and attention of the individual outside the workplace. Maintaining the cohesion of the family, its assets and its reputation requires the skilful handling of individuals and the judicious application of authoritarianism. The nature and complexity of these disputes are illustrated by transactions among members of the CAS and Y families.

One long-running dispute which gathered momentum over a period of several months before culminating in a split, began with DY's decision to renovate and then sell an old Toyota jeep given to him by his father. The proceeds were to be used as a down payment on a new car and to help his mother finish the construction of the family's new house in The Community. The Toyota was renovated by ACAS (a member of RY's *barkada* and the nephew of RCAS, RUCAS and TY) whose business was supported by his mother, JCAS, whose business was, in turn, supported by DCAS. The estimate for the job was P80 000. Once the work was completed a buyer was found through another son of DCAS. The buyer agreed to pay P270 000 for the renovated Toyota—some P70 000 more than DY had expected. ACAS then presented DY and TY with a bill for P110 000. Not only did this exceed the original estimate, but it also included charges for parts which DY had given to ACAS. ACAS's creative billing was motivated not so much by greed as by the opportunity to pay back a debt to his mother. At this point TY learnt that her mother had, on her previous instructions, managed to sell a ring belonging to TY for P25 000. TY asked DCAS to give the money to JCAS. By repaying ACAS's debt to his mother (DCAS's daughter), ACAS would be paid off indirectly, and TY and DY would not have to draw on their savings. DCAS, however, chose this moment to recall an old loan which she had made to her daughter, JCAS, and kept the proceeds from the sale of the ring for herself. JCAS and ACAS redirected their anger at TY and DY who now considered their payment for the Toyota made in full.

The rising tension between DCAS and DY was exacerbated by a second incident. DY had bought a plot of land from DCAS at the market price of P38 000. Eighteen months later he sold the land for P120 000. DCAS then suggested he pay 'interest' on the difference even though there had been no loan. Effectively, DCAS was asking DY to regard the profit he had made from the transactions as an advance on which interest should be charged: he would not have been able to make such a profit had she not agreed to sell the land to him at the time and at the price that she did. Not surprisingly DY refused to make any money over to DCAS. DY's successful deal had been preceded by another land transaction which had proved costly for DCAS because of DY's indecision. DCAS had bought several large plots of land which she wanted to register in DY's name in order to mitigate her taxes. DY agreed, but was then advised by his father that his own tax burden would be increased. DY pulled out of the arrangement with DCAS at the last moment, much to her chagrin.

At the root of these complex dealings was an attempt by DCAS to foster a degree of dependency in DY and TY upon herself: after failing to manufacture a financial obligation, she had then sought to exercise a moral claim over them. Both DY and TY had shown independence at leaving Davao and setting up a house, a medical practice and a pharmacy in a neighbouring province. Not only did this deprive DCAS's family of immediate and free medical services, it also signalled to her the beginning of the family's disintegration. This she would find impossible to accept for she had been intent on weaving a web of dependence around her children, her grandchildren and her great-grandchildren who were all instrumental in her promotion of the family's reputation and prestige. They had been integrated into her business empire; they were given neighbouring plots of land on which to live; DCAS had set up RUCAS with his own private surgery and had funded his unsuccessful political ambitions; and she had established extensive lines of credit (which she had underwritten) for those of her offspring who had ventured into business; and she had funded the education of the rest and provided them with introductions into professional institutions at the start of their career. Weekend gatherings were as much an occasion to ask for financial help or for additional contacts: these were gifts and services which had to be distributed with care if jealousy was to be confined.

Given the symbolic as well as practical importance of land, it is not surprising that the tensions which underlie, or are generated by, land sales within a family reflect many different considerations, only some of which are 'economic'. When XY bought land from his mother MY, he did not pay the full market value. On his understanding he was given a discount precisely because this was transaction within the family—an understanding which MY made no attempt to disabuse him of. While overseas XY sent back additional funds to begin the foundations for a house on the plot of land he had bought from his mother. But since XY had not paid the full amount, MY used the funds to pay back her debt to the Credit Union, to finance the construction of the family's new house in The Community, and to pay for FY's medical expenses. XY was furious, but his brother, DY, supported their mother: XY should pay for the land in full before laying the foundations. The dispute was further complicated by TY who implied that MY was attempting to elicit a proportion of her own and DY's savings in order to complete the Y family's new house.

From XY's point of view, MY was asking him to pay the full market price for land she had already paid for in full simply in order to cover miscellaneous costs which would not add to the value of the land. Moreover, she was asking him to do so while he was attempting to secure an investment and a home for his wife and child. That he had been excluded from a joint investment among two of his brothers and his sister in a farm added to his unhappiness. From MY's perspective, a wider range of matters had to be considered. To MY the land she and FY had accumulated was many things, including a form of savings to help them cope in their old age. Yet here, too, the issue of control and dependency is pertinent for the land was also the symbol of her family's prestige as well as a crucial base for its cohesion. Should *de facto* ownership of the land fragment, a key element of dependency and security and therefore cohesion would be lost. Although most of the land had been distributed among her children who were now its legal owners, all these holdings (except for a few lots which had been paid for in full by three of their children) were still, in fact, controlled by MY and FY. MY was determined that if large sums of money were needed, and if there was no alternative, it was better to sell land within the family to the most reliable of her children (over whom some measure of influence could be exercised) for fear that it would otherwise be sold off for easy money. In addition, MY brought a range of moral issues into the dispute: she had given XY life; she had educated him; and money had been needed to pay for his father's treatment—precisely the eventuality for which she had originally bought the land. Why did he not trust her? Moreover, the 'full' market price, she maintained, was no more than 60% of its true market value, and this was for a plot of land which would rise substantially in value over the next 10 or 20 years. She was only taking something which she deserved in return for all her sacrifice.

One further, though perhaps rather melodramatic, example which illustrates the many different aspects of reciprocity and its soft-framed institutions within the extended family, is the murder of PT. PT was the aunt of V whose grandmother was the sister of MY's grandfather. PT had, for many years preceding her death, managed a small farm owned by MY. Unknown to MY PT had sublet part of the farm to the X family who gave a portion of their produce to PT as rent. PT kept all of the rent from her sub-tenants and paid only the agreed percentage of the produce from the rest of the farm to MY. Tension grew between PT and her sub-tenants when she accused them of encroaching upon too much land, and then stopped their water supply by damming up the stream which first ran through her part of the farm. Both parties then turned to the *barangay* captain to mediate in the dispute. The *barangay* captain advised them to take the matter to MY since it was she who owned the land. However, it was only when VY and DY (surprised by the consistently low return from the land) visited PT that the sub-tenants and the dispute were revealed. MY informed PT that since the sub-tenants had farmed the land for a number of years, they could remain, but they would now pay the rent (either in cash or produce) directly to MY. PT would continue to pay rent to MY for her portion of the farm. PT angered by this arrangement, continued to harass MY's new tenants. In separate negotiations MY agreed with the X family's eldest son that after his wedding, he and his wife could move to another of MY's farms. PT, feeling she had won a victory, teased the eldest son: he was leaving, she was not. That same night, as PT answered a knock at the door, a long knife was pushed through her body, killing her instantly.

A short while after the murder, the mother of the suspected killer (the eldest son of the X family) arrived at a party being given by SY. The mother, MX, informed MY that she wanted to sell the tenancy to another family, but first she wanted payment from MY for the crops and labour she had put into the land. If MY did not pay, she would be reported to the Ministry of Agrarian Reform for illegal eviction. MY retorted that the tenancy was not strictly legal, she had let the X family stay out of kindness, and, not least, PT, a relative of MY, had been killed, allegedly by the eldest son of MX. She then implored MX to persuade her son to make amends on this earth, and made it clear that the X family could stay or leave: the choice was theirs, but MY would make no payments.

PT had been playing a dangerous game: she had betrayed MY's trust by subletting the land, and by evading the rent; and she had toyed with the livelihood and emotions of her sub-tenants. MY had been extremely practical and balanced about the whole affair, though it may be that her attitude to these events was to some extent shaped by the manner in which PT and her family had treated V (who often helped MY in the house). At the age of 19, V had become pregnant. PT, with whom V lived, regarded the pregnancy as an insult to herself and to her family. V was beaten on the arms and legs, and her head was banged against a wall. EY (whose great-grandmother was also MY's great-grandmother)—who later came to work with V in MY's house, and who managed another of MY's farms—was afraid V's unborn child would be harmed and arranged for her to be brought to MY. MY persuaded the father of the unborn child to pay for a maid to help V; MY would pay V's medical expenses which included treatment for severe depression. After delivery, the baby was removed from V on medical advice and left in the care of the father's parents.

The 'turning' of reciprocity

It has been suggested that although common-sense, pragmatic responses to particular circumstances, and the transmission of ideas may give rise to institutions and values which are similar in their external signature and in atmosphere, as well as in substance and patterns, there is no reason why similar patterns or signatures should inevitably emerge among different societies. This may well make it difficult to appreciate that differences in etiquette, mannerisms, facial expressions, and symbolism, as well as in institutions, values and exchange, represent a unidimensional perception, presentation and interpretation of phenomena rather than profound expressions of distinct and unique 'cultures' which are presumed to exist. Yet if the observer is aware of, and is prepared to encompass within analysis, different aspects of institutions, values and exchange, then the similarities among groups begin to emerge, and the significance of differences in the nuances of behaviour, in values and in atmospheres and the external signatures of institutions, begin to recede. The analysis of material progress among the Overseas Chinese, then, should focus not upon the cosmetic details of ethnicity and a presumed deterministic 'culture', but upon complementary and conflicting sets of wealth values, forms of exchange and their associated institutions and values which not only cut across unidimensional perceptions and presentations of ethnicity and culture, but also comprise such perceptions and presentations.

'Filipino' social networks are, in many respects, similar to those created, for instance, by communities in Bethnal Green, East London, by British and American elites, by the Creole in Sierra Leone, and by the Chinese of Singapore. It is, then, perhaps not surprising that there are 'Filipino' institutions, values and forms of exchange which could be turned towards the prosecution of trade, and which, if so directed, would then create business operations and other supporting institutions and values closely resembling those often regarded as specifically 'Chinese'. The transfer of land, and spontaneous credit institutions, may help illustrate the different aspects of reciprocal networks and their potential for redirection towards the support of trade.

The Community, formerly a squatter settlement, is now an NHA (National Housing Authority) site in which a community mortgage programme operates: that is, squatters are to buy the land they occupy from the government. It is noted elsewhere (R.N.W. Hodder, 1991) that there are a number of problems facing the implementation of this programme: to date (summer, 1994) the majority of squatters have not yet begun to purchase the land. However, sales of land among the squatters and within families, do occur, and in some parts of The Community adjacent plots of land have been bought up by several members of the same family, creating unwalled compounds. The complexity and tensions which are often inherent in these transactions or sales have been described above. Sales between different families, however, have been smoother, even though these sales have remained informal transactions based on trust. For example, the SD family sold a plot of land over which they had claim to the DG family for P60 000. The sale was made purely on trust, for there was no legal agreement and, indeed, legally, the land still belonged to the government. The DG family had not in fact bought the land: they had only bought the moral right to buy the land through the community mortgage programme from the government at a cost of a further P30 000. The total cost of the land, then, was P90, 000—cheaper than a plot of similar size and quality outside The Community, but nevertheless a substantial investment.

Another illustration of the multidimensional nature of reciprocal networks is the rotating credit institution which, it was noted earlier in this book, is indigenous to many societies throughout the world. One very common arrangement in Davao requires each individual to place an equal sum of money into the pot at an agreed interval (perhaps every week or two, or every month); each member then takes his turn to claim the pot. Since there is no interest, each individual repays the others (if he is one of the first to claim), or build up their credit (if they are one of the last to claim), by continuing their regular contributions. Such an arrangement relies entirely upon trust. There are few naïve enough to believe that circumstances and human nature will never allow one individual to run away with the pot, and it is for this reason that rotating credit institutions tend to be organised within the workplace, and that most will prefer to be the first to claim the pot.

More complex, formalised credit institutions also exist. Yet these, too, are often established and run initially by individuals with little or no administrative or managerial training or experience. One such institution is the Credit Union (formerly the Santa Ana Cooperative) which was established in 1967 with a handful of members. The Union now has a membership of almost 5000, a number of whom live in The Community. Each prospective member is required to attend a seminar designed to instil the individual with 'correct' values and objectives that enable the Union to function effectively. To join, an individual must deposit at least P500. The amount deposited—or 'share capital'—determines the amount a member can borrow. Six months from the date of membership, an individual can borrow up to 150% of his share capital; up to 200% of his share capital may be borrowed after another six months; and then up to 250% after a further six months. Two years after the date of membership an individual is permitted to borrow up to a maximum of 300% of his share capital. For all loans which exceed the member's share capital, some form of collateral (which may include anything from electrical appliances to land) must be provided. The interest on all loans is 1% a month, and the maximum period for repayment is 12 months. The interest is deducted from the loan before it is paid to the borrower who must repay the loan within the period agreed with the Union. Although an individual may increase his share capital at any time by depositing more funds, 5% of every loan (after interest has been deducted) is retained by the Union and then added to the borrower's share capital. These forced savings are regarded by the Union as an important method of teaching its members financial discipline. In addition to the interest and forced savings, borrowers must also pay a small, fixed sum to cover services and a loan protection scheme. The maximum absolute sum an individual may borrow is P150 000. Loans ranging between P75 000 and P100 000 must be approved by the Credit Committee; and loans above P100 000 can only be approved by the Board of Directors.

The Board of Directors comprises nine individuals, including the Manageress (the head of the Union) who acceded to the post in 1991 after a string of appointments which followed the death of her brother—the founder of the Union and its Manager from 1970 to 1983. In addition to the Board of Directors, the Credit Committee and 19 administrative personnel (accounts officers, loans officers, cashiers and collectors who are directly responsible to the Manageress), there are four other committees (Audit, Supervisory, Election and Education) whose members are elected by the Board during the General Assembly. Committee members are Credit Union members who are mostly housewives, labourers and very small businessmen, and the occasional professional. The administrative personnel are, on the whole, graduates of commerce or accounting and are usually recruited through Union members to ensure a trustworthy and reliable workforce. No member of the Board or of a committee is allowed to sit on another committee. This regulation, combined with the centralisation of power over administrative personnel in the Manageress, and with the authority of the Board (chaired by the Manageress) to elect committee members, provides the Manageress with an effective means to prevent a complete loss of direction consequent upon the friction and rifts generated when individuals and cliques attempt to secure greater influence and status within the Union.

Despite the problems which these organisational measures were introduced to mitigate, and even though the Union prefers to replace the word 'profit' by the careful phrase 'gross income minus expenses' (which is then translated into the term 'net savings', for 'profit' is regarded as a dirty thing in an organisation whose purpose is to serve), the assets of the Union are nevertheless substantial. The share capital, the savings derived from interest payments by members, the interest derived from funds deposited in a number of bank accounts, and the building which houses the Union's offices, are valued at around 56 million pesos.

Arguably, however, desultory economic transactions such as these, and the ambiguity with which profit is often regarded, merely highlight the difficulties of 'turning' reciprocity and its associated institutions and values towards the extension and institutionalisation of trade. Reciprocity and its soft-framed institutions which comprise relationships appear to be inherently unstable. It is an instability which derives partly from the maze of obligations, and partly from the play of emotions (such as friendship, affection, loyalty, trust and altruism) which are regarded not as ideals in their own right, but as supporting and legitimising institutions for reciprocal exchange. For if such emotions were to be treated as moral absolutes, then the construction and use of reciprocal networks would be severely constrained. This intricate complex of relationships, prior obligations and uncertain emotions makes almost impossible the establishment of reliable and predictable relationships and the observance of agreements which are necessary for the effective conduct of trade.

The manipulative (and sometimes grasping) nature of relationships and the heavy, suffocating atmosphere of emotional blackmail which may be created, are often as unsavoury to those involved as they are to the observer. As already suggested earlier in this book, the imposition of a parochial morality that requires the writer to explain away such relationships would succeed only in distorting their nature. There are 'rights' and 'wrongs', but these vary with circumstances and priorities. In this Filipino Community, emotions, concepts and behaviour such as friendship and kinship, or envy and jealousy, rivalry and perfidy, are openly talked about and recognised, often dispassionately, for what they are. The expression of such emotions and the conduct of such behaviour are treated not as a transgression of an unspoken code of moral absolutes, but rather as a tool. Yet because they are used to institutionalise and legitimise reciprocity, and because there is no attempt to realise such ideals as friendship, loyalty, trust and altruism as virtues in their own right, there is, in consequence, no satisfying or effective sanction against perfidy and manipulation. The absence of such moral absolutes implies acceptance of envy, suspicion, jealousy, bitterness, resentment and of other base emotions—all instruments of the politic individual. Without the presence of a strong individual or some external discipline, emotions may suddenly turn upon their puppeteer. Individuals may erupt with frightening displays of anger and violence against those whom they believe have done them wrong but against whom no effective political, social or legal retributive action can be taken, for there is no righteousness in the abuse they have suffered, no publicly recognised immorality in the other's deceit. 'Face' offers a degree of satisfaction but only by validating envy, jealousy, resentment and bitterness: rather than critically examine his own feelings, an individual duped or merely outperformed may salve his indignation by interpreting the other's behaviour as a deliberate attempt to harm his own standing. Sanctions may be brought against those who commit adultery, who are sexually promiscuous, or who indulge in some other form of behaviour which may disrupt the family's extensive reciprocal networks, threaten its stability and soil its prestige; but the manipulation of reciprocity and its soft-framed institutions is regarded as a fact of life and, indeed, these relationships constitute the channel though which sanctions are implemented.

Formed as an end in themselves, then, the very nature of reciprocal networks creates too much uncertainty for the conduct of trade. There may even be a case for arguing that Filipinos may create highly sophisticated and fluid reciprocal networks when reciprocity is pursued as an end in itself, but when directed towards trade their networks, in comparison to those constructed by Chinese, are less certain and less complex. Figures

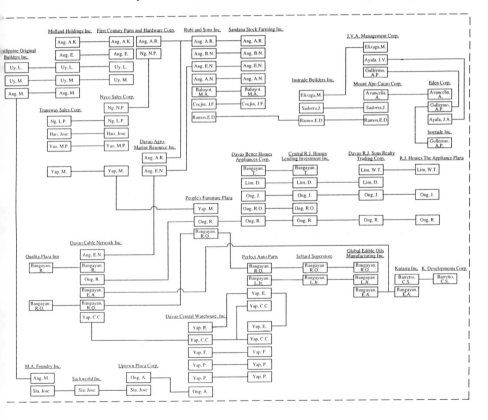

Figure 6.5a Merchants' Interests in Davao: Cells and Webs. Note: (i) Only enterprises whose shareholders also hold shares in other enterprises registered in Davao have been included. So, too, only those merchants with shares in more than one enterprise registered in Davao are shown

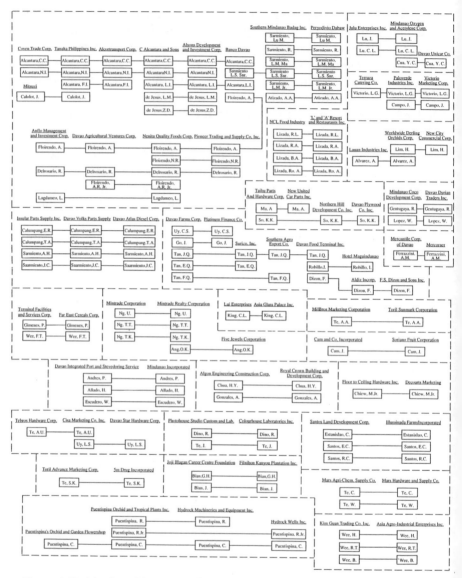

Figure 6.5b Merchants' Interests in Davao: Cells and Webs. Note: (ii) of the 107 enterprises shown, only 19 are not included in the top 325 corporations (a figure which excludes solo proprietorships). See text and notes to chapter 6 for details

6.5(a) and (b) may perhaps be indicative of one common perception of Chinese in Davao—that they are more skilled and more thorough in establishing a firm base of relationships upon which, and through which, businesses may expand. Filipinos, on the other hand, develop more self-contained and defensive businesses within a fairly small group which is protected by rather superficial connections with political allies. In other words, it is thought that Filipinos are less able to translate their complex networks of reciprocity into effective institutions for the support and extension of trade. It is, then, tempting to envisage a continuum of societies which proceeds from the conduct of reciprocity for its own sake, to its redirection towards trade, and finally, to the complete separation of trade from reciprocity. Transition along this continuum would be marked by greater authoritarianism, bringing with it the stability and focus needed to enable the 'turning' of reciprocal networks towards trade. To further distance and focus reciprocal exchange and its associated soft-framed institutions, emotions would increasingly be regarded as something in their own right: trust, honour and reliability become crucial attributes for a merchant; his reputation indispensable for the conduct of trade. Eventually, emotions such as affection, romantic love, trust, loyalty, friendship, altruism and kinship become ideals which are regarded as definitively good in their own right; jealousy, envy, resentment and bitterness are seen to be definitively 'bad'. Individuals thus come to rely upon legislation and legal institutions as reciprocal networks are pulled away from trade networks, allowing greater predictability in behaviour and wider opportunities for economic transactions. Filipinos, it could be suggested, lie at the beginning of this continuum; Chinese lie somewhere along its second and third stages.

The materials for the construction of such a distinction between these two notional groups might perhaps be found among some of the common perceptions Chinese and Filipinos have of themselves and of each other. The Chinese, it is frequently claimed, are clannish and family oriented. They have an ability to view emotions, circumstances and relationships objectively. Their relationships are much more utilitarian. They see politics as a business, not as the exercise of power for its own sake. And they adhere strongly to certain ethics: trust only those whom you know well; save for future investment; live within your means; negotiate hard but keep to your word; protect one another; perpetuate business through volume not through large profits. Filipinos, on the other hand, are perceived as outgoing, friendly and undisciplined. They are good singers and dancers and enjoy life, but they are unable to work hard or with a clear focus. They only consider the short term and in business they are only interested in quick, easy and large profits. They depend upon political patronage and raw influence to achieve success in business. They are too romantic and kind to treat relationships as an instrument for business. And while they may be easy to negotiate with, they seem unable to keep to their agreements.

The line between generalisations, caricature and myth, however, is a fine one: generalisations are often little more than the sum of their caricatures from whose shadows of reality a convenient legitimacy is extracted. Perceptions such as those above could be drawn upon to support the cultural determinist's view that Confucianism explains 'the Chinese' and their economic success. Equally, it could be argued that such perceptions are indicative of a clash of wealth values: 'the Filipinos' world is filled by their relationships; 'the Chinese' are primarily concerned with trade, and it is to this end that they direct relationships. Yet even this latter argument is not 'true'. It would be pedantic to shy away from the words 'Filipino' and 'Chinese', but the individuals who engage in exchange, who comprise institutions and who hold values are neither one thing nor the other. For although a particular form of exchange may dominate for a time, individuals possess many different aspects: they may each be engaged in different forms of exchange; they may each participate in different institutions; and they may each within themselves hold conflicting ideas and values. They may thereby create a mosaic of forms of exchange, institutions and values with shifting aspects.

Interestingly, the individuals who voice such unidimensional perceptions of themselves and of others are well aware of the true complexity. Their stock phrases are often no more than a simple justification for their own feelings and prejudices which alter with circumstances and mood. When faced with envious criticism of his success a 'Chinese' merchant will, for instance, say 'we, the Chinese, unlike Filipinos, rely upon our word of honour in the conduct of business'. But he will, in the same breath, recognise that Filipino merchants depend upon their word of honour and operate as effectively by word of mouth in the conduct of trade, and he will extend lines of credit to other 'Chinese' and 'Filipinos' alike. The 'Filipino' which the 'Chinese' merchant creates for whatever purpose is different from the merchants and customers he actually deals with. Occasionally, it may also be that the stock phrases which an individual advances are merely a repetition of half-remembered and confused explanations suggested by others. Indeed, the views of 'Chinese' expressed by some individuals (whether Chinese or Filipino) often bear a striking resemblance to those enunciated by academics: the influence of Confucian ethics, 'Chineseness', an ability to 'network', clannishness and an emphasis on blood, are all trotted out in an unthinking routine. Either academics have been successful in drawing such platitudes into their analyses, or, more likely, their analyses have been drawn into the lay world: it is not unusual for texts on management and entrepreneurs to be found on a bookshelf at the office or at home. But unlike the academic text, or so it would seem, the individual is aware that his own maxims (or those which he has learnt) are no more than a convenient justification—a lazy 'shorthand'—for the complex reality on which his energies, efforts and abilities are necessarily concentrated. While the academic focuses upon the unidimensional world he imagines within the intellect, the individual which the scholar analyses is aware that he, and others, and the institutions they create and the values they hold, possess many different aspects. Owners, managers and workers (whether they be Filipinos or Chinese) will constantly change their minds: they will, as circumstances seem to require, adhere to one or more of the many different concepts of 'Chineseness' or 'Filipino' to lesser or greater degrees, similar to that chameleonic way in which they commandeer the common touch, the mannerisms and etiquette of the elite, the authoritarian persona, or warm, approachable friendliness. Attitudes, ideas, approaches and mannerisms are adopted which, if consciously rationalised, appear contradictory. Yet this is all no more than commonsensical and intuitive reactions to different and changing circumstances which seem to require different responses. Much also may depend upon the varied experiences and the personality of the individual concerned. An owner or manager may be authoritarian for the simple reason that he is inexperienced in the exercise of power; or it may signal drive, intelligence and a corresponding impatience and intolerance of lethargy in others. Moreover, the many sides of his own personality and of those individuals with whom he must deal, and the unforeseen circumstances and complex networks of relationships he must negotiate, will mean that his 'shorthand' is often contradictory. Filipino merchants will lament the inability of 'Filipinos' to succeed in business; Chinese merchants who turn to Confucian explanations of their success see no differences between the way in which Chinese and Filipinos conduct business; to be 'Chinese', it is claimed, is to be an extension of being 'Filipino', and yet 'the Chinese' are inherently different because of their culture; a farmer will not trust 'the Chinese', and yet he will associate with, trade with, and rely upon, close Chinese friends rather than another Filipino to buy his crops; a Filipino whose great-grandfather was Chinese will complain about the clannishness, secrecy and dishonesty of 'the Chinese' and yet he is proud to bear a surname which is Chinese; 'the Chinese' are fundamentally different from 'the Filipinos', it is claimed, and yet there is no point in trying to make a distinction between the two.

The different and contradictory perceptions of 'the Chinese' and 'the Filipinos' are expressions of these concepts' different aspects. Presentations, interpretations and perceptions of values, exchange, institutions, ideas, mannerisms, facial expressions, language, food, smell and all the other details of being are seized upon to create multidimensional concepts which are the basis for unidimensional perceptions and presentations. In other words, whether he be 'Filipino' or 'Chinese', that individual's perception of 'Chinese' and 'Filipino' is a device with many aspects: it is a lazy, explanatory shorthand; it is a vehicle for anger and frustration; and it is a focus for the individual and for groups. It is also a negative kernel, or opposite, against which a more cosmopolitan elite can evolve. By creating 'the Filipino' who is not competent at business and yet good at dancing and establishing relationships for their own sake, and by creating 'the Chinese' who is a cultural automaton skilled at a limited form of commerce that is peculiarly 'Chinese', individuals (whether they be Filipino or Chinese) thereby conjure up an imaginary, unidimensional society which they are distinct from. By implication, the individual places himself into a separate group—the successful, cosmopolitan, merchant elite. In one sense this elite is a fabrication: it is no more than an expression of that mix of multidimensional values, exchange and institutions. In another sense, however, there is a recognisable, if fluid and very imprecise, group (or, more accurately, series of shifting and changing groups) signalled by their ability to redirect successfully reciprocity and its soft-framed institutions towards the prosecution of trade.

Apparent differences created by the presentation, interpretation and perception of 'Chinese' and 'Filipino' and 'cosmopolitan' are consciously and deliberately used by these merchants as kernels for the establishment of relationships among one another to facilitate the conduct of trade. In the same way that a Neuva Ecijan or Ilocano will use their place of origin and language as a common point from which to begin a relationship (or to consolidate one already established), so Chinese will use whichever nuclei of 'Chineseness' seems appropriate to improve their position and opportunities. One young Chinese businessman, whose grandfather had adopted the name Ramos, initially found his name a drawback. By refusing to exchange information on prices and customers, several individuals whom the young entrepreneur had targeted as possible suppliers made it difficult for him to establish the beginnings of a relationship with them, until he brought his 'Chineseness' to bear. He did not meet Somers-Heidhues's criteria for ethnic Chinese, for he did not read or write Chinese, he did not belong to any Chinese association, he had not been educated at a Chinese school, and he spent more time with Filipinos than with other 'Chinese'. But he spoke Chinese, and this was sufficient to overcome the suspicions of the few who seemed willing to limit their opportunities to their own definition of ethnicity.

There is, then, no real continuum which proceeds from reciprocity conducted for its own sake through to its gradual distancing and eventual separation from trade—it is only an imaginary device. Certainly this notion of a continuum should not be used to illustrate or analyse presumed cultural differences. Both Chinese and Filipinos distance reciprocity and focus it upon the conduct of trade by emphasising trust and reliability, and by adopting the multidimensional concepts 'Chinese', 'Filipino' and 'cosmopolitan'. An individual's reputation and honour are crucial for the prosecution of trade in Davao. Regardless of whether they be Filipino or Chinese, individuals deal, obtain capital and secure credit and supplies on the basis of their reputation. Their reputation is built up, not through socialising, but through proving reliability in business: agreements must be stuck to, supplies must be delivered on time, and payments must be made on or before the date due. If an individual breaks his word, if cheques bounce, if payments are late, then the individual quickly becomes known as unreliable and untrustworthy. A Chinese may, perhaps, be more open with another Chinese; but if either has a poor reputation, then the other will not deal with him, or will be very careful in doing so. A Neuva Ecijan may be willing to accommodate another Neuva Ecijan, but neither will automatically trust the other merely because they are both from the same part of the Philippines. Fictive or genuine kinship ties with another of high standing may, initially, open doors, but it will not ensure survival, let alone success. The son of a reputable Chinese businessman in the milling trade began his own business with a strong network of contacts inherited from his father. But this network, and his business, soon crumbled: he was late with his deliveries, his cheques bounced, and he fell behind with his repayments. His credit lines were reduced and in one instance suspended by a Chinese supplier who, at the same time, was happy to provide a reputable Filipino merchant with a credit line of hundreds of thousands of pesos.

Reliability and trust, and therefore predictability, is of far greater importance than ethnicity, place of origin or kinship. Yet while considerable importance is attached to these values, they are not moral absolutes: they facilitate the conduct of trade by making relationships more predictable, but deals and profits will not be sacrificed for them. Reciprocal exchange and its soft-framed institutions, distanced and focused upon trade, are built around some convenient kernel in the hope that a strong, predictable and useful relationship can be established or strengthened. If there is no value in these ambiguous relationships, they will be cut or allowed to wither, or used to open up new opportunities.

The 'turning' of the family

It is in the utilitarian nature of relationships that the 'turning' of reciprocity is most apparent. In giving pre-eminence to trust and reliability, and thereby building these emotions and this behaviour into expedient absolutes, reciprocity and its other soft-framed institutions are distanced from and focused upon trade. The willingness to introduce greater stability and predictability into relationships is also crucial if the family is to be turned towards trade's extension and institutionalisation.

Maintaining the unity of the family despite tensions and rivalries depends largely upon the emotional resilience of the parents—especially the mother—and perhaps the eldest child. Whether it is a consequence of fashion or necessity that the woman remains at home or works, the ability to weave networks of reciprocity is an essential skill. Independently, or through her children, she may build up such networks separate from or entwined with those of her husband, and thereby endow herself with considerable power. The father may be nominal head of the family and treated symbolically as such: he is presented with food and drink; when he enters the house his hand will be taken by his children and grandchildren and raised to touch their bowed foreheads; his views, however opinionated, will be listened to with quiet respect, and he will be allowed to win any argument of no consequence. But should he falter in his loyalty to wife and children, the network of relationships in which he nestles will fall apart or be turned against him: innuendo, allegation, distrust and sullied reputation will soon intrude upon his professional life.

Whoever holds the true power, the cultivation of dependency which the matriarch or patriarch constructs independently or through other members of the family, the liberal use of religious dictums, and affected expressions of emotions mixed with genuine compassion and affection, are set within a base of authoritarianism which may be called upon to override the instability inherent in reciprocity and its soft-framed institutions. All this allows the head to operate with speed and a wide latitude for hypocrisy and contradiction. In the interests of the family's unity, preservation and prestige, and in the perceived interests of the individual, actions and behaviour not tolerated in other family members become, in the head's instance, acceptable and unquestionable. There is little room for sentimentality although religious platitudes and claims of goodness, charity and selflessness are brought into play. Compassion and affection exist, but take equal place, or are subservient to, the preservation of the family and its reputation and, by extension, one's own standing.

Authoritarianism, the focusing of reciprocal networks, and the manipulation of relationships are patterns of behaviour which could be redirected towards trade and which might thereby give rise to institutions and means of operation very similar to a company frequently regarded in the literature as specifically 'Chinese'. The potential for redirection is clearly seen in the economic transactions conducted by Filipino families. Yet although the shift from rather disparate concerns to a coordinated and directed enterprise is possible, in many instances the energies, goals and diversions of the members of the family are not conducive to the establishment of such an enterprise. Quite apart from the financial risks such a transition entails, family members would have to 'turn' their relationships and re-examine their values if they were to create a successful company geared to the pursuit of profit.

The Y family of The Community, for example, are engaged in a number of loose projects. At Binugao, a little way southwest of Toril, the family has a farm (owned by SY) which is run jointly by the father and two of his sons—VY and DY. The family, and their two labourers, practise intercropping, creating a very complex pattern of land use which requires considerable attention to timing and the organisation of supplies and marketing. The logistics of this project are made still more intricate by the growth cycle and pattern of the banana trees which are to generate most of the farm's income for a number of years until other crops, including lumber, mature. The bananas are bought, packed and shipped to Manila by SM—a Chinese merchant with a small business about which more will be said later in this chapter. In addition to the farm, DY has a medical practice and owns a pharmacy which is supplied in part by XY who works as a hospital nurse in the Middle East. MY manages her various properties in Mindanao, Cebu and Manila through complex—though, as noted earlier, somewhat unstable—networks of reciprocity.

A second example is BB who used her husband's remittances from overseas to invest in a jeepney. She secured a licence and a route for the jeepney which was driven by her brother and his wife who also lent money to salesgirls working in a large department store owned by the Gaisanos. BB had also run a food store—a business she had taken over from RY's common-law wife—selling hot lunches and snacks to the salesgirls. The business had collapsed when she was abandoned by her cook who objected to serving both as her domestic and as cook for her business only to be paid a domestic's salary which in any case he often received several months late. She also bought three trucks and attempted to gain a foothold in trucking. This business also ended in failure, for the drivers proved extremely difficult to handle and she found herself being constantly undercut by the larger trucking companies. This enterprise was then followed by a mahjong parlour which she set up in an extension to her house.

Many of the problems which BB had faced were those of circumstances beyond her control. But underlying all her business concerns were disputes between herself, her workers and her husband's family. She had been accused of hostility towards her father-in-law, and of siphoning off her husband's remittances to her own parents in Manila—funds which should have been used to pay back the loan from her father-in-law which she had used to buy the house from which she now ran a business. Moreover, she had not been able to keep either her cook (the pillar of her catering business) or her drivers. Consequently, she had allowed others to build up an unflattering impression of her own character—an impression which could then easily be strengthened by prejudice: she was a Tagalog, not a Bisayan, and this was enough to explain, and to confirm, her parsimony and inability to handle other people. The very fact that she had engaged in trade made her unpopular, for she would have to place profit before the demands and obligations required by her relationships. That she had been unable to combine firmness with the continued successful management and reorientation of her relationships made matters still worse.

A third example is that of SM—a Chinese merchant who purchased bananas from the Y family and a large number of other small farmers in Davao. Although of Chinese descent, SM does not speak, read or write Chinese, nor is he a member of any association, nor does he socialise primarily with Chinese, nor does he in any way appear to be Chinese. SM's wife, a Filipina, is the granddaughter of DCAS's brother. It was through this connection that SM became the purchaser for ACAS's crops, and then for FY's produce. SM employs two labourers who pack the bananas. Once packed, the bananas are shipped by sea to Manila where they are purchased by SM's brother—SCM—who owns two stalls in a public market in Quezon City. Both SCM and SM's sister were originally sent to Manila by their father to stay with their mother's cousin from whom they learnt the retail trade. SCM then married the daughter of a stall-owner in the market in which he subsequently established his own stalls. SCM has worked in the market for 15 years and is now the market association's president.

The price at which SM purchases bananas from small farmers in Davao is determined by the retail price in Manila, by his brother's profit margin, by shipping costs, and by the price offered by his competitors. SM's profit margin is around one peso per kilo. His business and that of his brother are kept entirely separate, though each has occasionally lent the other money free of interest. When SM started his business he would include his load in shipments made in the name of the Dabaw Panabo Banana Growers' Association—an organisation designed partly to bypass the middlemen. The technique was expensive and, in any case, it proved difficult to identify which of the crates arriving in Manila belonged to SCM. The problem was later solved by another of SM's brothers—SCMD—who worked for the Davao Central Chemical Corporation. SCMD arranged for SM's crates to be included with the corporation's shipments and for SM to be issued with his own notice of loading.

Although there is a level of cooperation among family members, their individual business or professional concerns remain disparate, and there is some tension among the brothers. After leaving school, SM's oldest brother—SCMD—paid for SM's training as a mechanical engineer, and also his fees to the Board Examination which SM passed. SM was then sent to Manila where SCM arranged for him to work for his friend. There, SM worked to gain experience but without pay. SM returned to work in a small shop in Davao owned by a fourth brother, SCW, who manufactured wooden pallets for TADECO—a Floirendo Company (see Figure 6.5). After a heated dispute with SCW over pay and the division of responsibilities, SM then took a position in a large furniture company owned by the family into which his sister had married. Here, too, pay was low and he was required to work in a number of different capacities, including shop assistant and supervisor, as well as engineer. Driven by the need for more money, and for greater independence, SM eventually moved into wholesaling. Nevertheless, tensions between himself and other members of the family continued. SM agreed to take on as a packer the cousin of the eldest aunt of SCM's wife. The new recruit, however, proved careless. By the time the first shipment for which he had been responsible arrived in Manila, the bananas—which had been packed badly and too late after harvesting—had begun to rot. SCM would not pay for the shipment nor bear any portion of the loss—a sum which amounted to SM's average monthly income.

The company

It is clear from such examples that the redirection of a loose association of desultory activities towards a more focused economic organisation is no easy matter. Beyond the all-important desire to trade there must be a preparedness to distance reciprocity and its soft-framed institutions from trade by assigning pre-eminence to trust and reliability. In this way, the intricate complex of prior obligations is simplified and marginalised, and the tensions and heated emotions generated by the play of relationships which attach no absolute morality (nor even a strong code of right and wrong) to altruism, loyalty, friendship, trust, envy, jealousy, affection, resentment and bitterness, are either avoided or at least prevented from impinging upon the organisation's external connections and its internal operations. Again, however, trust and reliability, despite the preeminence with which they are endowed, remain expedient absolutes. If mutual self-interest is removed, or if one party believes itself sufficiently powerful, then all that is left is a purposeless relationship. In large organisations, most especially, there will be those who are unable or unwilling to distance and focus their relationships upon trade. The skilful handling of family members' and employees' rivalries, resentments, ambitions, egos and jealousies, as well as their varied abilities and multidimensional personalities, requires prescience and determination if tensions and disputes are to be defused, or used constructively, or prevented from spilling over into the company and thereby obscuring its vision and dulling its effectiveness. There must, then, be a judicious mixture of authoritarianism and paternalism to ensure that individuals maintain a distance between trade and reciprocity; there must be additional procedures and organisational methods of control (which breed dependency among workers upon the owner or manager); and there must be the promotion of an ethos which is critical of jealousy, envy and politicking and of other emotions and actions which supersede the interests of the company or divert it from its aims.

Within companies, then, it is not surprising that while key personnel are those in whom the owner (or owner and manager) have considerable trust (a value, or emotion, which overrides any kernel for the establishment of a relationship), this trust is often supported by an organisational arrangement which helps ensure control. These methods differ according to individual preferences, decisions and circumstances. One example is a large plaza constructed in Buhangin by a prominent Chinese merchant family—LS. The company may be thought of as comprising three sections—management (operations), finance and administration. The management section is organised into four divisions—supermarket, department store (food), mall, and the main department store. The departmental manager of the supermarket ran the original 'Park 'N Shop' in Buhangin before it was bought by LS from Mehang—another Chinese merchant. The departmental manager of the department store (food), who has considerable professional experience of managing stores in Manila, was recruited by LS through a third party. The manager of the mall is a high-school classmate of the president—RLS. The management of the main department store is handled directly by the president. His assistant manager, now in his sixties and with long experience of managing stores in Manila and London, was hired initially as consultant. The

president's operations manager within the department store is a gradu-
ate of La Salle, Manila, and a member of another prominent Chinese
merchant family (the Bangayyans) who are also shareholders in the
plaza. Shadowing the management section is the finance section which
is also headed by the president. All cheques signed by the management
and administration sections must be countersigned by finance. Vice-
president of finance is an old friend of the LS family, the wife of the merchant So King Co, and the daughter of the lumber merchant Go Tam Go
who is a partner in RLS's plywood company and a shareholder in the
plaza. Departmental comptroller of the supermarket is a schoolmate of
the vice-president of finance. Departmental comptroller of the department store (food) is a friend of the LS family. Departmental comptroller
of the main department store is another long-standing friend of RLS,
and, in common with the departmental manager of the department
store (food), is also a former manager of the original Park 'N Shop.
Departmental comptroller of the mall is another close friend of the president. The latter three departmental comptrollers are all Filipinas.

The administration section—which includes personnel, internal audit,
market research and general support—also falls under the direct control
of RLS but it does not form a tier between the president and management and finance, and has no authority over these latter two sections of
the company. The responsibilities of the administration units are partly
as their titles suggest, partly to provide ideas for improving operations,
and partly to keep an eye on the other two divisions. There is no hierarchy among the administrative units, for each is directly answerable to
RLS, though the director of personnel (N, the nephew of MW's son-in-
law) does have a crucial role about which more will be said later.

The close ties with these key personnel and the organisational arrangements provide the option for strong, direct control, but they also allow
the individual to operate with a fair degree of autonomy. Over the long
term, decisions are centralised with the president and the Board. Once
decisions have been reached and plans formulated, key personnel are
then left on their own. They are able to hire and fire employees of the
rank of supervisor or below (although their immediate subordinates are
determined by the president and the Board), and, for the most part, they
are responsible for the organisation of merchandising, though some
goods from Hong Kong (where the LS family have properties) and
China are routed through a sister of RLS in Manila.

Overseas connections and the organisation of supplies are an important
means of control in another 'Chinese' company (referred to here as the
Q company)—a successful and growing enterprise that imports and sells

construction materials, fixtures, wallpaper and carpets for homes and offices. There are nine branches—three in Davao (including the main office) and one each in Tagum, General Santos, Zamboanga, Cagayan de Oro, Cebu and Manila. In most cases the stores are run by professional and qualified Filipino managers. Some were hired through agencies, some through personal contacts, and some had become acquainted with the president through his previous dealings. Relatives of the president are placed in each branch. The relatives are not involved in, nor are they supposed to interfere in, the store's operations, but they will report on the general atmosphere in the branch and on any actions or decisions which they regard as damaging, and in this sense they act as overseers. Every two months or so the president or his internal auditor or his operations manager will tour all nine branches.

Each branch manager's duties are well defined and fairly limited. Decisions concerning the extension of credit and the appointment and dismissal of employees must be referred to the president who is also responsible for arranging supplies of goods. Most items are imported from Europe (mainly Italy, Germany, Austria, Greece, Spain and the United Kingdom), Brazil, America, Japan, Singapore, Malaysia and Taiwan. Of the imports, 60% originate from Southeast Asia—a proportion which reflects, not personal networks (for supplies are generally purchased through impersonal contacts, trade fairs and promotional literature) but lower prices, a quality and style of products which are more in tune with local tastes, and international arrangements on tariffs. It is these supplies which enable the president to deal with the problem which he views as the most difficult and most important to overcome—control over a growing and expanding company. Branches are not allowed to hold stocks and must buy goods through him: in this regard each branch is treated as a separate company. If the value of goods held in a branch is greater than money paid for those supplies, or if the money returned to head office after sales is less than the money paid for goods plus profit margins, or if they are unable to buy the same or a greater quantity of supplies, then something is clearly wrong unless it can be shown that growth has been faster than projected, and has outstripped the collection of credit within the time limits agreed between head office and the customers.

Another dispersed company which supplies technical personnel for the installation, operation and maintenance of machinery for processing wood, faces a similar problem of control. The company provides its services in Malaysia, Indonesia and Japan as well as in the Philippines. The tour of duty overseas for its personnel is around four years. The majority shareholder and the director of the company, a Filipino, brought most of his key personnel with him when he left the Alcantara, a Filipino merchant family, to set up his own company: it is, as the director puts it, 'employment based on trust'. In addition, duties are closely prescribed, and information and contact remain centralised with the director.

One organisational method of control born out of luck and circumstance is the adaptation and institutionalisation of the feudal relationship which has at its core the landowner's and estate workers' feelings of mutual responsibility—a relationship very much akin to that which still exists between the old landed families and their workers in England even today. GZ, the owner of a popular and successful beach resort in Davao, employs around 200 workers. Most of his employees either lived on his family's estates at one time or are the children or grandchildren of those labourers who worked on the estates first built up by GZ's father and grandfather. In common with many of those landed families who were unwilling or unable to diversify their business interests, large tracts of land had to be sold off. The estate's labourers, however, were moved on to lots reserved for them in one of a string of private residential subdivisions between the Poblacion and Tibungco built on land formerly owned by GZ's father, grandfather and cousins. The labourers were sold the lots (together with a house) for around P4000—an insignificant sum, but one which had to be paid to GZ's family over 20 years. Around this organisational method of control which itself institutionalises reciprocity, GZ weaves relationships between himself and his employees and their families—relationships which, in many instances, GZ inherited from his father and his grandfather and on which he spends a good deal of time maintaining and strengthening. He is known to all of his employees in the subdivision where he will often eat: he is treated by them 'as if a King', and they are his *bata-bata* (or, very loosely, 'runner'), willing to do him any kind of favour because it is an honour for them to be asked. In return, he will help them in an extraordinary variety of ways: he will use his influence if they or their kin have problems with the police; if kin or close friends need a job, he will help them find one; and he will mediate in disputes with neighbours. Indeed, he must set aside several hours each week to handle the deluge of petty disputes and problems in which his workers and their families become embroiled. Nevertheless, this investment in time and energy is important. Reciprocal exchange, cemented by the soft-framed institutions of trust, loyalty, friendship and affection provides him with a sound and effective labour force over which he may, if he so wishes, exert considerable influence.

Clearly, organisational methods of control—which are, in essence, designed to breed in employees dependence upon procedures and arrangements focused upon and manipulated by the company's head and his key personnel—are dependent for their efficacy upon the decisions and actions of the individuals who comprise the company. The clarity of direction given to the company and the certainty of its control are very much rooted in the nature of the relationships established. To ensure the smooth operation of whatever organisational arrangements exist and the fulfilment of the company's objectives, a careful balance is struck between a relationship that is close enough to allow the development of expedient trust and of effective cooperation, and one that begins to distract individuals from the task in hand and to disrupt the prosecution of trade. Thus the emphasis given to trust, and the implementation of organisational arrangements designed to shadow the decisions of individuals, may themselves be supported by additional safeguards which ensure that reciprocal relationships are distanced from, and carefully focused upon, trade. The merchant thereby creates an intricate lattice of varied personal relationships comprising reciprocal networks, each strand of which supports, monitors and controls another.

In return for loyalty and performance, key employees may be plied with a variety of gifts, including subsidised or free housing, cars, medical benefits for their families, and free trips overseas, perhaps to Hong Kong or Singapore. The give and take of material items and services ranges way beyond any job description or contractual obligation and is accompanied by plenty of opportunities to strengthen emotional bonds: regular meals are taken together at a restaurant or at each others' homes; an employer may take a key employee on a casual, informal, fishing trip; there may be regular bowling competitions; and an employer may become closely involved with his employees' families and their personal lives. The establishment and fostering of such relationships with key employees are as true of the large company as they are of the small company, whether it be 'Filipino' or 'Chinese'.

At these extremely informal occasions, employer and employee will joke and make fun of each other. The traffic in glances, the practised touches on the arm, the changes in tone of voice, the relaxed smile, the gentle ribbing of each other's ego, the interest shown in the other's family, and the willingness to help and to listen, convey to the employee that he is an important and integral part of the company. At work, although the atmosphere remains cordial, a slight formality is introduced. Both sides understand that their crafted relationship is useful to the realisation of their shared goal and, as the owner of a Filipino company put it, for the protection of the company. Such a relationship makes the give and take of orders easier; it enables a greater appreciation and, crucially, a better anticipation of the other's wishes and expectations; and, by making it unnecessary to constantly refer back for confirmation, it allows the individual to act with more speed and flexibility. Further down the hierarchy, too, reciprocity and its soft-framed institutions may be carefully distanced from and focused upon trade. Even in the larger companies there is often direct communication between managers and the lower rank. The experienced manager replaces harsh demands with a stable, calm, relaxed, confident and untroubled appearance—the common touch underlain with cold, unquestionable authority. Staff must feel they can open themselves up to their superiors, but they must not be given the impression that the relationship established can be misdirected.

Relationships are also played against one another: at all levels individuals are vying for influence and attention. Within the plaza these shifting cliques serve as a means to keep individuals slightly tense and on a competitive edge. It is here that the director of personnel has an important role to play, for he acts as an insider in most of these groups and keeps the president informed on the state of relationships among his employees. It is only when politicking, rivalry and ambition become an end in themselves or begin to interfere with the operation and direction of the company that these cliques become harmful.

The owner of the Q company who, as noted above, places his relatives in branches around the Philippines to monitor his professional managers, will also deliberately use tensions between his kin and employees. It is made quite clear to both managers and relatives that the latter have no say in the operation of the store. Yet it is also expected that tensions will arise between the two: the manager resents being monitored, and the relative's zealousness overcompensates for his own impotence. Much of the owner's time during his periodic visits to his branches is spent over a few drinks patching up arguments and rifts between the manager and his unwelcome shadow.

Reciprocity and its soft-framed institutions may also be distanced from and focused upon trade by the explicit codification of behaviour. In the plaza this code is set out in a mission statement which carefully plays both to the pursuit of profit and to the community or, from the individual employee's point of view, to his relationships outside the company:

> 'We believe that business is an important element of modern society and must be managed not only to earn profit and sustain itself but also to promote the dignity of persons, the growth of the community and the improvement of the quality of life.

> It is therefore our public responsibility to utilize all our resources with maximum efficiency, recognise the value of the contributions of those who work with us and those who support us with capital, quality goods and services. Most of all, it is our key objective to provide prompt and honest service with utmost courtesy to those who patronise our prod-

ucts and services—the customers whose total convenience, comfort and happiness we are committed to always pursue with our best effort.'

This is an ethos which influences the selection of rank-and-file employees, and it finds symbolic expression in the National Anthem (a long and complicated affair) which is played at 10.00 o'clock every morning, and in the cardboard and wooden signs, which pepper the corridors, loading bays and storage areas behind the shops and public walkways of the plaza, enjoining staff to smile: 'A Smile is the Best Thanks we can give our Customers'; 'Smile and Feel Better'; 'Feel Stressed or Tense? Try Smiling!'

More often than not a code of behaviour remains unwritten but nevertheless influential. Within another 'Chinese' company employees are made aware that certain standards and patterns of behaviour are expected from them in order to ensure that a single direction is maintained by the company. Relationships and behaviour which interfere with the smooth running of the company will not be tolerated. It is clearly understood, and made quite explicit, that relationships must either be suspended or redirected for the good of the company. So, too, in another company, which supplies machinery for mills, a clear line is observed between the director (who is also one of the owners of this family enterprise) and his employees even though he knows all of his employees well (for many are the sons and daughters of the former employees of his father) and will usually eat and drink with them after work and during the weekends and on the occasional public holiday. Neither he nor his workers appear to feel that 'He is their boss and They are His workers.' Yet all know that employees are *expected* to inform the director of any problems at work or at home which affect their performance: for despite long-standing relationships bequeathed from one generation to the next, employees who begin to fail in their duties will be given three warnings and then be removed. There is no contradiction here between the friendships and the pre-eminence attached to performance. The relationship is founded upon reciprocal exchange: its supporting emotions, such as friendship and trust, though genuine, are not regarded or treated as absolute virtues in their own right. The principal aspect of the relationship is to facilitate trade. As the director put it: 'I am good to them, but I see to it I get something in return.'

The owner's personal involvement in the affairs of his employees at work and at home, and the help which is extended to their families, may also carefully be institutionalised in order to ensure that relationships are distanced from and focused upon trade. In one 'Filipino' company the owner knows his employees and their families well; he sets aside a proportion of his time outside the company to be with them; and their emergency needs are met by the company. Nevertheless, help, though given personally, is governed by company policies (which the owner himself drew up) in order to avoid conveying the impression that one individual is being singled out for special treatment. Other companies (both 'Filipino' and 'Chinese') run cooperatives which provide benefits, services and goods in a relatively impartial manner, but in some instances with the personal involvement of the owner.

Despite the attempts to 'turn' reciprocity and its soft-framed institutions towards the prosecution of trade, reciprocity, pursued as an end in itself, does invade the company. A common problem faced is that of *pakikisama* and *utang na lo'ob*—terms which describe, among other things, informal groups or cliques. Such groups are an expression of the reciprocal exchange of favours, support, tasks, services and material items. As already noted earlier in this chapter, these groups can be 'turned' to the company's advantage by creating an atmosphere of constructive, if sometimes tense, competition. They may also form the basis of team-work by allowing greater flexibility in the assignment and performance of duties. In one of the Aboitiz companies, supervisors, though expected to keep *pakikisama* in bounds, attempt to direct the loyalty of these groups towards 'the management' and the company. As a branch manager put it: 'my loyalty to you [the group] ends where my loyalty to Aboitiz begins'. Moreover, by identifying leaders and allowing them to participate in the formation of decisions, it may become possible to fit this group into the company's organisational structure. In an Alcantara company, a supervisor of 3500 men had attempted to use these informal groupings to organise 'quality circles'—groups of front-line workers who were expected to identify problems in production, and then find and implement solutions to those problems. In this case, however, it appeared that the individuals comprising these informal groups all thought alike and, consequently, there was no useful exchange of ideas. Individuals who create these groups are primarily concerned with their own status (as measured by reciprocity's associated wealth values), and with the use of the group, either as a reservoir of emotional support and companionship, or as protection against uncertain and badly paid employment. Rather than appear to put one's own concerns above those of the group, or to vie for greater prestige within the group, the individual will merely toe the line, provided it suits his purpose so to do. If he were to appear to jostle for higher status within the group or to turn his back on it, then 'face' (that all-purpose self-justification of one's own feelings of envy, jealousy and hostility born of mediocrity), accusations of selfishness, ostracism and pressure to demonstrate unthinking loyalty are used to safeguard and strengthen reciprocal exchange within the group and thus to preserve its unity. All this well serves the existing leader's crude philosophy of leadership.

It is for these reasons, too, that these informal cliques are perceived as a nuisance at best, or, at worst, as a threat. The formation of cold, exclusive groups creates tension and feelings of uncertainty, envy and jealousy in other employees. Loyalties are directed away from the company, and the company's objectives are obscured. Team-work and a focused attention on profit are replaced by individuals who, vying for prestige within groups which may, in turn, be scrambling for influence among each other, now refuse to cooperate, feeding each other false information, and finding ways to ridicule. Worse still is when these cliques become strengthened by the soft-framed institution of the *compadre*. The fact that individuals are co-workers is a kernel around which relationships may be established outside the workplace. Individuals may become *compadres*—that is, the godparent of the other's children or grandchildren, or the sponsor at the wedding of the other's children or grandchildren—or they may become the *compadre* of a workmate's kin or friend. Groups within the workplace become increasingly oriented towards the conduct of reciprocity for its own sake, and the preservation of the group's interests. As these groups crystallise and consolidate, members of different groups begin to link together and inveigle other individuals from higher or subordinate levels within the company's hierarchy or from parallel units originally designed to monitor and control. For example, a superior may become a member of an informal group and thereby render his subsequent decisions questionable. Moreover, favouritism—real or imagined—may build up envy, resentment, suspicion and tension within the workforce as a whole.

From the perspective of the owners and managers, the greatest threat from these informal groupings is their potential to merge with or transform themselves into unions, for it is at this point that reciprocal exchange and its soft-framed institutions which comprise the relationships of the group become entrenched. It has already been suggested that one aspect of *pakikisama* and *utang na lo'ob* is mutual protection. It may be, then, that once the financial stakes are raised (by improving pay, for instance) the group might weaken, for it is no longer so important to the individual to remain part of it. However, this is only one aspect of these informal groups. Whims and demands, legitimate and fabricated, reasonable and unreasonable, sensible and ludicrous, become as one: vital economic issues and moral debates used to support relationships and relative status within the group and among different groups, are merged. Inevitably, the company becomes snarled up in bickering, politicking, rivalries and the struggle for prestige.

Owners and managers, then, will frequently act to suppress these exclusive cliques. As soon as the owner of one 'Chinese' company notices groups developing he will remind individuals of their loyalty to the company—objectives must be realised, and clarity of purpose must be maintained. An employee who continues to behave in a manner which runs counter to, or obscures, those objectives, will be removed. In the larger 'Filipino' company of the Alcantara, supervisors spent much of their time identifying and breaking up these informal groups; and, in their turn, *padrino* managers (or supervisors) were immediately dismissed. In common with a Floirendo company, rank-and-file employees were shifted around every six months as a matter of routine.

But perhaps the greatest threat to the company is the family itself. The distancing of family relationships and their focusing upon trade are both difficult and crucial, for should disputes, tension, envy, jealousy and rivalry spill out into the company's operations, the consequences may be disastrous. GZ is, at the time of writing, only the temporary owner of the beach resort: although the company is his, the land is owned jointly by his parents and his brothers and sisters. His family have demanded that, in two years, his company should be turned into a family corporation. From GZ's point of view, the rest of his family are not interested in, or have no competence in, business; their claims merely reflect his success and their failures; his brothers and sisters are envious; and his father has neither the skill nor ruthlessness nor willingness to take the risks necessary in making and running a successful company. Clearly, such tensions, if not controlled, are potentially ruinous.

The enterprise owned and managed by members of the CAS family referred to earlier in this chapter is another example of a 'Filipino' company whose head is acutely aware of the potential destructiveness of family tensions. After escaping from an unhappy stepfamily by marrying a suitor with the largest wage packet, DCAS started her business in 1937 with the purchase of an army-surplus jeep. She renovated the jeep and sold it in Manila for a profit which she then used to buy more jeeps. She continued with this circuit until she had built up her own fleet of 60 units. DCAS then diversified into gasoline stations. The profit margins from the transport business and from the gasoline stations were small, but there were few competitors and her turnover was substantial. Always diversifying as competition intensified or as other circumstances changed, and always maintaining a high turnover, she created a large business empire which eventually came to revolve around a real estate company. She and her husband have an equal part of the majority of shares in the company: the remainder are divided equally among their children. But it is she alone who makes all the decisions concerning this core asset. The remaining gasoline station was handed to her son and daughter-in-law (RCAS and LCAS), but, as noted earlier in this chapter, DCAS retains indirect influence over their affairs, and those of her other children and grandchildren, by establishing and guaranteeing lines of credit on their behalf, by extending help and advice on business and personal matters, and by providing invaluable contacts with other merchants, politicians and professionals. The monopoly of power she has effected over her core assets, the reciprocal networks she has focused upon herself, and, through these networks, the influence she wields over her family's various business, professional and political interests, have been deliberately engineered, partly in order to limit and redirect family squabbles which may spill over into the core of the family's trading interests.

The dangers stemming from relationships within the family are also well-recognised within the smaller companies. One example is a 'Chinese' company owned and managed by seven brothers and sisters. Together they run a mill which processes corn, and two shops which sell mechanical parts for rice and corn mills. The mill, run by the oldest brother, and the shops are managed as separate entities, although the mill and shops will occasionally grant each other interest-free loans, and the oldest brother will buy spare parts and new machinery from the shops provided they are able to supply the best equipment for the best price. Tensions exist, but all members of the family are keenly aware that disputes and destructive emotions must be kept out of their business affairs, for they have seen a number of 'Chinese' enterprises collapse precisely because jealousy, envy, rivalry and consuming ambition were allowed to invade the company. However, it is not by will alone that relationships are distanced from, and focused upon, trade: organisational methods of control and codes of behaviour have been institutionalised to help safeguard the company. The father, who is now semi-retired from his own business (which is kept entirely separate from his children's company) and who advises his children on long-term strategy, is sufficiently removed from their trading operations to act as a mediator and arbitrator, and all are prepared to accept his decision as final. Another institutionalised pattern of behaviour (which operates more by happy accident than by conscious design) is the constant switching of tasks among family members and employees. All members of the company—family and employees—must be flexible. Family members must work as manager, accountant, collector, salesman and driver; and employees are expected to work as collector, driver, labourer, salesman, shopkeeper, warehouseman, and, in the mill, as plant operator. The lack of clear-cut duties reduces labour costs: the longer each individual is kept productively occupied, the better. It is a technique which also requires less planning. Some may find it a rather dubious solution, and indeed it may be a rather crude response which will not prove to be cost effective as the company grows, especially if individuals demand that pay should be linked to responsibilities. Nevertheless it is a practice which is also perceived to encourage team-work, and, as in some of the larger companies such as those owned by Alcantara and Floirendo, it helps prevent the emergence of reciprocal groupings within the company.

In order to distance and focus family relationships upon trade, other companies may define and specifically limit the authority of family members. To avoid giving the impression that decisions are coloured by nepotism, the owner of one 'Filipino' company excludes all of his own kin and those of his key personnel. In another large 'Filipino' company, professional managers were introduced partly in order to reduce the confusion produced by family members who did not clarify their own areas of responsibility, and partly to act as an emotional buffer among family members and between the family and their immediate subordinates. Within the Aboitiz companies, family is regarded as an effective tool because of the speed and flexibility with which decisions can be made—decisions which are often necessarily intuitive. Nevertheless, the problem of succession from one generation to the next is recognised, and the turmoil which can be created by emotions and disputes spilling out into the companies' operations is guarded against. In-laws may provide useful contacts: for example, DOM, who married into the Alcantara family, is a Presidential Advisor (see Figure 6.4), and his brother is an influential figure in the Department of Trade and Industry (Manila). On the other hand, in-laws may create extremely damaging tensions: for example, the DL family, which married into the Floirendo dynasty, is now trying to wrest control over a number of companies from the Floirendos. The Aboitiz family, however, specifically excludes in-laws from the Board, management and operations; once chosen, a future spouse may be carefully vetted; and the cult of professionalism is strongly propagated. Family members are well-groomed: they are sent to the best schools and universities and are given experience in various companies and at all levels from clerk upwards. But if they do not measure up they will be, and have been, sidelined. Members who do not pull their weight or who are incompetent and inefficient are removed. The ability to build relationships and exhibit the common touch remain important considerations: inexperience in the exercise of power and over-compensation for an introspective preoccupation with shyness are to be ironed out. Yet trade is conducted on the basis of contracts: in their relationships within the company and in their dealings outside the company, trust is seen merely to be one of the requirements of professionalism.

Professionalism, professionals, government and corruption

The cult of professionalism may not be incongruous with a family company, but it is difficult to square the perpetuation of dominating family lines within a professional institution even if it is run as a private corporation. One example of such a private institution is Davao Doctors—the largest and most respected hospital in the city. Five major shareholders (all doctors and all founders of the hospital) have at least one son or daughter (and, in some instances, a daughter-in-law or son-in-law as well) who are undergoing or who have completed residency training and are now practising in the hospital. The chief of hospital administration is the daughter of one of the hospital's founders who also sits on the Board of Directors. There is also at least one other large and respected family with three members who work in the hospital in some capacity.

Membership of this elite core within an elite institution undoubtedly constitutes a useful advantage, particularly if one is to obtain a surgery in the hospital from which to build up a practice. Securing rooms in the hospital requires a 'good-will' donation of P75 000; rent is set at P75 per square foot; and utilities and local tax must be paid by the practitioner. If the practitioner is able to buy, or is given a certain number of shares, then anything from 10% to 50% of the rent and running costs will be shouldered by the hospital. To some extent the law which requires individuals to obtain a professional qualification and to pass the National Board Examinations before they are allowed to practise medicine, and the tendency to select applicants from the better medical schools, helps prevent reciprocal networks from interfering with the hospital's attempt to acquire competent and professional staff. Moreover, it might be suggested that the children of serving doctors and members of the Board are imbued with a subtle blend of values and experience which provide them with an important and inimitable advantage which should not be dismissed simply because of fears that others may cry 'nepotism'. Clearly, however, were it to become apparent that family lines were strongly supported and institutionalised, abuse, bitter politicking and mistrust would inevitably affect the running of the corporation. Indeed, the instability and uncertainty introduced by the conduct of reciprocity for its own sake, or by its use with the intention to support the operation of a professional institution, may well render that institution ineffective. One striking illustration are the institutions and procedures for planning in Davao (R.N.W. Hodder, 1991).

Given the pervasive importance of reciprocity, its practice within professional institutions, and its use in the conduct of trade, it is of no surprise that business and politics are closely entwined. There is no point in attempting to represent the link between merchants and politicians as a noble alliance. Nor is there any point in putting together elaborate and indulgent justifications of these relationships. However, wining and dining politicians, making donations to opposing candidates simultaneously (one payment 'under-the-table', the other public) during elections, and securing the protection of these power brokers, are pragmatic actions. If the merchant does not spend both time and money establishing a solid base of relationships he will find continued success and further expansion quickly undermined. What is acknowledged, even by the merchants themselves, as corruption from one perspective is often a necessity when viewed from another. Corruption is little more than a form of mercantilism—a mercantilism derived from and strengthened by the use of trade and the material progress it creates to support the prosecution of reciprocity for its own sake and to provide symbols of prestige and worth as judged by reciprocity's associated wealth values. In short, corruption often represents a symbiosis between the mercantile interests of the trader and the prestige of the politician. The issue is not one of morality but of the effect which corruption has upon the general conduct of trade. If the exercise of political ambition is the establishment and maintenance of introspective economic monopolies to further strengthen the power of the politician, then trade and sustained material progress will be stifled. To what end are reciprocity and trade being directed?

Although the kudos derived from political connections, an interest in a variety of social contacts, genuine friendship, or perhaps the cultivation of political ambitions are considerations which are, or may become, of greater significance, political connections are established, first and foremost, to protect the merchant from other politicians or bureaucrats acting independently or perhaps as proxies for a competitor. The incidents which arise are generally petty, *ad hoc* and opportunistic—such as the use or manufacture of some infringement of laws and procedures which are built upon by an official hoping to extort a bribe or perhaps to ridicule. DCAS, who has also served as a *barangay* captain for 17 years and has established an extensive web of political allies, which includes a gaggle of former mayors and serving congressmen, many of whom are now her *compadres*, found herself in dispute with a local military commander. In one instance, an officer arrived with a number of his men and a bulldozer to remove a squatter settlement from her land in her *barangay*. The action would have been politically unacceptable, most especially since the fate of the squatters was, at that very moment, waiting upon the decision of the court. A few weeks later DCAS was wanted for questioning after she had given money to a man who had appeared at her house asking for a Christmas donation for his family. According to the military the man was a Communist guerrilla. In both instances a telephone call to her *compadre*—a congressman in Manila—was sufficient to prevent any further escalation of the dispute.

Another example is GZ who has been fortunate enough to inherit a long-established network of political connections from his grandfather. GZ's beach resort is on an island which lies just off the coast of Davao City. The first two years of operation were extremely difficult. There was no electricity or fresh water, and the enterprise barely managed to break even; it was also during this time that the local government on Samal Island threatened to implement a 30% amusement tax on his entrance fee—a laughable idea had it not been for the crippling effect this would have had upon his business. It was at this point that his aunt, the Vice-Mayor of Samal, whose campaign GZ had financed, was able to elicit sufficient support from councillors to block legislation, although the threat of further attempts to implement the tax still exists. At the same time GZ made a point of establishing a good relationship with the Mayor of Samal by donating money to his political campaigns and by ensuring that he was well attended whenever he visited the resort.

It is difficult to know, even for the merchants who believe they have been targeted, whether or not such incidents are simply bad luck or manufactured by vindictive and greedy officials or by competitors. Rumour and slander, and the unstable and unpredictable nature of reciprocal networks, make it easy to slip into a state of paranoia. Nevertheless, this particular form of mercantilism is not only used defensively: political connections are also used aggressively. They are, for instance, often used to facilitate smuggling operations. Large shipments of automobile parts from Japan and elsewhere in Southeast Asia, paid for by *ad hoc* groups of merchants, are brought into Cebu. The coastguard and customs officials are paid off with a 'gift' comprising part of the shipment which is later sold back to the consortium once they have received income from the sales of the goods shipped legally into Mindanao from Cebu and distributed through corporations such as Rose Motors, Gaisano, Altrac, Toyo and Yi Kian Gwan. Second-hand clothes (known as *oki-oki*) from other countries in Southeast Asia are brought into Mindanao through similar channels and then distributed legally

through retail outlets—both Filipino and Chinese. Another similar operation is Mindanao Textiles—a company with a substantial export quota
to the United States. The company, originally 'Filipino', was bought up
by a Hong Kong manufacturer who shipped garments made in Hong
Kong through the company to the United States. Violations of trade
marks are commonplace: the Gaisanos, for example, import jeans made
in Shanghai through their Hong Kong office, and then sew an interna-
tionally recognised label on to the garments which are then sold in their
stores in Davao for a substantial profit[4].

The main threat to trade, however, comes from the evolution of large
monopolies institutionalised and protected by government. Evidence is
hard to come by, although such arrangements do appear to be common
knowledge and would seem to be confirmed by recent government
reforms (see, for example, Tiglao, 1994b, c). Certainly, when government
does not question actions and arrangements which are clearly inimical
to competition and thus to trade, then allegations of collusion are bound
to surface. For example, the shipping routes between Davao, Cebu and
Manila are dominated by five companies—Sulpichio (with 37% of the
market share in the summer of 1994), Williams (17%), Lorenzo (17%) and
Gothong (9%) and Aboitiz (with 20% of the market share). Sulpichio,
Lorenzo and Gothong, which originate from one line, are owned by the
Go cousins; Aboitiz is owned by a Spanish–Filipino family mentioned
earlier in this chapter; and Williams is owned by Chiong Bian—a congressman. All companies are members of the Conference of Inter-island
Shipping (CIS) which is overseen by the Maritime Industrial
Administration—a government body. Although some progress has been
made recently to open up ports to other operators, CIS pegs and regu-
lates prices. Banana exports, too, are dominated by a few large merchant
families, including the Floirendo, Alcantara, Sarmiento, Ayala and
Elizalde who, it is alleged, were granted export rights as political favours
or as part of an attempt by Marcos (who also used Chinese merchant
families in an attempt to break the power of the Filipino merchant elites)
to divide the country up into development areas headed by merchants
with a strong incentive to oversee economic progress. These merchants
are also capable of indirectly affecting domestic prices by unloading
excess and rejected crops on to the local market. Similar allegations have
been made against grain merchants. The price of corn in Davao, it is
claimed, is heavily influenced by General Milling—a large corporation
owned by Oi Ting Sing; the price of other grains is said to be fixed by a
group of seven merchants in Manila headed by Antonio Roxas Chua.
Investigations have been conducted by government agencies which
have yielded no evidence.

It was noted at the beginning of this chapter that there is now an attempt to marry policy formulation with the needs of business. If it is to be successful, relations between merchants and politicians will have to be clearly focused upon trade: in other words, within the domestic economy those economic imperatives which facilitate the successful prosecution of trade and sustained material progress will have to take precedence over partisan mercantilism. Operations within the international economy, however, may find advantage in strong alliances between the merchants and politicians.

The associations

The terms 'Filipino' and 'Chinese' emphasise not inherent, distinct or deterministic cultural differences, but rather mixes of wealth values, forms of exchange and associated institutions and moral imperatives. 'Filipino' and 'Chinese' are not, and should not be taken as, analytical categories; instead, they should be viewed as concepts which constitute part of reciprocity's soft-framed institutions.

As already made clear in this chapter, there are extensive networks among Filipinos and Chinese, and there are striking similarities both in the way in which greater stability and predictability are brought to behaviour and relationships, and in the way in which institutions are used to facilitate, safeguard and legitimise trade. The redirection of institutions, exchange and values towards the prosecution of trade illustrates that apparent differences among groups and societies are not a consequence of unique and deterministic 'cultures', but of unidimensional perceptions and presentations of exchange, values and institutions.

These connections among merchants—Chinese and Filipino—find expression in informal meeting places such as Josephine's Cafe at the AV Hotel. Here, merchants and politicians—who, as members of the cosmopolitan elite, adopt a characteristic offhanded, almost lazy, manner, and wear casual, but usually expensive, branded clothes, watches, glasses, jewellery and shoes—meet to conduct transactions, to talk about possible deals, to exchange information, or simply to cement existing relationships. Another, though for some less respectable, nexus for their connections is the Casino.

More formal expressions of these nexus are the Davao City Chamber of Commerce and the Filipino–Chinese Chamber of Commerce. The Davao City Chamber of Commerce which was established in 1968 is staffed by half a dozen or so young, energetic Filipinas. The chamber's membership numbers around 300 merchants—Filipinos and Chinese, the prominent and the relatively unknown. The chamber is regarded as more thrusting, cosmopolitan, outspoken and effective than the Filipino–Chinese Chamber of Commerce and a more likely source of national and international contacts. Correspondingly, the Filipino–Chinese Chamber of Commerce—with a membership of 200 Chinese merchants—is regarded, even by many of its own members (a number of whom would meet some of Somers Heidhues's criteria of 'ethnic' Chinese), as parochial, secretive, introspective, narrow, limited and anachronistic.

Yet, whether members or not, merchants rely far more upon the establishment of their own networks for the prosecution of trade than upon any chamber. And while the Davao City Chamber of Commerce may seem more focused upon the establishment of national and international links, there are merchants who join for no other reason than to enjoy the social life it offers regardless of any incidental, though welcome, business contacts which may result. The Filipino–Chinese Chamber of Commerce is an even more ambiguous institution. In addition to administering the Filipino–Chinese Fire Brigade, to settling disputes between the members of the chamber, and to routeing information on social events, on laws passed by government, and on donations required for some cause or other in the Philippines or in China, the Filipino–Chinese Chamber of Commerce is also an important source of gossip about who is doing well in business and who is in difficulty. It is perhaps not surprising that the chamber is seen as introspective and parochial, for it may also serve as a useful symbol, presented even by many of Somers Heidhues's 'ethnic' Chinese as an antitype against which the 'cosmopolitan' may be created.

Membership of international clubs—such as Rotary, Lions', JC's, Kiwanis and Masons—is very mixed, and connections established within the Philippines or overseas are very extensive. The main focus of these organisations appears to be welfare. In the case of the Lions' Club, the income raised through members' contributions and social events—such as cultural shows or concerts—is spent on medical and welfare projects. However, members are not solely concerned with the administration of such activities. Incidental, but useful, contacts are sometimes made; and membership may provide a more effective arena for influencing the decisions of politicians and civil servants. A quiet word with a politician in a cosmopolitan organisation publicly devoted to the welfare of all, may prove to be a far more effective technique than aggressive lobbying by an organisation representing special interests. The Davao City Automobile Association, for instance, was dominated by Chinese merchants who used the organisation to lobby against, or water down, legislation: the association now rarely meets, for its members' interests are now pursued through a variety of other, more effective channels. Motivations for membership of the chambers and clubs, then, are extremely varied, and this is also true of those other institutions such as the non-governmental organisations (NGOs) and the universities in which merchants regularly participate. There may well be useful, practical advantages to participation: a member of the Gaisano family joined the board of PITO-P (Private Investments and Trade Opportunities)—an NGO which was attempting to attract investment in Davao, and to stimulate trade between Mindanao, Indonesia and Malaysia—and, in so doing, provided himself with a range of additional contacts overseas. But merchants may regard their own participation as no more than an interesting hobby. Participation may reflect genuine interest in the services which such institutions provide; it may be that a merchant who has decided to limit his own material ambitions, is now looking for a new stimulus; or it may be nothing more than a welcome escape from loneliness or from some unhappiness at home.

Another focus of reciprocity is a Taoist group based in Taiwan which has set up a branch of its organisation in the home of a Chinese merchant who was also the former president of the Ong family association in Davao. The Taoist group in Davao comprises an odd mixture of mysticism, palm-reading, fortune-telling, the supernatural, faith-healing and the propagation of vegetarianism. The Davao branch meets once a month on the third floor of the merchant's home (the ground floor is a loading bay, the first a warehouse, and the second his family's living quarters) in a room which originally held the ancestral shrine. The new altar—a collection of large, heavy and ornate rosewood furniture—was given to him by the sect's Taiwanese members. Business is an important aspect of the group's activities, and the holder of the shrine, the merchant, has been offered various deals with members in Taiwan who frequently visit Davao. But this is only one aspect of the group. The merchant has limited his ambitions for yet more money, and now he, like some of those who enjoy taking part in the activities of NGOs, the Church or the universities, desires a new stimulus or sense of worth. The Taoist association also gives him a good deal of psychological comfort, reinforced·by a reassuring sense of destiny. His selection as a member and as leader in Davao had come at a time when he and his wife and children were being moved out of a small house rented from the Villa Abrille. The merchant's father—the original founder of the Ong family association in Davao—had been a blood-brother of the head of the Villa Abrille family. It was understood that the merchant's father would be given the house and land, but when the head of the Villa Abrille family died, his son would only give them the option of buying a nearby lot. The purchase of this land and the construction of a house required a substantial loan and entailed considerable risk. The merchant was advised by the Taiwanese who recruited him into the Taoist group to take the risk—a course of action which eventually proved successful: the house, which also served as a loading bay and warehouse, substantially reduced the merchant's overheads during a period in which his particular line of business (cosmetics) boomed.

Given that a Chinese merchant may participate in the Davao City Chamber of Commerce, in the Filipino–Chinese Chamber of Commerce, in an international club and in a family association, it might be suggested that some merchants are operating in, or view themselves as operating in, two different worlds: one is the limited and unidimensional world of 'the Chinese', used to help in the construction of the other more active, outward-looking world of the cosmopolitan. The attitude of Chinese merchants who are not active in, or members of, any chamber, club or family association, may give some weight to this interpretation. While they may see the advantage of membership of one or other chamber of commerce or international club, they may have no invitation to join. In any case, it is the relationships they establish in the course of business which are vital to their companies. The prosecution of trade leaves little or no time for socialising. As for the family association, these are, as one Chinese merchant put it, 'only for relationships with society'. In contrast to those who feel pressure to contribute time and money to the family association as their success grows (especially since individuals within the association are likely to have a reasonable idea about the state of another's business), the 'unsociable' merchants either feel no pressure to legitimise their activities or to demonstrate altruism, or, if they do, they regard that pressure as no more than moralistic, judgemental, unjustified and institutionalised self-righteousness.

Here, then, is perhaps another aspect of the association. By establishing relationships around some common, if fabricated, kernel such as name,[5] or city, or language, and thus to admit, symbolically, that they are as one, it becomes easier to pressurise individuals into following certain lines of behaviour and to adopt certain values prescribed by a few activists who thereby wield and demonstrate their social and political strength. Within the Chinese association, then, it may be that reciprocity becomes an end in itself. In practice, very few members may be active. In the case of the Song Te association any individual with the correct surname may register: if they are poor they have no obligations; if they have money all that is required is the occasional donation. For the most part, it is only the officers who attend the meetings, and within this and other associations it is they who tend to confine their relationships to a small group—largely other Chinese. In the words of one Chinese merchant: 'if they don't owe you anything, they won't have anything to do with you'. Here, then, the notion of 'the ethnic' Chinese is reinforced. Moreover, the family associations—which are usually housed in large, two-storeyed or three-storeyed buildings with their ground floor rented out as a shop or warehouse—often stand separate from, and rise above, the surrounding buildings. These exhausted shells, daubed with Chinese characters, occasionally guarded by high walls, and always with a locked metal grill for their front door, do indeed look mysterious and imposing: an obvious subject for speculation and rumour. Yet in many instances little goes on within these buildings: if there is secrecy it is because of embarrassment with their own stagnation and lethargy. For the most part, activists comprise a few old men (loyally shadowed by their tolerant sons) gossiping about each other. The minor issues, which spawn the intrigue and the slander, the criticism and the bitterness, and the envy and the dogma, and which now consume their thoughts and their energies, are retained only as convenient nuclei around which emotions collect, magnifying these old men's petty concerns and pedantry into matters of great import. But these intensely parochial and cloistered institutions are not peculiarly 'Chinese'. 'Ethnicity' and 'Chineseness' are convenient fabrications—explanations of something which is perceived to be anachronistic and restrictive, and which therefore serves as a useful opposite against which the 'cosmopolitan' may be created.

These are the three aspects of Davao's Chinese family associations: a nexus of reciprocal networks established for their own sake; a means of legitimising material success by focusing and publicising contributions of time and money for welfare activities (such as providing for the poor, for funerals, for education and for the unemployed); and a unidimensional presentation which helps to define the 'cosmopolitan'. The Chinese family association may also serve as a nominal focus for individuals who feel they ought to be doing something to help others in less fortunate circumstances. The Ong family association, which has around 100 members, is small but its members are almost all successful merchants. They provide a few scholarships, but the association has little work to do since the building pays for itself (the ground floor is rented to a Chinese merchant who sells crafted, nara-wood furniture to the few who can afford it in Manila, India and the United States), and most of the association's members are financially secure. The association is supposed to meet once a month, but in practice its officers rarely gather. Meetings with other branches of the association in Manila, Cebu, Baculod and

Iloilo used to be held twice a year (once in Manila and once in each of the provinces by rotation so that each province and city, including Davao, would host the meeting once every four years), but are now held only every other year in Manila and once every eight years in each provincial branch, partly in order to reduce costs, but mainly to reduce the effort required to organise such gatherings which many officers find extremely tiresome affairs. When the Davao branch does meet, its officers must cajole as many of the association's members as possible to attend the meeting or else the turn-out is certain to be rather thin. Appointments for posts in the association are decided upon with a good deal of humour: no one likes taking on responsibilities. When the time comes for elections to be held, telephones are left off the hook, and merchants are difficult to find. The solution has been to rotate the office of the president among the oldest members (who then discharge their duties in a resigned but generally good-natured manner), and to promote officers automatically through the hierarchy. There is supposed to be a secretary manning the association's office, but he is now too old and stays at home: and since everyone else is too busy, the building is usually left empty. If members need to contact each other, they will visit or telephone each other at work or at home.

The business activities of the Chinese family associations in Davao would appear to be limited to arbitration between merchants who are members. Nevertheless, disputes and rivalries among Chinese may degenerate into kidnappings, violence and murder—crimes into which Communist and Muslim guerrillas and hired thugs are sometimes drawn as proxies.[6] Additional business contacts may be established through the association, and such contacts are welcome; the association may also prove useful to some merchants who need credit references; and the association will distribute welfare in one form or other to support those individuals and their families who have fallen on hard times. However, there is little direct involvement in a member's business. Occasionally it may be that some members in the same family association will help keep another individual's business afloat, or will provide enough capital to enable him to start afresh, but all this is kept quite separate from the affairs of the association. The association's distance on such matters is based on pragmatic considerations: aid extended to help the business interests of one member would create jealousy among other members who would probably then demand assistance for their businesses; there would be no means of coping with the financial liabilities incurred by the association; and few would be prepared to pay for funding the mistakes, poor judgement or ill-luck of another. The Ong family association is, in theory, able to help a merchant. But first it must be decided whether bankruptcy was a result of incompetence, of poor judgement or of circumstances beyond the merchant's control. The distinction between incompetence, poor judgement and bad luck, however, depends largely upon perspective. In practice individuals very rarely apply for help, and, indeed, no member has ever been given assistance.

Philippine town in Chinatown

In addition to Chinese associations, there are also a variety of Filipino associations in Davao based upon religious affiliation, occupation (such as the vendors' associations of the markets), and local community. Indeed, there is evidence that more than 80% of the urban poor belong to some kind of association (Amoyen, *et al.*, 1992). There are also Filipino associations, such as Boholanos, Nueva Ecijan, San Pablosanos, Pangasinan Ilocanos, Tagalogs and Pampangueños, formed around language and place of origin (province or city). Organisations such as these seem to have existed at least since the first quarter of the twentieth century when there was little mixing between the different groups of immigrants from Cebu, Bohol, Capiz, Antique, Ilocos and Bicol (among other places) which comprised Davao's small population. Moreover, it was only logical that the Filipino associations would reflect the interests of their founders—a particularly dominant or high-profile family or group of families. One example is Club Dabawenyos (formerly Hijos de Dabaw, and then Hijos de Mindanao) which was originally founded by immigrants who had arrived in Davao under the Spanish during the late nineteenth century. The association was formed around families that had been resident in Davao for more than 10 years, and its members represented some of Davao's prime-movers: at first the Suazo, Bastida, Pelayo and Lizada; later, the Cervantes, Cabaguio, Santos and Gempesaw; and then, after the Second World War, the Palma Gil, Castillo, Rodriguez, Cabreros and Garcia. All these families were, or still are, notable landowners, merchants or political dynasties, and are, in many instances, also related to each other by blood or by marriage.

The associations formed around language and place of origin were still active up to the declaration of martial law. In common with the Chinese associations, the Filipino associations had connections with other parts of the Philippines; they were partly an expression of defensiveness and partly a focus for elites; and through their welfare activities and their members' chameleonic networks, they provided an opportunity to display, and, at the same time, to legitimise their material progress and status. The former president of the Neuva Ecijan (an association which had had a membership of around 7000 strong) shamelessly declared that her motivation for running the association was her enjoyment of dressing-up in expensive clothes and jewellery: that it could also open up new opportunities for additional deals was welcomed, but it was the social life and the pomp which were, for her at least, the main attraction.

The imposition of martial law, the operations of the Communist and Muslim guerrillas and the kidnappings and murders of merchants and their families made the display of material success and status a luxury that was dangerous—politically as well as physically. In any case, with the rise of the international clubs during the 1960s and 1970s, the Filipino associations came to be seen as more and more parochial and anachronistic. A few of these associations still exist as chains of individuals occasionally meeting at each others' homes, perhaps to discuss who should be supported in an approaching local government election.

The Filipino associations overseas, however, are stronger and more active institutions. In Hong Kong, for instance, there are well over 100 such associations (though by no means all are registered with the Philippine Consulate) which are formed around dialect, place of origin, religious affiliation and occupation. They include, among many others, the Iloilo Association, the Bantanga Association, Magafel (Mindanao and Muslim), Ayacib (Cebuano), Bicol Sanao (Bicol), Hinunangan Leyte Hong Kong Overseas Association, Hong Kong Musician's Union and the Filipino Catholic Association.

The similarities between the Filipino associations and those of the Overseas Chinese extend beyond the use of kernels of dialect, place of origin and profession. One example is Associacion Negrense which is organised according to a written constitution and series of by-laws. In theory at least, these procedures and rules govern the actions of the association's officers and members. In addition to the chairman (political correctness seems to have little place in these institutions run entirely by women), there is a secretary, an assistant secretary, a treasurer, an assistant treasurer, an auditor, an assistant auditor, a press-relations officer, a sergeant-at-arms, and a business manager. Under the association there are four sister organisations—Himwa, Bamwah, Silaynon and Pontenedra. Each of these associations is centred around members who originate from a particular city or town in Negros. All four are, at the time of writing, independent from Associacion Negrense, for unlike some of the other associations in Hong Kong (which are based on smaller geographical kernels from within the same province or same language group), they do not pay dues to their parent organisation. Nevertheless, relationships between the five associations are close, particularly between Associacion Negrense and Bamwa: the chairman of Bamwah is the sister of the advisor (an experienced grandmother now in her late sixties) of the chairman of Associacion Negrense.

As in the Philippines, these associations—as all the others in Hong Kong—exist largely as chains of individuals who meet for a few hours each week (usually on Sunday) in a park, on the steps of City Hall, or in shopping centres where strings of stalls and shops (some owned and run by Filipinos) catering to Filipino tastes and budgets have appeared in Central and Hung Hom. Prompted by complaints over the large numbers of Filipinos milling about Central and Tsim Sha Tsui, the Hong Kong government has recently allocated seven schools to be opened at weekends to domestic helpers of any nationality, much to the dismay of local residents. These centres are operated by the Bayanihan Trust—an association of Filipino merchants—which has raised HK$2 million. A further HK$1 million was raised by the Royal Hong Kong Jockey Club, and another HK$10 million is to be raised from corporate sponsors.

The chairman and officers of the Associacion Negrense are elected every two years. Once elected a chairman may, in consultation with the other officers, replace an officer if she should for any reason leave her post, but the Philippine Consulate must be notified of any changes. Any member of the association can stand for any post, although if it is thought by the incumbent chairman that an individual who wishes to present herself as a candidate is unfit for the post, the chairman may attempt to garner enough support within the association to prevent that individual from ever standing for election. Individuals who wish to stand for a post, then, must drum up support for themselves among the membership (which numbers around 100 in Associacion Negrense). The result is the creation of small cliques, each pushing its own preferred candidate. A chairman requires not only commitment, but also the maturity to ignore intrigue, rumour, gossip and allegations of misconduct, and the ability to bring these groups together. Another crucial attribute which a chairman and her officers must possess is the ability to build up contacts and support outside the association, not only with other associations, but also with merchants and with the consulate. The dues paid by each member is no more than 10 dollars (Hong Kong) each month for an association with a turnover of between HK$40 000 and HK$50 000 or more each year. Donations from business and the attendance of local dignitaries (Hong Kong Chinese, Filipino, British and American) at fund-raising and other social events provide an important boost to confidence and to the members' self-respect, and they also help secure the position of the chairman and her officers. One recent event organised by Associacion Negrense was the 'Mother of the Year Award'. The event was held in the hall of a church school which the association hired for five hours at a cost of HK$3550—a sum equal to the monthly wage of a domestic helper. Candidates for the award included members from Silaynon, Himwa, Pontenedra, Bandiangnan, Iloilo, Tigbauan and Janiuaynons associations, and the event was attended by the Vice-Consul and the Labour Attaché of the Philippine Consulate, and by members of the Philippines Sunday Tonight Show. Sponsors for the event included the Philippine National Bank, United Bank of the Philippines, Metrobank, Indio (a Filipino insurance firm), *Bayang Himig* (a Filipino magazine published in Hong Kong), the Cabaya Computer Centre (which donated three scholarships for computer courses), and *Diwaliwan* (another Filipino magazine) which printed the programme for the event free of charge. For each advertisement which appeared in the programme the association received HK$1000. The merchants' participation in the event is motivated partly by philanthropic sentiments, but participation also makes good commercial sense, for it provides the opportunity to demonstrate interest and concern about the problems and difficulties faced by their potential customers.

Part of the money raised at such events is used to extend help to members whose contracts are broken or terminated, who become ill, who must return home following a death in the family, who are either charged with or are a victim of crime and abuse; and if a member dies in Hong Kong, the association will help pay for and organise the return of the body to the Philippines. Money is also remitted to the members' home towns following a calamity or if there is a request for help with a worthy project. For instance, after Typhoon Bissing struck the Visayas in 1994, the Associacion Negrense raised P10 000 for those families whose homes had been destroyed in Valladolid (Negros Occidental), the home town of one of the association's officers; another P20 000 was donated to help with the construction of a church in the chairman's home town. To ensure that money is received by and used as intended, contributions are usually made to, and through, individuals known personally to members; and to avoid any suspicion of fraud, letters of thanks duly signed by local government officials (such as the City Mayor or Provincial Governor) are circulated among members for inspection.

Another important aspect of the association is as a salve for loneliness. Many of the women are mothers and, in some cases, grandmothers, or the eldest child of a family, who have by choice or necessity worked, or intend to work, in Hong Kong for many years. Knowledge that the money they remit home will enable their family to enjoy opportunities, and a better standard of living, which they could not otherwise have afforded, provides the immigrants with some comfort, as does the knowledge that their experience overseas may enhance their status when they eventually return home.[7] But the work is hard, their status and pay in Hong Kong could hardly be lower, and holding their families together is not easy. It is true that many Filipinas do not join associations, at least partly because they are a forum for gossip, rumour, criticism and intrigue, especially if they know members of each other's families and wider social networks in the Philippines.

Nevertheless, communion with others sharing the same hardships, who speak the same language, who enjoy the same foods, who share the same sense of humour, who may know the same people at home, is essential, both emotionally and in a practical sense, for communion is also a focus for constructive energies which help keep at bay the all-consuming moroseness that nostalgia and sentimentality breed. The associations, then, are often merely institutionalised expressions of reciprocal groupings which already exist.

The association is also a public symbol of self-respect. 'Being' Negrenese, or Ilocano, or a native of Iloilo or Baculod, not only provides a kernel around which relationships can be formed: it is also an expression of pride. Even if an individual is not actively involved in running the association and organising events, the fact that these organisations exist, that events are held, and that help is given to others in Hong Kong and in the Philippines, is a source of pride: the Ilocano or Negrenese and by extension 'the Filipino' are not dependent servants without ability, but rather individuals with initiative and drive who are prepared to stand up for themselves.

Conclusions

The arguments developed in this book, and most especially in this present chapter, are likely to attract criticisms not only from cultural determinists, but also from those who feel themselves to be 'Chinese', 'Filipino' or 'African'. To many Chinese, Filipinos and Africans (and, for that matter, British), the idea that fundamentally they are little different from each other would be difficult to accept and would seem to run counter to their own experiences and emotions. Indeed, when all is said and done, it is, perhaps, the perceptions which individuals have of themselves, rather than the interpretation which another gives to their lives, that really matters in both a practical as well as in an intellectual and emotional sense. Yet it is also true that when it suits their purpose individuals will adopt a unidimensional perspective of themselves and other people. It is in this regard that those who play to a deterministic 'culture' feed the darker side of human nature; and it is in opposition to such ideas that trade, particularly when released from the bonds of mercantilism, is of such vital importance.

Notes

1. My thanks to Ernesto Corciño, Consultant (Historical, Cultural and Protocol Affairs), Office of the City Mayor, and to officials of Agdao market, for this information.

2. Rankings are based on gross sales and profits estimated from tax paid. Quite aside from the difficulty with identifying who is, or is not, Chinese (see text), there are a number of other problems confronting an attempt to estimate the strength of Chinese companies in Davao. For example, government and businesses are often extremely reluctant to release information on sales; and overseas investors (including Chinese) may use Filipino 'dummies' to circumvent legislation restricting their ownership of a company to around 40%. It is a little easier to obtain information on taxes paid. From this information it is possible to work backwards to provide a rough-and-ready indication of the relative strength of enterprises. Again, however, there are a number of problems with such information which should be mentioned.

First, there is the complexity of the tax system. There are two sets of taxes—a national tax (administered by the Bureau of Internal Revenue (BIR)) and a local tax. Both rely on the declaration of income and expenditure. It is not clear how the BIR is able to cross-check names of individuals and companies against records of additional ownership, employment and share dividends. The ownership of corporations and partnerships may be registered with at least three different agencies—the Department of Trade and Industry (DTI), the Securities and Exchange Commission (SEC) and the Business Bureau (an organ of local government). All such enterprises are supposed to register with the SEC (a national-line agency) in Davao. The SEC, however, has no computer system, and records are split among Davao and Manila and the other provinces where an enterprise may be registered. Some companies may also choose to register with the DTI (a national-line agency), but are only required to register their initial capitalisation. The BIR in Davao relies on computer print-outs which give the name of companies, the individuals or the company responsible for paying the tax, and the tax paid. Presumably, the BIR in Manila is able to cross-check declarations from records sent in by different agencies in Davao and in other parts of the Philippines. The uncertain logistics of collecting accurate records is compounded by a fairly complex series of taxes. For the national tax, a tax-payer registered as 'non-VAT' pays 3% on sales of less than P200 000; 'VAT' taxpayers must pay 10% unless they are producing for export in which case they are exempt. In addition, partnerships and corporations pay up to 35% on their profits. Income tax for individuals is divided into a compensation (or salary) tax—which has a maximum rate of 35%—and a tax on single proprietorships which has a maximum rate of 30%.

The local tax is divided into a number of different categories based on type of business. Some of the most important categories include: manu-facturers, producers and processors; wholesalers, distributors, dealers and retailers; contractors; hotels and motels; restaurants, soda fountains and bars; *carenderias* and other food dealers; and cafés, cafeterias, ice-cream parlours and other refreshment parlours. In each category a flat rate tax is applied to gross sales over a fixed amount (which varies from between 2 million and 6.5 million pesos depending on the category). Under these fixed amounts there is a sliding scale which specifies the amount of tax to be paid if gross sales fall within a certain bracket. For the first two categories listed above (production and distribution) the sliding scale is divided into two parts: for those goods deemed non-essential the amount of tax paid is about twice that paid on sales of goods classified as essential.

Secondly, owners will use a variety of measures to evade or reduce their tax burden. For example, corporations may be registered as single pro-prietorships. The major stockholder (or stockholders) will thus pay no more than 30% on their share of the profits, and no more than 35% of any salary which they receive from the corporation. Alternatively, a cor-poration or partnership or single proprietorship may be registered in the name of another individual who is paid a retaining salary by the owners of the company which pays no more than 30% of its profits to the BIR. Merchants may also doctor their books or they may be too incompetent or too lazy to keep proper records. Different sets of books are kept by the organised swindler, but there are many honest merchants who do not keep a proper account of their transactions. Some businessmen will sim-ply open two bank accounts: one for costs; and the other for income realised above costs. Provided the latter account grows, and the former remains relatively stable, the merchant is happy. Other businessmen will wait until the end of their own peculiar financial year at which time they will make a very rough estimate of their profits: for the rest of the year they live by applying Mr Micawber's recipe for happiness to each trans-action. According to serving and retired BIR officials these rather infor-mal and chaotic, and yet reasonably effective, financial methods are practised even by some of the larger firms, both Filipino and Chinese.

(See also Limlingan's (1986) description of similar financial techniques practised by Chinese entrepreneurs in Southeast Asia).

Thirdly, there is a significant level of corruption within the BIR at all levels. More is said about the nature of this corruption later in the text of the present chapter.

For all these reasons, then, it is not possible to be certain either that records of the business interests of merchants in Davao, or the estimates of the relative strength of their businesses, are complete or accurate. Notwithstanding these problems, however, it is possible to gain a rough indication of the strength of individual enterprises by examining tax and ownership records held by the BIR, the Business Bureau, the DTI and SEC.

3. These echelons are based upon records of local tax paid by individual enterprises. Once the local tax paid reaches 30 000 pesos, a flat rate tax takes over from the progressive rate. Any company paying between 30 000 and 99 000 pesos must be realising gross sales of at least 4 million pesos; those paying between 100 000 and 499 000 pesos must be realising at least 20 million pesos; those paying between 500 000 and 999 000 pesos must be realising at least 100 million pesos; and those companies paying over 1 million pesos must be realising gross sales of at least 200 million pesos. Although it is not possible (in theory) for an enterprise which is paying less tax than the figure used to help mark the lower end of each echelon to be included in a higher band for gross sales, it is possible (because of variations in the flat rate tax among sectors) for an enterprise with higher gross sales than another enterprise in a different sector to be included in a lower tax band than that other enterprise. In other words, the bands for gross sales overlap. However, in combination with the fact that these estimates begin for companies with higher gross sales, this downward shift means that each echelon will include some particularly strong companies, at least as measured by gross sales.

The collection and interpretation of these records would not have been possible without the help of several individuals connected with the BIR, SEC and the Business Bureau. Some of these individuals are named in the preface of this book. However, the author remains solely responsible for the gathering and interpretation of this material.

4. My thanks to Vice-Governor Turan of Cotabato who gave me permission to name the companies involved and to give his name as source.

5. It was not unusual for Chinese immigrants to take the name of other immigrants who had died on their way to the Philippines, or to take the name of children who had died in the Philippines whose deaths were

not registered, or to claim kinship with individuals in the Philippines in order to circumvent restrictions on immigration. When an association is formed around five or six names and around a common ancestry which is traced back more than 3000 years to northern China (as in the case of the Uytong Family Association), the kernel for the association can only be regarded as expediency or extraordinary naïvety.

6. My thanks to a serving officer of the 5th Infantry Battalion, 1st Infantry Division, Lanao del Norte and Lanao del Sur.

7. Some returning workers, however, find others, including family members, extremely jealous of their time overseas and likely to claim as much of their savings as they can. These experiences may constitute an important catalyst for changes in values since demands on hard-earned money are often strongly resisted and resented.

7

Conclusions: Greater China and Southeast Asia's Chinese economy

Introduction

Considerable interest in the notion of a 'Chinese-based' economy focused on a 'Greater China' has been stimulated by the huge investments which Overseas Chinese have made in China, and by the marked improvements in China's economy which have occurred during the 16 years following the implementation of the reforms in 1978. Clearly, the notion of a strong and specifically 'Chinese' economic network reaching across the Far East is likely to find ready support from cultural explanations of the relative economic success of Overseas Chinese in the region. In view of the obvious implications such ideas have for the Far East as a whole, and given the arguments which have been ranged against cultural explanations of economic success in this book, this final chapter considers the significance and legitimacy of the arguments and beliefs underlying the idea of a Southeast Asian Chinese economy centred on a 'Greater China'.

The chapter begins with an analysis of China's political economy. It is suggested that China is now faced with a three-way tension between trade, reciprocity, and state procurement and distribution. In this sense, the economy has become trimorphic in nature. The reforms and the initial conflict between trade, and state procurement and distribution, created opportunities and stimuli for the establishment and strengthening of reciprocal networks. Bureaucrats, their subordinate economic units and private traders, found themselves operating within increasingly vague and fluid circumstances. However, the promotion and protection of their own particular interests by the manipulation of reciprocity and its soft-framed institutions have added to the instability and uncertainty of China's economy and polity. If China is to realise sustained and rapid material progress, then the authorities must privatise domestic enterprises, both state and collective. This will atomise the economy, divest provincial and sub-provincial authorities as well as central government and the military of their direct economic concerns, and enable the central authorities to institutionalise trade in a manner which focuses loyalties upon Beijing, distance reciprocity from the conduct of business, and thereby enable merchants, whatever their nationality, to operate more freely and effectively. It is possible that the atomisation of the economy and its domination by trade could take a generation or even longer to accomplish, but China's leaders will have to act far more quickly than that. The alternatives will be a return to centralised planning or, more likely, either the break-up of China (an event which would be facilitated by trade with foreign merchants), or the appearance of a strongly nationalistic military government intent on creating and maintaining a Greater China.

Reform: the evolution of China's trimorphic economy

Although a degree of material progress is consistent with any form of exchange, state procurement and distribution not only hinder sustained and rapid material progress: they may also cause a general deterioration in the economy. In an economy dominated by state procurement and distribution, it is the state that buys, sells and supplies the goods which the state itself produces. Under this arrangement all economic activities are controlled and coordinated by the state; economic units are mere appendages of state administrative organs. It is a cloistered system in which there can be no exchange for profit among independent economic units and individuals. Terms such as profit, wages, tax, subsidies and price merely describe the different means (or economic levers) used to help effect the transfer of goods. The nature and use of such levers are determined by administrative plans and commands.

Clearly the fundamental principle of a cloistered arrangement such as this is to balance production with the use and deterioration of goods. However, as a number of writers observe, balance is nothing more than a theoretical expedient. More importantly, any drive for material progress requires imbalance as a prerequisite (contrary to the logic at the very heart of this unitary economy) and creates two striking phenomena which militate against sustained and rapid material progress—these are investment swings and an obsession with constant increases in production (R.N.W. Hodder, 1993b).

The two key components of the reforms implemented in China have been to draw a distinction between state administrative organs and economic units; and to replace the administrative control and organisation of economic activities with a less rigidly organised system for procurement and distribution (see, for example, Yang Chengxun, 1984; Chang Xiuze, 1990; Zhou Diankun, 1986; Ye Zhenpeng, 1991; Gao Shangquan, 1984; An Zhiwen, 1985). The reforms, then, represent an attempt to imitate certain characteristics of an economy dominated by trade. These imitation characteristics, it was hoped, would act as flexible and responsive mechanisms to handle more effectively the central problem of the economy—that of maintaining balances and of controlling imbalances deliberately brought about in order to enable designated enterprises and sectors to achieve a degree of material progress. However, the reforms and the unintended effects they produced, revealed to the authorities a tension or conflict between different forms of exchange and their associated institutional patterns, moral imperatives and wealth values. If sustained material progress was to be achieved, clear choices would have to be made.

Problems of reform: imaginings incarnate

A peculiar twist of Marxist and Keynesian thought is that the interpretation of an atomistic, 'external' economy as a unitary economy—and, therefore, the belief that the notion of balance is relevant to an economy dominated by trade—lead to measures designed in part to restore balance and equality. This inevitably requires the centralisation of economic power, domination by procurement and distribution, the formation of aggregate bodies and the creation of a unitary economy. This is to make intellectual imaginings incarnate. The criticisms levelled at profit and other 'market mechanisms'—criticisms quite misplaced in an economy dominated by trade—become precisely applicable in an economy dominated by procurement and distribution. In China, this misinterpretation of the fundamental nature of an atomistic, 'external' economy caused the Chinese authorities to misjudge the effect which their attempts to imitate such an economy would have. To reform, they believed, was to introduce 'market mechanisms'; to reform, they reasoned, was therefore to enable a more flexible, responsive and targeted flow of goods; thus, to reform was, by definition, to provide a better way to maintain balances and to control imbalances. In practice, however, imitation created severe imbalances and near chaos (see Wei Jie, 1991; Chang Xiuze, 1990; Ye Zhenpeng, 1991; Wang Zhihua, 1990).

Huge investment swings turned in favour of manufacturing (light industry), the processing sectors of heavy industry, and construction, particularly in eastern China. Investment in agriculture, heavy industry (including energy and the extraction and supply of raw and semi-processed materials) and transport suffered an absolute or relative decline. The movement of rural labour (both surplus and productive) to non-agricultural activities in urban and rural areas exacerbated a number of problems. Industrial enterprises, many of them already grossly overstaffed, came under pressure to take on more workers or at least to keep their existing workforce; public utilities were unable to cope with the flood of migrants; land was taken out of cultivation; and water conservancy, irrigation and drainage facilities deteriorated.

The weakening of vertically structured administrative command routes used to control economic activities and the movement of funds, and the greater rein given to indirect policy levers, also led to an increase of the collectives' share of economic activities and industrial output, and to a rise in extra-budgetary revenue and expenditure without a corresponding fall in state expenditure. While the state and collective systems of ownership remained intact, the devolution of power to make decisions, the growth of extra-budgetary funds, and the emphasis on horizontal linkages required different sets of economic units and their administrative overseers to compete and at the same time to cooperate with each other. The existing framework of self-sufficiency and self-contained industrial systems, although heavily criticised as a fundamental obstacle to modernisation, was nevertheless adapted to meet the new pressures on economic units and their administrative overseers. Provincial and sub-provincial authorities and those administrative departments and ministries responsible for particular economic sectors now acted to promote and protect their own economic units. The duplication of economic activities; areal blockades to the movement of goods, materials, technology and enterprises; and the establishment of 'open', special economic zones by individual provinces, municipalities and even counties: all served to intensify competition for goods, materials, energy, technology and funds, thereby creating unnecessary waste, increasing pressure on transport, sending confused and confusing signals to both domestic and foreign enterprises, and setting up still more complex imbalances within the economy (Hong Changshi, 1984; Zhang Fubao, 1984; Hu Bing, 1987; *SHJJXX*, 23 May 1987 and 6 September 1986; *JJRB*, 18 October 1993, p. 2; *ZHGSSB*, 18 April 1990, p. 2; Zhao Qingyi, 1994).

State enterprises—still obsessed with constant increases in production, still carrying enormous welfare responsibilities, and yet unable to rely on the much-weakened bureaucracy which had previously supplied their needs and procured their output—were left floundering by the edge of swirling streams of goods, materials and funds set in motion by the reforms. Managers of these enterprises attempted to claw back materials, goods and funds through political manœuvring. State and commercial banks (for neither were independent entities) continued to grant loans; and enterprises borrowed money from each other. In 1994 these triangular debts amounted to a sum equivalent to one-third of the state-owned industrial enterprises' output value. A proportion of loans and subsidies flowed into collective and private enterprises which supplied, or acted as outworkers for, state enterprises, or which borrowed from state enterprises directly at high rates of interest.

The rise in subsidies and loans represented a rise in the quantity of goods rerouted with the intention of redressing imbalances. The effect, however, was indirect and untargeted expenditure which acted to shelter enterprises from the consequences of their own actions and from those of the state, enabling those enterprises to fritter away funds and goods on non-productive activities, to increase wages and bonuses, or to invest wildly and excessively in manufacturing and processing activities. Moreover, this expenditure was financed partly by running up debts while structured and properly directed investment fell. Consequently, output in some sectors declined, stagnated or failed to keep pace with that achieved in other sectors. It was clear that the level of expenditure could not be sustained. The burden of subsidies combined with other demands on state budgetary expenditure made it more and more difficult to maintain the material fabric of China's economy. If allowed to go unchecked, it was argued, the rise in subsidies could lead to the collapse of the fiscal system (*ZHGSSB*, 3 March 1990, p. 2; *JJRB*, April 1990, p. 2; Wang Zhihua, 1990).

These fiscal problems had already had worrisome repercussions for those state organs specifically concerned with the procurement and distribution of China's material produce, and which had already been weakened by the removal of many items from their exclusive preserve and by the rent or sale of their lower echelons to collectives and private entrepreneurs. In order to shore up inadequate or deteriorating facilities, or in order to diversify into more profitable activities, funds for the procurement of materials were diverted to pay for capital construction projects and for other purposes at a time when funds for the procurement of certain goods—especially non-essential agricultural and sideline goods—were already being restricted by the banks. Whereas previously it had been possible to obtain further subventions from the state, by the late 1980s the demands on the state had reached a point at which extra funds were no longer forthcoming (Yahuda, 1990). The result was that state organs of procurement and distribution were frequently compelled to issue procurement bonds to be redeemed by the vendors for cash at a later date, or simply to reduce altogether the amount procured (*JJRB*, 24 March 1989, p. 3; *ZHGSSB*, 7 February 1990, p. 7).

It is true that attempts have been made periodically to control spending. At the end of 1994 all new investment in fixed assets classified as 'medium' or 'large' were banned, and the discount rate was raised; a proportion of state enterprises were already by this time lying idle, their workers at home on minimum pay; some use had been made of China's bankruptcy laws; and there are plans to improve unemployment insurance for workers (although this still leaves the state with bills to pay unless some form of private insurance is introduced). However, it would seem that the authorities are not yet prepared to end their profligacy by closing down the weaker state industrial enterprises and by privatising the remainder. Nor do they yet seem willing to dispense with the state organs of procurement and distribution, for although many key goods and materials leak constantly into the collective and private sectors of the economy, the state continues to dominate the procurement and distribution of these items. Indeed, more often than not, the state will print money rather than refuse to buy goods and materials from production units and farmers.

No doubt the authorities fear that throwing millions of workers (110 million workers are employed by state industrial enterprises alone) into the ranks of a vast surplus labour force, which already amounts to something between 50 million and 100 million, will only lead to social unrest. It may also be that although inefficient, state industries still produce more than 40% of China's industrial output value, and, together with the state organs of procurement and distribution, they constitute some sense of a cohesive economic system without which China might shatter into a kaleidoscope of merchant empires. Collective-owned industries, which now produce around half of China's industrial output value and, perhaps more importantly, soak up much of China's huge labour force (*ZHGSSB*, 5 October 1994, p. 5), effectively lie outside the central authorities' control.

By the mid 1990s the number of state industrial enterprises had risen from around 90 000 in 1985 to more than 100 000, and yet their share of the nation's total industrial output value has fallen from 65% to under 50% over the same period. They absorb a substantial proportion (close to a third) of the nation's fiscal expenditure in the form of subsidies, and about two-thirds of the nation's investments in fixed assets; and yet, in 1993, state industrial enterprises were responsible for 80% of all losses incurred by all industrial enterprises above the level of the *xiang* (that is both state and most collective industrial enterprises). In 1994, at least half of all state industrial enterprises were still losing money. This deterioration in their performance to some extent reflects changes in the way statistics are gathered: many formerly 'hidden' costs—such as rates of depreciation, technical developments, and bonuses—have now been 'revealed' (*ZHGSSB*, 31 October 1994, p. 2). Clearly, however, to argue that there has been no deterioration, only a failure to improve, appears to be little more than semantics: private enterprises, domestic and foreign, are responsible for an increasing proportion of industrial output value and exports because those enterprises are growing so much faster; increases in output by state-owned and collectively owned industrial enterprises remain investment-led; and inflation was left running at more than 20% by the end of 1994.

The central problem with state-owned enterprises (and, for that matter, collective-owned enterprises) is not simply that they are inefficient. Nor is it that they ignore profit and prices and are geared to bureaucratic orders. 'Soft-budgets', subsidies, 'resource constraints', a lack of incentives—and all the other phenomena described by Kornai with such terms—derive from the *fact* of a unitary economy and from the *fact* that enterprises are publicly owned. Inefficiency, bureaucracy and inattention to profit and prices derive from the practice of a unitary economy—an economy which requires either stagnation, or vast increases in output in one unit or sector at the expense of others, or widespread deterioration. The imbalances and near chaos which occurred after 1978 were a consequence of the absence of 'true market mechanisms' (Wei Jie, 1991). In particular, enterprises were not yet truly independent economic units. The economy was such that the question of balance and imbalance dogged all strategies for material progress. Gain in one sector was a consequence of loss in another. This phenomenon, Hu Chuangnuan (1981) asserts, was rarely acknowledged in the late 1970s. Many had mistakenly believed that balance was relevant to an economy dominated by trade, and that to imitate such an economy was sufficient to bring about sustained material progress. The reforms, however, initiated severe imbalances and set in motion a train of mutually reinforcing problems. By their own misinterpretation and by their actions the authorities had unwittingly acknowledged and demonstrated the inability of the prevailing wealth values and form of exchange to bring about sustained material progress; they had revealed a fundamental tension between different forms of exchange and their associated wealth values, institutional patterns and moral imperatives; and they had begun to lay bare another logic for sustained material progress. Private trade organisations and trade networks developed rapidly, challenging and weakening the key institutions and values of state control: state and collective ownership; the vertically structured administrative system; the state system of procurement and distribution; self-sufficiency; the emphasis on constant increases in production; self-sacrifice; and the pursuit of equality (R.N.W. Hodder, 1993b).

Understandably, the attitude of the central authorities towards private trade has been ambivalent. Private trade was a useful supplement to the state and collective sectors of the economy; yet its dangers were all too apparent. However, even after the Tiananmen Square Massacre in 1989, there was no clamp-down, and, despite obligatory loyalty to the notion that state ownership would continue to form the 'mainstay' of the economy, it would seem that by the mid 1990s the vital importance of private trade has been recognised, and its eventual domination of the economy has been accepted.

The rise of private traders[1]

During the period 1978–94, the growth in the number of private trade organisations (a term which is used here to exclude private enterprises) in retail sales and in industrial output value has been dramatic.[2] The number of private trade organisations rose from about 100 000 in 1978 to 14 million by the late 1980s. By the middle of 1993 there were some 15.5 million private trade organisations which together employed 25 million workers and possessed licensed funds amounting to 70 billion yuan (ZHGSSB, 29 September 1993, p. 6). A year later, there were nearly 19 million private trade organisations registered in China: their combined workforce numbered 32 million, and their licensed funds amounted to 103.2 billion yuan (ZHGSSB, 8 September 1994, p. 1). The growth in the number of private enterprises has also been impressive. Huang Dejun (1989) gives a minimum figure of 115 000 for the end of 1987, to which could be added another 50 000 private enterprises registered as collectives, and a further 60 000 registered as cooperatives, making a grand total of 225 000. Other sources, however, put the number of private enterprises at around only 40 000 for the end of 1988 (JJRB, 21 October 1989, p. 2); 66 500 for the end of June 1989 (JJRB, 18 October 1989, p. 1); and 90 790 for the end of 1989—a fall of 14% over the end of 1988 (ZHGSSB, 21 March 1990, p. 1). By 1991 there were some 99 000 private enterprises (a figure which probably excludes those enterprises registered as collectives and cooperatives) which together employed 1.7 million workers and possessed licensed funds of 12.1 billion yuan (*Dalu Siying Qiye Jingxuan, 1991*). By 1993, the number of private enterprises had risen to 184 000 with a combined workforce of nearly 3 million and licensed funds of 45.3 billion yuan (*ZHGSSB*, 10 October 1993, p. 1). In the late 1980s, many private enterprises were said to be joint-funded partnerships with funds of 50 000 yuan or less, employing between 15 and 30 workers, although very large enterprises with funds of more than 10 million yuan and over 1000 employees also existed. By the mid 1990s each of the seven largest manufacturing private enterprises registered sales of between 115 and 830 million yuan, and net profits of somewhere between 10 and 137 million yuan (*ZHGSSB*, 4 October 1994, p. 5).

Across China, the growth of domestic private traders has been somewhat sporadic. Available provincial data for China on private trade organisations provide no clear evidence for any regional or any other static or structured geographical pattern.[3] When each province is ranked according to each available indicator—numbers of private trade organisations (excluding private enterprises), numbers of people employed in those organisations, aggregate turnover and funds, the average share of aggregate funds and turnover for each organisation—provinces show considerable movement in their ranked order for each indicator from one year to the next (GGJGTS, 1987), and by the mid 1990s all provinces were registering growth in the number and strength of private traders (both private trade organisations and private enterprises). For the year 1992–93 the fastest rate of growth (of 20% or more) in the number of private trade organisations occurred in Beijing, Anhui, Ningxia, Heilongjiang, Shanxi and Jilin. The rate of growth in the number of private enterprises was even more startling: in Guangdong the number rose from 11 743 to 39 398, while in Beijing, Hainan, Anhui, Heilongjiang, Shanghai, Hubei, Xinjiang and Jiangsu, numbers more than doubled.

Significantly, the majority of private traders (70% of private enterprises and just over 50% of private trade organisations) were to be found along China's eastern seaboard. Moreover, they had appeared to shift from rural to urban areas. Although most private trade organisations (around 70%) still operated in rural areas, the rate of growth there was only 6.6% compared to 17% in the urban areas (*ZHGSSB*, 29 September 1993, p. 6). The proportion of private enterprises operating in rural and urban areas had almost been reversed by the early 1990s. In the late 1980s the vast majority of private enterprises operated in the countryside. About 20% of private enterprises were to be found in urban settlements, predominantly in the suburbs of the municipalities: only a mere handful operated in the urban districts of the cities proper. The number of private enterprises within the administrative jurisdiction of China's largest and most important cities reached only a few thousand: Tianjin contained around 8000 by the end of 1986 (Huang Dejun, 1989), perhaps rising to over 9000 by the beginning of 1988 (*GGJ*, Tianjin, 1990); Guangzhou held barely 1700; Beijing had just 1300 by April 1988; Shenyang and Chongqing held between 1000 and 2000 each; in Shanghai there were fewer than 1000 private enterprises operating within the municipality by the end of June 1987; and by June 1988 there were fewer than 500 in Chengdu (Huang Dejun, 1989).

The growth of private enterprises in rural and suburban areas partly reflected the diminishing influence of administrative will. Administrative restrictions (some formulated locally and others embellishments of national guidelines) governing the hiring of labour, the provision of welfare by enterprises for their workers, the availability of sites, and the issuing of licences, were more likely to be enforced near or in the city. But there were also economic reasons, such as inflated rents, which forestalled any attempt by private enterprises to operate at close quarters with subsidised, large-scale, state-run and collectively run enterprises. Moreover, it was generally accepted that private enterprises had evolved naturally from the gradual concentration of land and the specialisation of households brought about through the transfer of contracts allowed for under the responsibility system (Huang Dejun, 1989) In Wenzhou, for instance, the implementation of the responsibility system, the concentration of land and the specialisation of agricultural activities led to the emergence of small family industries dependent upon the waste goods and materials shed by the large state-run industries. At the same time, these families sent their representatives all over China, establishing contacts for the marketing and supply of materials and goods. These networks were regarded as the key to success (Huang Dejun, 1989). By 1991, however, the proportion of private enterprises found in China's rural areas had fallen to 60%; and by mid 1993 around 53% of all private enterprises registered as such were operating in urban settlements where (as would be expected) the rate of increase in the number of private enterprises was almost three times that in rural areas (*ZHGSSB*, 10 October 1993, p. 1).

Given these figures it is not surprising that the importance of private traders within China's economy has also grown markedly. Between 1978 and 1994 the private traders' share of retail sales grew from less than 0.5% to more than 30%; and their share of industrial output value rose from almost nothing to somewhere between 7 and 14% by 1992. All these figures probably represent substantial underestimates, in view of the difficulties in collecting data on the private sector, the growth of markets, the various techniques which are used to avoid taxation (one source suggests that more than 90% of private trade organisations are evading a proportion or all of their taxes [ZHGSSB, 18 August 1994, p. 6; JJRB, 2 August 1989, p. 1; ZHGSSB, 21 March 1990, p. 7]), the huge floating population of rural migrants, and the complex interactions among state-owned, collectively owned and both domestic and foreign privately owned enterprises.

In some parts of China, certainly, private traders had become important, even dominating, economic concerns. In Wenzhou for example, around 70% of all rural enterprises were likely to have been private enterprises: about 45% of all rural enterprises were known to be private concerns, and at least half of all registered *xiangzhen* enterprises were also thought to be owned privately. As early as 1987, private traders as a whole generated between 38 and 62% of Wenzhou's total industrial output value, depending on the degree of infiltration accepted (Huang Dejun, 1989; see also Parris, 1993). In Tengzhou, Shandong, 20.3% of the municipality's revenue for 1992 was derived from private enterprises. In addition, these enterprises had stimulated the flow of more than 1000 types of goods though the city's 116 markets, and they had identified and occupied particular niches for development, including the manufacture or marketing of reproductions of antique furniture, spinning machines and machine tools which were exported to, or imported from, Russia, America, Indonesia, England, Germany and Hong Kong, earning around US$30 million (ZHGSSB, 18 July 1994, p. 6) in foreign exchange. In Fu'an, Fujian, private traders generated around 38% of the municipality's total industrial output value, and somewhere between 30 and 50% of the municipality's tax revenue, for (as in Wenzhou) around half of all enterprises registered as collectives were in fact owned by private traders (ZHGSSB 25 October 1994, p. 5). Even in Shanghai, 3.4% of the municipality's tax revenue was collected from private enterprises which, by 1994, numbered around 12 000 (ZHGSSB, 20 September 1994, p. 5).

The institutionalisation of trade

The growth in the numbers and significance of private traders has been accompanied by the gradual institutionalisation and legitimisation of trade. This has been, and is likely to remain for the time being, a somewhat haphazard and uncertain affair, but it may prove to have far-reaching implications.

Article 1 of the 1982 constitution, which purportedly safeguards the interests and rights of private traders (including the right to own and inherit property), merely legitimised the continuation of what had already begun—the practice of trade, and the rise of merchants. More importantly, merchants were drawn into direct competition with, and were even encouraged to infiltrate, the state and collective sectors of the economy: smaller units of state-owned and collective-owned industries and organs of state procurement and distribution were leased, contracted out, and, in many instances, sold to private entrepreneurs; centralised control over planning, prices and procurement and distribution was relinquished for many categories of goods and materials; and free markets were allowed to flourish. And in 1988, private traders were permitted to sign contracts with foreign enterprises.

Furthermore, merchants have been presented in a more favourable light. It is now commonly accepted that they constitute a significant channel for the absorption of surplus labour; they help strengthen links between urban and rural areas; they represent a welcome competitive stimulus for state-owned and collective-owned enterprises;[4] they are an important source of tax revenue; they purchase national treasury bonds; they make frequent contributions to worthy causes; and individual entrepreneurs are named and praised for their courage, vision, willingness to take risks, and strength to rise above and ignore gossip (ZG, Tianjin, 1988; ZHGSSB, 29 September 1993, p. 6).

Indeed, even many criticisms of private traders are essentially constructive. The failure of private traders to move further towards the realisation of their full potential is seen largely to reflect the administration's failure to provide an institutional framework within which private enterprises can freely operate, a comment which often implies criticism of the inefficiency, ignorance and hostility of local authorities. Private enterprises, it is said, are unfairly and heavily taxed; the creation of two tracks for prices and supplies forces merchants to buy goods and materials at excessive prices through both official and illegal channels; credit is more difficult to obtain, and more expensive, for private traders than it is for state-owned and collective-owned enterprises; and merchants often have to suffer under a barrage of resentment, envy and hostility from local officials and from their peers.

The solutions frequently proposed are a change of values, and the construction of supporting institutions. First and foremost, it is argued that the necessity for private traders must be recognised by the general populace and, most especially, by local officials. Moreover, the view now put forward is that the antagonism between 'socialism' and 'capitalism' and, therefore, between public ownership and private ownership, is arbitrary and misleading. During the initial stages of socialism, a public-ownership system by itself is not effective; it requires the help of private ownership. A reasonable level of exploitation, so it is argued in one report by the Communist Party, is quite acceptable at this evolutionary stage. Worries that the rapid development of private enterprises will undermine and topple the public-ownership system are said to be misplaced. Moreover, private enterprises in China are, in the main, joint-funded organisations and, therefore, very similar to those of China's publicly owned enterprises which issue shares. In this and in many other respects, it is now claimed that public and private systems of ownership are converging.

Similar arguments form part of the debate over the notion of a 'civil economy'—another common vehicle used to justify the rise of merchants under the control of the central authorities.[5] Although interpretations of this term vary, state-owned, collective-owned and private-owned organisations are often placed into two categories. One, which includes private traders and collective-owned enterprises, is described as 'non-state'. The other category embraces state-owned enterprises which are, or should be, run and managed by individuals, or by groups of individuals, under the contract responsibility system or under share ownership. Indeed, it is suggested that the terms 'private ownership' and 'public ownership' are losing both their significance and their meaning. Comparisons with Japan and the NICs in the West Pacific Rim are made, for it is these successful mixed economies which demonstrate the potency and efficacy of one important strand of 'Chinese' cultural influence— the notion of an economy and society founded upon the actions and efforts of the general population (*JJRB*, 14 October 1993, p. 5).[6] Moreover, even if the state-owned enterprises' share of China's total industrial output value were to decline still further, this would not affect their ability to guide the economy.

In order to foster the necessary values and to institutionalise trade, it is now commonly argued that the legal standing of private traders must be given far more precision; clear and more detailed procedures and regulations for the assessment, collection and payment of taxes, and for the extension of credit, must be formulated and implemented; advice on managerial, technical and legal matters should be provided; and administrative organs should cooperate and coordinate with each other to enable private traders to expand their operations across areal and sectoral administrative boundaries. Furthermore, government organs should facilitate both domestic and international trade links among private traders.

Perhaps one of the most interesting developments has been the re-appearance of associations. By 1989 there were over 2000 national-level associations—a term which is used here to include chambers of commerce, federations, societies, guilds and other similar bodies representing specific groups such as rural marketeers, specialised households, private enterprises, exporters, and Overseas Chinese investors (Gold, 1989; Pearson, 1994; *GDQB*, 24 December 1993, p. 1). To some extent these associations are little more than a formalised, public expression of existing relationships among private traders or between merchants and the administration. In many cases, however, membership is compulsory, and officials from administrative units sit on the associations' boards.

The stated activities and goals of the associations are wide ranging. They implement and disseminate government laws and policies; they promote business attitudes, exchange information, liaise with government, establish trade contacts (both at home and, as in the case of the Overseas Chinese associations for example, abroad), moderate competition, and propose new policies and institutions; and they mediate in, and settle, disputes between members, and between members and the government. If the All China Federation of Industry and Commerce (ACFIC—one of China's nominally democratic parties) can be thought of as an association, then it would seem associations may also decide against the administration,[7] for the ACFIC has accused the administration of prejudice against private traders. More specifically, it is claimed that state administration has been unwilling to extend credit to private traders—a charge followed up by a call for the establishment of private banks for merchants.

Predictably, the rise of associations is an issue which writers concerned with the question of democracy in China have been quick to draw into their analyses. Those who believe that, in the long run, democracy works 'best' for nations attempting to realise sustained material progress, must first create the sham argument that an authoritarian state works 'best' in order to strengthen the moral authority of their own position. Yet neither is 'best' for sustained material progress, for there is no specific political formula or institutional pattern necessary for the realisation of such progress. It is true that there are certain values and attitudes, such as equality of opportunity, anti-monopolism, and an acceptance of inequality and profit, which find sympathy with the ideal of democracy. But it is also true that certain moral positions which are regarded by some as indivisible from a democratic state—such as welfare, egalitarianism and hostility to profit—are also important for the institutionalisation and legitimisation of an authoritarian and centralised state. This point can be taken a little further. The notion implicit in much of these discussions that China's political institutions and supporting values have been imposed upon the mass of the people by an authoritarian elite may not be entirely appropriate. If institutions and values which some may regard as characteristic of a democratic society can be 'turned' in favour of an authoritarian state, then it may be that many restrictive attitudes, values and ways of behaviour, accepted and manipulated by individuals to create an authoritarian state, form part of wider institutional patterns and moral values of a democratic one. Introspective and bickering cliques which, for whatever purpose, manipulate values and emotions such as 'equality', 'contributions to society', 'being-one-of-us', group loyalty, envy, resentment and jealousy and all the other characteristic vices of the closed institution and the village mentality, are likely to find greater support within an authoritarian regime than within an atomistic, democratic one.

Reciprocity, then, may also serve state procurement and distribution. While it may be that individuals have strong sentiments for the opportunities presented by the conduct of trade and for liberation (political, economic and social), they may not view, hold or adhere to those sentiments as values in their own right, nor may they be willing to subordinate personal interests and personal relationships to democracy's imperfect institutions and procedures. Even the intellectual 'elite' of the pro-democracy movement seem unwilling to put united action before their own personal disagreements and ambitions. So often they appear to be more concerned with publicity and prestige than with democracy and justice as values in their own right. As Gao Xin, a former hunger-strike leader (and later a visiting scholar at Harvard University) has put it: 'People like Shen Tong and Chai Ling benefited from June 4th—they ate steamed bread soaked in other people's blood' (cited by Goldstein, 1994, p. 26).

The imposition of democratic political institutions and supporting values upon individuals who may feel some emotional attachment to these institutions and values, but who see no personal stake in them, and who are unwilling to accept them as self-defining and absolute, may lead only to chaos, or to the formation of powerful groups of special interests founded upon webs of reciprocity, and legitimised by the notion of 'democracy'. For similar reasons, White (1994) has argued persuasively against a sudden drive for democracy. Among a people whose attention and energies are still consumed by the day-to-day needs for material survival, and whose central government has loosened its control over economic affairs, a push for democracy would create innumerable factions of self-interested elites.

In an earlier paper, White (1993) makes the important point that there is a close relationship between:

'the spread of market relations and the differentiation of ownership brought about by the Chinese economic reforms on the one hand, and the rise of new forms of social organisation and the adaptation of existing social organisations on the other' (p. 67).

White further hypothesises that as an economy and society atomise, the association will serve as a link between state and society; associations will represent and protect in an increasingly atomised population; they will provide a means to coordinate and cooperate; and they will bring greater stability and security. In China, however, the state:

'has moved to . . . organise the newly emergent, dispersed sources of economic power by encouraging the establishment of social organisations to act as intermediaries. Though there has been some impetus for their formation from below, the dominant impulse has come from above' (White, 1993, p. 85).

And since these associations do not exhibit a clear distinction between 'public' and 'private', are not entirely independent, do not behave as pressure groups, and do not comprise an entirely voluntary membership, then the associations in China cannot be described as 'civil' organisations, though embryonic elements of a civil society can be detected.

'This relative weakness of "civil society" must be situated in the context of a semi-reformed command economy in which the state retains its dominant position in the economy. Its weaknesses must also be perceived within the context of the dynamics of reform, in which this dominance is gradually being undermined as the number of participants in the non-state sector increases. One can hypothesise, therefore, that to the extent the economic reforms continue . . . these socio-economic forces will grow in strength and a more powerful "civil society' will emerge . . . " (White, 1993, p. 86).

These associations, White (1994) suggests, may help with the transition to democracy—a transition which might be realised through a form of 'new authoritarianism':

' . . . the resulting political system would be a hybrid, comparable to counterparts in countries such as Singapore, Taiwan or Mexico where elements of single-party dominance co-exist with elements of institutionalised competitive democratic politics. Its political bases would be diverse, but at least this diversity would secure public acknowledgement and find more available channels of self-expression than the previous system of artificially imposed uniformity. . . . To the extent that political reform is accompanied by continued transformation of the economic system, the balance of power between state and society would continue to shift in a direction favourable to pluralist politics . . . '(p. 90).

This transition, and the final steps towards democracy, would require a more reform-minded leadership prepared to allow the formation of a true market economy; the willingness of the leadership and the political activists to make accommodations; the legal protection of civil and economic rights; a redefinition of the party's role; and the evolution of truly independent associations which could then be brought into the political framework.

Yet even this crafted argument (which ends in a wish-list) for the eventual realisation of democracy in China skirts around the central issue. The wishes, ideology or political faith of commentators, the categorisation of organisations as 'civil' or 'governmental', and the hypothetical formulae believed necessary for the evolution of a market economy and a democratic society, may produce unidimensional interpretations of 'what is', and limited prescriptions for 'what should be'. To be sure, associations possess many aspects, and these include the potential to be 'turned' towards the political objectives of the totalitarian state as well as the democratic one. Though White acknowledges this, he suggests that associations, rather than acting clearly and directly to institutionalise 'market relations', in fact provide the state with the means to perpetuate control over the economy. Only with further reform in China, the atomisation of society, and the retraction of state controls will associations be able to play their role within 'civil society' as prescribed by White.

All this seems to ignore the possibility that associations may nevertheless institutionalise and legitimise trade, and, as part of this, help target loyalties upon Beijing. Associations, as one expression of corporatism, may well strengthen rather than weaken. Certainly, it would appear that associations are intended as, and work as, coordinating and regulatory bodies which are both flexible and focused upon the prosecution, extension and safeguard of trade. This is a task for which existing administrative organs, whose priorities, outlook and areas of responsibility are rigidly constrained and narrowly directed towards the support of state procurement and distribution, are obviously unsuited (*ZHGSSB*, 11 August 1994, p. 3). The associations, then, may ensure that government is tailored to meet the requirements of private traders, and that the aims, desires and behaviour of merchants do not run counter to the concerns of the ruling party. For instance, trade associations (whose members are drawn from enterprises which would otherwise fall into separate areal, departmental and ownership administrative divisions) are regarded as more flexible, more pragmatic and more capable of separating managerial from administrative concerns than the existing compartmentalised administrative units such as the bureaux or general companies. They are therefore better equipped to coordinate activities, reduce competitive tensions, mediate in disputes, and ensure that all members work for their common good. To operate more effectively, however, they must be shielded from the interference of existing administrative units. The legal

position of associations must therefore be strengthened, existing administrative institutions must be eliminated gradually, and the quality of the
associations' personnel must be improved (*ZHGSSB*, 11 August 1994, p. 3;
see also Ye Zhen *et al.*, 1993).

Another related strategy for the integration of merchants with the
administration has been the admission of private traders into the ranks
of the Communist Party. Although doubts are raised about the wisdom
of this practice, it would not seem to be an uncommon one. If occasional
reports are any indication, demand for entry outstrips the ability or willingness of the party to accommodate new members. A recent survey of
the Private Labourers' Associations, the Private Entrepreneurs'
Associations and the Municipality Party Committee in Tengzhou,
Shandong, revealed that 20 of the 172 entrepreneurs questioned had formally applied to join the party. Another 46 were exploring the possibility
of membership, and the remainder were intending to join. In recent
years, 23 merchants had been praised as 'exemplary' taxpayers or
'advanced' workers by municipal authorities, and 7 had been selected as
provincial and municipal representatives to the Chinese People's
Political Consultative Conference. Eleven more had been selected as village heads or had been made responsible for developing local service
and manufacturing industries. In Shenzhen, too, around 1600 merchants
and their employees were made party members (*ZHGSSB*, 26 September
1993, p. 6; 18 July 1994, p. 6).

More generally, informal links between private traders and the administration appear to be fairly widespread. A proportion of workers
employed by private enterprises are former cadres, military personnel,
graduates, technicians and administrators, skilled at establishing and
maintaining complex networks of personal connections with members
of state administration, scientific research institutes and large state-run
factories (Huang Dejun, 1989; see also Parris, 1993). Many others are
drawn from *xiangzhen* enterprises and are either related to the owner or
have been introduced to the owner by the owner's relatives. In most private enterprises in Tianjin, for instance, production, supplies, sales, personnel, materials and finances are all the responsibility of the owner's
(or owners') immediate family, relations and close friends—the pivotal
connection often being a woman (ZG (Tianjin), 1988).

It is said that kinship ties and personal connections contribute to the stability and cohesion of private enterprises, enable administrative work to
be kept to a minimum or to be dispensed with altogether, and allowing
business transactions to be conducted by word of mouth without the
need for involved and cumbersome accounting techniques and contrac-

tual agreements. Kinship ties and personal connections also enabled private enterprises to overcome administrative obstacles to the supply of land, credit, technology and training. Most importantly, they functioned as channels for the flow of information on market conditions and on changes in demand for goods and raw materials, thereby forming the basis of trade networks that drew in many distant parts of China, and extended overseas (ZG (Tianjin), 1988; GGJ, 1990; Huang Dejun, 1989).

But clearly, then, it would appear that merchants have, understandably, established more extensive and much stronger links with provincial and sub-provincial governments than with the central authorities via their associations. This, together with the appearance of provincial and sub-provincial political elites formed around place of origin and kinship (Zang Xiaowei, 1991), has magnified concerns over localism. Not surprisingly, the reliance of merchants upon the establishment and strengthening of personal relationships with the local administration and its subordinate economic units is frequently portrayed as a response to uncertain and hostile conditions created by local officials, and as a traditional mode of operation that is unsuitable, limiting and damaging. Similar criticisms have been levelled at collective-owned units which have been rented or leased to their existing managers or to private individuals (ZHGSSB, 19 October 1994, p. 5). The central authorities may have good reason to feel uncomfortable about the current *modus operandi*. In a society which has not yet evolved clear and strong institutions for the extension, safeguard, legitimisation and prosecution of trade, it may be that the devolution of power, the decline of state procurement and distribution, and the conduct of reciprocity for its own sake (and the desire for the power and prestige thereby derived), will produce considerable instability and uncertainty both for merchants and the central authorities. This is particularly true among people who, until recently, have had little or no experience in the conduct of trade, and who may view the money and material items which trade brings as legitimate symbols of prestige and personal worth as judged by reciprocity's associated wealth values. Rather than pursue the ideals of democracy prescribed by external commentators, or align themselves with Beijing, private traders may well continue to develop inherently unstable alliances with members of the provincial and sub-provincial administrations, for both administrators and merchants share the same goals—the realisation of material progress, and the survival of these political elites. As Solinger (1992) argues:

'A return to a regime based purely on planning would undermine new-found opportunities now enjoyed by cadres, whereas a leap to fully open and unobstructed markets would deprive the most successful merchants of their special inside channels. As a result, there is an implicit pact to preserve the monolith and a stasis in the symbiotic tie between bureaucrat and merchant' (p. 130).

The interests of the central authorities would appear, therefore, to lie in preventing the dissolution of a distinction between merchants and the local political elites, and in ensuring that the loyalty, and taxes, of merchants are focused upon Beijing rather than on provincial and sub-provincial governments. It is unlikely in a country as large and as diverse as China, and in which the state system of procurement and distribution is being allowed to wither, that reciprocity and its soft-framed institutions will provide the necessary stability, cohesion and predictability for the prosecution of trade. China may find that it must soon privatise state-owned and collective-owned enterprises, allow private and independent banks to operate and to dominate, establish a strong legal framework for the conduct of trade, introduce private insurance schemes for welfare, unemployment, pensions and health, restructure its tax system, and promote the development of associations closely allied to the interests and policies of Beijing.

China's trimorphic economy

The investment-led growth of collectives, and the losses of state-owned enterprises were merely opposite sides of the same coin. And even though collectives had increased their share of China's industrial output value, the rate of growth of private enterprises (both domestic and foreign) was far greater. China's leaders have become more and more dependent upon merchants to deliver sustained material progress, and yet the leadership's preoccupation with the maintenance of the state and collective systems of ownership works against that very progress. State enterprises, and alliances between local government and their subordinate economic units, are dragging the economy down: policies designed to handle problems created by state-owned and collective-owned enterprises affect the operations of both domestic and foreign enterprises; crucial decisions such as lowering tariff and non-tariff barriers, which would allow China's entry into the World Trade Organisation, are difficult to reach and implement; and conflicts between central and local authorities create still more uncertainty and instability. The links between key individuals in the central administration and large state international trading companies help the central authorities exert control and influence over foreign trade, but these connections, too, give rise to as many problems as they solve (see, for example, Karp, 1993; Sender, 1994; and *The Economist*, 1994c). If China's economy were already dominated by trade, then large state enterprises, directed towards the international economy, and well connected with China's political elite, might do very well. But China's leaders are still wrestling with the unpredictable lurches of a unitary economy in which an increasing reliance on kinship and personal ties seems to indicate either an inability to control events in any other way or, perhaps worse still, the relegation of trade, trading organisations and material items, to symbols of prestige measured by reciprocity and its associated wealth values. The more vigorous approach to the protection of intellectual property rights, and the arrest of Zhou Beifang and the removal of his father Zhou Guanwu from the Shougang Company (merely one incident in a string of arrests and dismissals of ranking merchant-bureaucrats) might well indicate political infighting as one faction attempted to discredit another; yet it might also be part of a genuine attempt to imbue China with an impartial legal system.

Ironically, then, the gradual acceptance of private trade's eventual domination owes as much to political considerations as it does to a determination to realise sustained and rapid material progress. The growing material and fiscal power of provincial and sub-provincial authorities— bolstered by complex networks of reciprocity which span industries, administrative departments and ministries, and which embrace both domestic and foreign merchants—might end in the *de facto* privatisation of huge entities under the control and ownership of provincial merchant-bureaucrats or favoured representatives of political factions within central government, oriented towards the international economy, and loyal only to themselves and their merchant empires.

The central authorities, then, are faced with three options. Either they may continue with the implementation of an uncertain, indecisive and frequently contradictory hotchpotch of policies clouded by special interests and drawn up in response to events. Or they can recentralise and then watch the other economies in the West Pacific Rim and South Asia leap still further ahead. A third option is to allow economic units true independence; to develop institutions and values which distance reciprocity from trade, and which act to focus obedience, loyalty and tax revenues upon Beijing; and to ensure that any remaining state-owned enterprises are reoriented towards international trade. In this way, provincial and sub-provincial authorities would find the ground crumbling beneath them, and large state enterprises which have proved so damaging within the domestic economy could be turned to considerable advantage in the international arena. Moreover, the likelihood that border and coastal provinces (most especially in the southeast) could be pulled away from Beijing would be reduced, for as Segal (1994) has suggested:

'It may be that the only way to ensure that China does not become more dangerous as it grows richer and stronger is to ensure that in practice, if not in law, there is more than one China to deal with' (p. 64).

Clearly, all these matters are of considerable importance to Hong Kong. While tensions between trade, reciprocity and state procurement and distribution in China remain, there will always be some question over Hong Kong's future. Doubts have been raised about the survival of Hong Kong's currency after 1997, and it would seem possible, given attempts to strengthen the Shanghai clique in Beijing and concerns over the unity of China, that Shanghai's development may be promoted at the expense of Hong Kong. However, perhaps the greatest danger which the colony faces is not democratic uprising followed by dictatorial suppression, nor the vindictiveness of a ruling elite determined to settle old scores at any price, nor is it, for the time being, the collapse of the Communist state into chaos. Hong Kong's future depends upon the realisation that the colony has always been and remains little more than a trade settlement. Provided that future administrations are prepared to accept this fact and continue to extend and institutionalise trade in any way that is expedient, then Hong Kong has little to fear. Lee Kuan Yew may be quite correct in his observation that Overseas Chinese have one important advantage over their competitors in the conduct of business in China—their experience of forming and manipulating social relations specifically in order to further trade in the absence of effective and sophisticated legal frameworks. His observation that the rule of law is vital to Hong Kong's continuing success is also correct. Reciprocity may work both for and against trade. If the administration and business community becomes drawn into the parochial and narrow politicking from which Hong Kong has been shielded for so long, and if trade is thus once again turned to serve other forms of exchange and other agendas, then Hong Kong's economy will gradually be choked by political intrigue tinged with envy, resentment and bitterness.

Greater China and the merchant princes

If it is believed that Overseas Chinese economic success has something to do with their 'Chineseness', then clearly the extension and institutionalisation of trade in China would seem to lead naturally to the conclusion that an extremely powerful and extensive 'Chinese' trade network is likely to develop across East and Southeast Asia and perhaps beyond. After all, estimates of the economic strength of Overseas Chinese (see Chapter 1) appear to support this idea; and Overseas Chinese have been at the vanguard of the push into China by foreign merchants. Between

1979 and 1991, contracted foreign capital in China amounted to US$121.47 billion, of which US$79.627 billion was realised (Lin, 1993). Two-thirds of realised capital took the form of loans. The remaining US$26.885 billion is described as foreign direct investment. Overseas Chinese were responsible for two-thirds of this investment: Hong Kong and Macao Chinese, the Taiwanese, and the Southeast Asian Chinese provided 51.82, 9.29 and 5.57% respectively. The remainder was account-ed for by other foreign merchants. The exact source of funds, however, is probably impossible to pin-point, for it is not known how much originates from merchants who are not Chinese, or from China's own companies reinvesting funds through foreign dummies, for Mainland Chinese companies have been establishing operations in Australia, Peru, the United States, Singapore and Malaysia, as well as in Hong Kong which holds the most important concentration of Mainland Chinese companies overseas. Nor is the scale of Mainland China's investments in Hong Kong easily determined: Qi Luo and Howe (1993) maintain that Japan and the United States continue to dominate foreign investment in Hong Kong, whereas Ho and Kueh (1993) suggest that, with 3000 main-land companies now operating in Hong Kong, the total value of China's investments (which they estimate at between US$15 billion and US$20 billion) in the colony greatly exceeds that of Japan and the United States.

After 1991, annual realised foreign direct investment rose again to more than US$10 billion in 1992, and then still further to US$27.5 billion in 1993, and to US$28.8 billion in 1994, to reach an accumulated total of just over US$90 billion. Hong Kong Chinese and Taiwanese were responsible for around three-quarters or more of this investment. The United States (6.5%) and Japan (4%) were the next largest investors. Most investment was concentrated in labour-intensive manufacturing industries, and around half of all foreign direct investment in China during the late 1980s and early 1990s was absorbed by Guangdong and Fujian. By the mid 1990s, foreign direct investment was being attracted to other parts of China, most especially along the eastern seaboard, though Guangdong's and Fujian's combined share at just under 40% remained disproportionately large. Over the same period, foreign direct invest-ments were targeted at a much broader range of activities. Manufacturing activities (mostly labour-intensive industries which used second-hand machinery to produce goods for export in small enclaves) remained the largest single investment category, but real estate, public utilities, infrastructure projects, and services together attracted as large a share (if not larger share) of new foreign direct investment (*JJRB*, 8 October 1993, p. 1; *ZHGSSB*, 8 September 1993, p. 6; Lin, 1993; Qi Luo and Howe, 1993; Ho and Kueh, 1993).

The idea of an interconnected and distinctly 'Chinese' network, partly submerged in the cultural, social, economic and political structure of the indigenes, is certainly not an uncommon one. In his discussion of Greater China's foreign relations, Yahuda (1993) acknowledges the possibility of such a web:

> 'Greater China refers in the first instance to the close economic ties of trade, technology transfers and investment that have emerged since the second half of the 1980s linking Taiwan and Hong Kong with the rapid development of southern China. But it also suggests that the economic links are buttressed by familial, social, historical and cultural ties of *a peculiarly Chinese kind*'[8] (p. 687, italics added).

Sung (1992) expresses a belief in a more cohesive and significant 'Chinese' network which explains the characteristic pattern of foreign investment in China:

> 'The importance of cultural affinity is quite evident. People in Hong Kong have their ancestral roots in Guangdong, and Guangdong has received the bulk of Hong Kong's investment in China. Taiwan's investment is similarly concentrated in Fujian . . . ' (p. 7).

Moreover, 'cultural proximity' and networks of kinship allow merchants to evade formal barriers to trade, and to obtain favourable concessions from local authorities in Guangdong. Leung (1993) also emphasises the crucial importance of kinship for Hong Kong merchants wishing to subcontract production and services to entrepreneurs in China.

Weidenbaum (1992) expands upon the idea of 'Chinese' cultural networks within Greater China, and suggests that the 'Chinese-based' economy which extends from Guangzhou to Singapore and from Kuala Lumpur to Manila, is the backbone of the East Asian economy (p. 22). This 'Chinese-based' economy, Weidenbaum argues:

> ' . . . is rapidly emerging as a new epicentre for industries, commerce and finance. This strategic area contains substantial amounts of technology and manufacturing capabilities (Taiwan), outstanding entrepreneurial marketing and services acumen (Hong Kong), a fine communications network (Singapore), a tremendous pool of financial capital (all three), and the world's largest endowment of land, resources and labour (China). Examples of economic integration abound. A talking doll, to take a modest example, may be designed in Hong Kong and assembled in China, and contain a computer chip made in Taiwan . . .' (1992, p. 22).

Kotkin (1993) also portrays a regional economy which is specifically Chinese and which is of similar or even greater economic power. The proposal of a Greater China Economic Zone comprising China, Taiwan, Hong Kong and Singapore:

' . . . would boast combined foreign currency reserves roughly two-and-a-half times those of Japan, and would rank second only to that island nation as Asia's leading industrial power' (p. 197).

Kotkin goes on to suggest that:

'If it is willing to capitalise upon global expansion, the Chinese tribe has the potential to develop a worldwide presence—in terms of permanent settlement and global networks—not seen since the hegemony of the British. With their enormous human resources, historic flexibility and entrepreneurial skills, the Chinese could prove to be the best positioned of all ethnic groups in the coming century, and their emergence the most momentous contemporary movement in the ongoing history of global tribes' (1993, p. 200).

Wang Gungwu (1993) provides a suitably scholarly and scientific basis to this idea of a regional economy which is specifically 'Chinese'. His suggestion that there are various 'types' of Chinese which exhibit varying degrees of intensity of 'Chineseness' clearly implies that there is an underlying 'Chineseness' common to, or inherent in, all of them—a 'Chineseness' which includes the ability to borrow and adapt. Moreover, while he acknowledges that the term 'Greater China' can imply a desire for expansionism and a misleading or boastful grandiosity, Wang goes on to argue that:

' . . . if the emphasis is on the broader picture of cultural China, the term [Greater China] may refer to the traditions of Chinese civilization and what has transformed them in modern times. In this context, it may be said that Greater China has become Greater because its store of cultural values has been enhanced by modern borrowings, influences and adaptations. As a result, millions of ethnic Chinese now residing abroad might find it possible to identify with it '(1993, p. 926).

The implication that there is a unique and possibly deterministic Chineseness is reinforced by Wang's otherwise unaccountable[9] and didactic insistence that the use of the term 'Chinese Diaspora' is unsatisfactory 'because its association with Jewish history is unacceptable in South-east Asia' (1993, p. 927).

Sun Qian (1993) also believes that the ability to borrow and adapt is an inherent feature of 'Chineseness'. This is an argument which has proved useful in China. It was noted earlier in this chapter that 'Chineseness' is used to justify the notion of a 'civil economy' and privatisation, and to claim credit for the economic success of China's neighbours in the West Pacific Rim and thereby to provide both a precedence and a reason for the establishment of a mixed economy guided by the state. Zhu Rongji has similarly instructed his audiences at home and abroad that China's economic reform has been successful because it is rooted in its 5000 year old culture and tradition: the reforms are therefore irreversible.

Interestingly, Sun Qian, like Wang, also demonstrates a preoccupation with definitions of Overseas Chinese, but he unashamedly and explicitly provides his definition of Overseas Chinese and his understanding of 'Chineseness' with a strong racial foundation. Perhaps, then, it might be suggested that the desire to categorise 'Chinese' derives from a belief that Chinese individuals are all essentially variants of a true 'Chineseness' which includes the ability to adapt and change while retaining certain inherent racial characteristics. This appears true of Sun Qian, at least, who makes a distinction between those individuals born in China, those who have received a 'Chinese' education, and those who have not received a 'Chinese' education. Sun Qian goes on to use this distinction to make another—this time between Chinese culture and specific racial characteristics which Sun believes are peculiarly Chinese. Those who have not received a 'Chinese' education nevertheless retain a special 'Chineseness', while those who have received a 'Chinese' education are also capable of retaining their Chinese culture. These racial and cultural characteristics include: the ability to grasp opportunities; a desire for unity, and a desire to cooperate with each other and with the indigenes in Southeast Asia; an emphasis on racial harmony (for Chinese perceive man as the centre of all living things under heaven); a willingness and an ability to endure hardships; a single-minded approach to economic matters; and flexibility. Sun also claims for 'the Chinese' a natural grasp of economics; a strong locality consciousness which finds expression in Chinese individuals' desire to establish links with their place of origin, to provide succour to their home towns, and to establish trade organisations (such as the *bang*) on the basis of kinship and locality; the fusion of economic and moral values to the extent that diligence, trust, honesty and truth become crucial attributes for the conduct of business; a belief in competitive coexistence which is so strong that Chinese individuals would rather lose money than avoid helping each other; and a perception of society as an integrated, harmonious, organic whole in which the individual is controlled by the group, and in which the potential of the individual is realised only in so far as this advances the group.

To emphasise the uniqueness of these racial and cultural characteristics, Sun Qian draws comparisons with 'Westerners'. 'The Chinese' will grasp opportunities when they arise, but they are also willing to cooperate with other peoples, including the indigenes; the Europeans, however, are essentially hegemonists. Whereas Chinese stress 'relational' capitalism, Westerners are individualistic. 'The Chinese' will take considered risks, whereas 'the Westerners' see no distinction between adventurism and management. Whereas 'the Chinese' believe that competition, coexistence and harmony are complementary, Westerners can only envisage a completely closed economy or a completely open and ruthlessly competitive one. An attachment to one's place of origin is a phenomenon found in many different cultures, but it is an unusually intense sentiment among Chinese.[10] 'The Chinese' are not the only people to work hard, but it is extremely rare to find as much emphasis placed upon selfless effort and moral improvement as it is among 'the Chinese'. And whereas the 'Chinese' educated Overseas Chinese will work hard because they believe hard work and moral improvement are values in their own right, the 'Western' educated Chinese are only really interested in how much they are paid for the hours they put in.

Such interpretations, though crude, contain echoes of the cultural determinists' more sophisticated expositions: the views of this academic from Mainland China merely represent the darker side of that potent mixture of 'science' and 'culture'. If, however, one rejects the assumption that there is a unique, unidimensional 'culture' which is peculiarly Chinese and capable of explaining or even determining Overseas Chinese economic success, then a very different perspective emerges. While certain values, institutions and patterns of behaviour may be presented as unidimensional phenomena and thereby serve as a useful kernel which may facilitate the prosecution of trade, the fact that an individual is of Chinese descent becomes of little relevance, both to an understanding of Overseas Chinese economic success and to an appreciation of their links with Mainland China.

It is well to remember that the United States and Japan are the NICs and ASEAN countries' most important source of imports and the main destination for their exports. Indeed, with the exception of Hong Kong and China, by far the largest proportion of the West Pacific Rim's merchandise trade takes place with Japan, North America, Europe and the Middle East. And, with the exception of Taiwan and Hong Kong, goods originating from China comprise no more than 10% of imports of any country, and no more than 4 or 5% of their exports are bound for China. In the case of Taiwan, no more than 10% of its total international trade is conducted with China; and the strong relationship between Hong Kong and China merely describes the colony's entrepot status. Some 80 or 90% of Hong Kong's imports from China are re-exported; and three-quarters of Hong Kong's exports to China are re-exports from other countries. Moreover, the bulk of Hong Kong's domestic exports to China are destined for Hong Kong's own industries, which are geared to the production of goods for export. Among the other countries of the West Pacific Rim only Japan and Singapore remain in the list of Hong Kong's 10 leading markets for domestic exports: by far the largest proportion of goods are bound for Japan, North America and Europe. Much the same can be said about China: Japan, Hong Kong, Singapore and South Korea are the only countries in the West Pacific Rim among China's 10 leading trade partners—the bulk of China's exports are bound for Japan, North America and Europe.

Although the dominance of 'Chinese' investment and its concentration in south China are facts which easily lend themselves to a 'cultural' interpretation of economic relations, it would be fanciful to suggest that Overseas Chinese trade with China is driven by anything other than a desire for profit. Hong Kong's primacy in merchandise trade and foreign direct investment merely reflects its geographical advantages and its exceptionally favourable climate for merchants—both 'Chinese' and 'non-Chinese'. Certainly the most common 'reasons' for Overseas Chinese investment in China cited in both English-language and Chinese-language sources are pragmatic: the appreciation of domestic currencies; cheaper land and labour in China; rich natural resources (agricultural as well as mineral); and China's potentially huge market. The circumvention of export quotas, and loose and poorly enforced laws and regulations governing safety and conditions, the payment and use of labour, and copyrights, are additional attractions (Qi Luo and Howe, 1993; GDQB, 21 December 1993, p. 1; Lee-in Chen Chiu and Chin Chung, 1993; FJJJB, 13 March 1992, p. 3; GDDWJM, 1992, 2, p. 45). Reciprocal relationships formed around kinship or some other kernel are merely techniques of convenience. Overseas Chinese investors are attracted to the *qiaoxiang*, not because of any sentiments they may have for their home towns or for their ancestors, but simply because these set-tlements are located along the coast where transport is comparatively efficient, the flow of information is fairly quick, policies are sympathetic to the needs of merchants, and where the Overseas Chinese associations (established by Mainland Chinese) have been working to attract Overseas Chinese merchants and to facilitate their operations (Liang Yingming, 1993; GDQB, 24 December 1993, p. 1; FJDWJM, 1992, 4, pp. 2–3). Indeed, not only are the significance and value of reciprocal rela-tionships played down, but their disadvantages are emphasised. The problems derived from a reliance on reciprocity and its soft-framed institutions for the prosecution of trade have already been discussed in earlier chapters, and the implications of these problems for the rise of trade in China have already been commented upon briefly in this chap-ter. The distancing of reciprocity from trade, and the evolution of institu-tions and values which will extend, institutionalise, safeguard and legitimise trade are essential to both domestic and foreign merchants, and a political and economic necessity for the survival of China's central authorities. While it is easy to find those whose confidence in China's ability to attract foreign merchants is such that they are happy to spend time ranking nationalities according to the sums invested,[11] others are less sanguine. If trading opportunities are to be widened, if the scope and quality of investment and economic activities are to be improved, if China is to offer more than just cheap labour, land and promises, and thereby to remain attractive to foreign investors in a very competitive international arena, then a more transparent, organised, stable and predictable institutional, legal and regulatory framework must be created, and better services and infrastructure must be provided (Liang Yingming, 1993; Huang Xunping, 1992).

So, too, according to Liang Yingming (1993), has the importance of the Overseas Chinese in Southeast Asia been overemphasised. Estimates which put Southeast Asian Overseas Chinese capital assets at US$400 billion (a hundredfold increase over 1969) are little more than guesswork. Chinese merchants frequently operate in alliance with the indigenes whose economic power is often much greater than assumed. Based on figures covering the period 1967–80, Liang cites the example of Indonesia where 29.65% of enterprises formed out of domestic capital were owned by non-indigenes, 11.25% by the indigenes, 3.1% by cooperatives and 58.95% by the state. Of the shares issued by foreign enterprises, 9.7% were owned by non-indigenes, 12.77% by the indigenes, 9.25% by the state and the remainder by foreign concerns. Such evidence is dated, and easily influenced by definitions, political considerations and concerns about hostile reactions to the presence and success of Chinese minorities. But it makes the point well enough: a preoccupation with the identification of ethnic groups would merely limit China's opportunities. As noted above, even analyses which are based upon notions of racial or cultural determinism emphasise both the desire for interaction with the indigenes, and also the belief that Confucianism underlies economic success in East and Southeast Asia. The message seems clear: however different and unique Mainland Chinese may feel they are, such feelings should, for the moment at least, remain subordinate to the desire for trade.

It is further argued that Overseas Chinese enterprises are usually small and their capital dispersed (Liang Yingming, 1993). L. Lim (1992) also suggests that Chinese merchants and their enterprises are

'highly individualistic in their operations, prefer competitive to collusive modes of business behaviour, and eschewing the tight monopolistic corporations and networks which have helped the Japanese and Koreans to build large-scale, world-class enterprises' (pp. 45–46).

Overseas Chinese industrial technology is still heavily dependent upon

Japan, North America and Europe, and the accumulated stock of foreign investment in the West Pacific Rim is owned by individuals and groups who are not of Chinese descent. Most of Singapore's manufacturing sector is founded upon investments by companies which are not Chinese (L. Lim, 1992); by the late 1980s Japan had become the most important source for foreign direct investment in Indonesia (even investment by Nigeria and Liberia was only marginally less than the Taiwanese share at that time (Thee, 1991)); by 1992 Japan's trade surplus with the West Pacific Rim surpassed its surplus with the United States; and in 1993 the West Pacific Rim overtook Europe as the second most important destination for Japanese bank loans.

The Overseas Chinese may be better placed than many others to operate effectively in China, but it seems unlikely that China will limit its opportunities to the Chinese overseas. Although Overseas Chinese provide most foreign direct investment at present, China has no illusions about their motivations or their economic significance. Hong Kong Chinese and Taiwanese merchants are doing no more than making good use of their opportunities to expand their operations and increase their profits. Hong Kong is shortly to be handed back to China, and Taiwan is unlikely to tie its economy to that of the mainland, and neither, while influential economically, can match the economic, political and military powers of Japan, the United States and Europe nor the opportunities which these members of the old industrialised world offer.

From the point of view of Overseas Chinese, too, China is another emerging market which offers both opportunities and risks: to concentrate both opportunities and risk in one place is unlikely to prove a popular or wise strategy. A greater realism has been introduced by unpaid bills and broken contracts; by piracy (behaviour which, it has long been recognised in Chinese-language sources, damages businesses (FDJYZ, 1987)); by uncertain attitudes towards foreign currencies and their conversion; by the absence of cooperation between local and central government; by the administration's inability or refusal to insulate decisions on economic matters from its own factional interests; and by the unreliability of bureaucrats and factory managers of state-owned and collective-owned enterprises who treat foreign investment as a status symbol, and who will promise anything to get it (GDQB, 16 April 1993, p. 3). New regulations, procedures and laws are drawn up and implemented without warning, and there are sudden and unpredictable changes in policy.

Some of these actions are designed to enable more control to be exerted over the economy by the central authorities in Beijing; some appear to be no more than expressions of the paranoia that closed institutions and ignorance breed; and some are heavy-handed reactions by officials attempting to mask their unwillingness, or their own inability, to handle or accept the complexities of a market economy. Foreign merchants are required to raise local funds if their factories are to produce goods for the domestic market. Yet this is made difficult by a tightening of the money supply, the necessity for which is blamed on almost everything (including the inflow of foreign currency) except the structural weakness of the economic system. Customs duties on televisions, videos, fridges, air conditioners, telephones, copying and fax machines, and general office equipment (goods which are also subject to VAT of 17%) have been raised, adding still more to the cost of business. Parts and supplies are difficult to obtain or expensive since foreign merchants must cover exorbitant charges for transport—just one of the many hidden costs which are conjured up at will by officials. The stock markets at Shenzhen and Shanghai have been described as 'experimental'. The central authorities have said they will not become involved in disputes between foreign creditors and those state enterprises which are refusing to meet their debts while at the same time diverting liquid assets into property speculation. Laws on copyright infringement have not been enforced properly, though it has been possible for merchants to buy official help to close down pirate operations. Limitations on foreign ownership still exist, and the rates of return on equity are subject to constant questioning by officials worried that foreigners are making too much money out of China. Beijing has been prepared to interfere with the commercial decisions of Japanese companies because Japan invited representatives of the Taiwanese government to the Asian Games. The long-standing threat to use force if Taiwan should formally declare its *de facto* independence is still reiterated even if the wider context of such threats may at times be more conciliatory. And it would appear that key components of Hong Kong's Basic Law which would help maintain the necessary institutional and legal framework for Hong Kong's continuation as a trade settlement will simply be ignored.

'Being Chinese', then, would appear to be no defence. Overseas Chinese merchants such as Malaysia's Guo Henian, Thailand's Xie Guomin, Hong Kong's Li Jiacheng and Taiwan's Cai Wanlin are treated with considerable respect in some quarters, their financial worth and the size of their investments in China competitively ranked, and their donations to their home towns or to some other worthy cause well publicised (*GDQB*, 2 November 1993, p. 1; *ZJRB*, 25 March 1992, p. 5). And it is often repeated in Chinese-language materials that Overseas Chinese, particularly from Hong Kong and Macao, understand China's concerns and problems better than others. But these same actions and perceptions may exacerbate tensions, for often it appears that Overseas Chinese are being asked to make additional contributions and allowances for the privilege of being in China and because it is felt that their very Chineseness gives Mainland Chinese a moral claim on their generosity and patience. Hong Kong entrepreneurs have been asked to cover accommodation, food and the entertainment costs of their workers during national holidays to prevent them from migrating. High-profile entrepreneurs such as Gordon Wu and Li Jiacheng have run foul of the authorities in spite of (or perhaps because of) their carefully nurtured relationships with political elites. Injudicious comments about the personality and intellectual competence of China's leadership have been enough to have other merchants declared *persona non grata*. And some commercial disputes between Mainland Chinese and Overseas Chinese have ended in violence, abductions or life imprisonment. Some Overseas Chinese not only find it difficult to operate in China; in some cases they have come to the conclusion that the effort is simply not worth the trouble.

*   *   *   *   *

Internal political squabbles and xenophobic, inexperienced, ignorant and parochial administrators obsessed with their own power and prestige are not peculiar to China. But so often the impression given is that foreign merchants are being used for the short-term convenience of local or central authorities. It is no wonder, then, that American and European commitments to China have been comparatively limited. Perhaps most interestingly, it seems as if the Japanese have only been testing the waters. It may be that the Japanese perceive that their market share is partly represented by Hong Kong and Taiwanese merchants whose manufacturing in China is often based upon Japanese machines, equipment and semi-processed materials (Ho and Kueh, 1993; Kueh, 1992). It might also be that only a few are convinced that China is worth substantial long-term investments. Popular lore has it that China's relationships with the outside world are conducted in a sophisticated and subtle manner. Yet so often China's leaders seem to behave rather like men emerging from years of social isolation, demanding both attention and obedience. It is unlikely that the Republicans, who now dominate the United States Congress, and who appear markedly unimpressed with Sinologists' over-intellectualised and romanticised interpretations of Mainland China and with notions of specifically 'Asian' forms of capitalism determined by 'Asian' cultures, will be sympathetic towards China's attempts to centre itself according to its own terms on the world stage.

This impatience and harder realism may reflect similar and growing sentiments among other governments and merchants. For the time being, merchants will continue to find China attractive despite the risks, although they will perhaps be a little more cautious. Foreign investment appears to be slowing, and may even have begun to decline—a trend which may be strengthened by the central authorities' desire to curb the power of local governments to attract foreign merchants. Actual foreign investment in 1994 rose by 11% to US$28.8 billion, but contracted foreign investment fell by almost half to US$69 billion from US$122.7 billion in 1993. At the same time, interest and commitments in other countries in the West Pacific Rim are growing. Indonesia, which allows 100% foreign ownership of local operations for the first 15 years and minimal divestment thereafter, is one of the largest recipients of Japanese foreign direct investment in the West Pacific Rim, and has been receiving more attention from Taiwanese and Hong Kong Chinese merchants—including some, like Gordon Wu, who found China too unpredictable. The Philippines, too, which continues to liberalise the economy, to change attitudes towards foreign investment and to achieve very respectable levels of economic growth, is becoming more attractive to foreign merchants, especially American, Japanese, Malaysian (both Chinese and non-Chinese) and Taiwanese. Trade between Taiwan and Malaysia has now risen to US$4.55 billion, and Taiwanese investments in Singapore are also increasing. Of equal significance has been growing interest and commitments in the Indian subcontinent. It is true that India has its own problems with state enterprises, that it is also subject to political upsets, that there are Hindu-revivalist parties who are hostile to foreign trade (for there are those who still believe trade is simply a way of exploiting the Third World), and that by the end of 1994 actual foreign investment in India had slowed to reach an accumulated total of only US$6.5 billion. Nevertheless, even the Hindu-revivalists are not unified in their opposition to the free market, and, in general, opposition parties appear to be embracing reform. Moreover, India has a much stronger, more transparent and mature institutional framework: democracy is firmly entrenched; it has a legal system adapted from the British, and courts that will enforce the law; there is a somewhat chaotic but long-established and technically modern financial system; there is a large and skilled labour force; and there is a comparatively prosperous middle class numbering some 200 million. Not surprisingly, India is now seen by many investors, such as Daimler-Benz, to represent a market as important as that of China.

*   *   *   *   *

There are two main issues which may affect the relationship between Overseas Chinese and China. The first—the extent to which, and the manner in which, China institutionalises trade—has little to do with 'culture' or 'ethnicity'. Neither China nor Overseas Chinese will restrict themselves to the opportunities and markets provided by the other. If China fails in its attempts to establish a stable economy dominated by trade, then the Overseas Chinese will no longer find their 'Chineseness' convenient, and they will look elsewhere. If, on the other hand, China is able to create a sound institutional pattern, to distance reciprocity and its soft-framed institutions from trade, and allows the evolution of values and attitudes sympathetic to the prosecution of trade, China may feel its commercial and political interests lie with Japan, South Korea, the United States, the Southeast Asian political elites and perhaps, some years from now, with India. And if a more self-confident, sophisticated and mature Chinese state is to be integrated economically and politically with the rest of the world, then the Overseas Chinese may, as Lee Kuan Yew believes, eventually find their own trading empires dwarfed by Japanese, North American, European and Asian businesses.

The second issue, however, is China's long-term political and military ambitions. More specifically, is sustained material progress viewed by China's leaders as an end in itself, or as a means to other ends—political and military, regional and global? Here the cultural explanation of Overseas Chinese economic success may have some relevance, not because it is able to provide an accurate and objective analysis, but precisely because its unjustified and unproven assumptions and its pretence at 'science' can be used to legitimise feelings and notions of superiority, or to give credence to fears of the inevitability of Chinese economic and political hegemony in the West Pacific Rim.

China lays claim to the entire South China Sea to within 50 or 80 kilometres of the Vietnamese, Malaysian, Philippine and Brunei coastlines. China is unwilling to discuss the basis of its claim, although it is prepared to enter into bilateral negotiations on the development of oil and other marine resources with each of the nations concerned—Vietnam, Malaysia, the Philippines, Brunei and Taiwan. China's position is strengthened by the failure of other nations to agree on a common stand and to confront China directly on the issue of sovereignty, and by its military bases on the Spratly Islands which are claimed by the same six nations—China, the Philippines, Malaysia, Taiwan, Brunei and Vietnam. The motivation for China's reach into the South China Sea may largely be economic and, as Garver (1992) suggests, it might also stem from internal parochial considerations: Hainan province would gain financially by the exploitation of oil and other marine resources by foreign companies, and the *de facto* control of the islands provides further justification for China's navy to strengthen itself. The desire for an aircraft carrier battle group has been justified in China partly by criticism of the other countries' claims, and by the need to protect the Spratly's maritime territory. China's extension into the South China Sea might also indicate much broader strategic designs. Strengthened by the reacquisition of Hong Kong in the north, a strong military presence in the South China Sea would place China right in the middle of the prosperous nations of Southeast Asia, and would leave China with the potential to control sea and air routes along both axes, from America to India, and from Japan to Australia.

There are also concerns about the extension of trade routes from China to Burma. The import and export trade between Burma and the Dezhong Daizu and Jingpozu Autonomous Prefecture in which the town of Ruili is situated, was valued at 1.7 billion yuan in 1992, and constituted the prefecture's most important source of income (*ZHGSSB*, 4 October 1993, p. 6). Taken as a whole, cross-border trade between Yunnan and Burma is valued at around US$800 million each year:[12] border towns, such as Ruili, are growing and transport infrastructure is improving quickly; Chinese merchants from Yunnan and Fujian roam freely as far south as Mandalay, which has a swelling Chinese immigrant population; and military ties between China and Burma are being strengthened. It is believed that electronic surveillance equipment, operated and funded by China, has been set up on Coco Island—a Burmese possession in Indian territory, just to the north of the Andaman Islands which run southeast to the Nicobar Islands, Indonesia and the opening to the Straits of Malacca; and Chinese survey and fishing vessels have been detained by the Indian Coastguard. An alliance with Burma would give China access to Burmese military bases, the Indian Ocean and the Straits of Malacca. Similar fears have been voiced about growing trade between Yunnan and Laos which has also been accompanied by the migration of Chinese, the flow of military hardware, and the establishment of Chinese signals-intelligence stations in the heart of Indochina. Moreover, China continues its military build-up. China's defence budget may now be the second or third largest in the world, and the PLA (People's Liberation Army) runs somewhere between 10 000 and 20 000 companies with combined profits of around US$5 billion—much of which is thought to be spent on the research and development of more sophisticated weaponry. Taken together, China's military build-up, and its movements in the South China Sea, Burma, the Indian Ocean and Laos could be viewed as the start of an attempt to place Southeast Asia (and thus the trade routes so important to Japan, Taiwan and South

Korea) between China's pincers.

On the other hand, such machinations may be driven by genuine fears about China's cohesion and the survival of the party. To a great extent, foreign investment is something of a double-edged sword. Foreign investors have become crucial to China's economy, and it is for this reason that China is very sensitive over its recovery of Hong Kong—the touchstone of China's attitude towards trade. China may also be sensitive for another reason. As Segal (1994) has suggested, trade with other countries could serve to weave parts of China into the interdependent world economy and thereby draw the teeth of the dragon. In Yunnan, too, the operations of Chinese merchants in Burma and Laos may concern Beijing for similar reasons: much of cross-border trade between the Dezhong Daizu and Jingpozu Autonomous Prefecture and Burma is, according to one source, technically illegal (*ZHGSSB*, 4 October 1993, p. 6). Local officials, hostile to private enterprises, make it difficult for merchants to obtain land and credit, and there have been calls to legalise cross-border trade and to improve the system of taxation. And yet drugs flow across the border, often with official complicity. It is not inconceivable that the illegality of cross-border trade provides local officials with the means to extract bribes. As Lintner (1994) notes, 'Beijing should worry as much about Yunnan's proximity to Southeast Asia as the region does about growing Chinese influence' (p. 26).

The strength of China's determination to maintain its cohesion should not be underestimated. There is as yet no effective institutional pattern which can end capriciousness and bring predictability to the conduct of trade, but this does not mean an effective pattern cannot emerge. And if all else fails, an intense nationalism backed by military force may provide the necessary cohesion—a regime into which the notion of Greater China's 'destiny' and the 'scientific' analyses of all things Chinese would fit easily. These same analyses and feelings would also help Beijing present itself as the natural leader of the West Pacific Rim—a distinctly 'Asian' civilisation which owes its economic success, its uniqueness and its oneness, to the historical and cultural influence of an essential 'Chineseness'.

It may be that all such discussions on Beijing's future designs and interests are an expression of an unnecessary and perhaps even damaging preoccupation with China. There are many possibilities which may be realised. At present it would seem that China is prepared to institutionalise trade and to allow material progress for its own sake. If this is the case, then trade with Japan, South Korea, Southeast Asia, the United

States, Europe and perhaps India will greatly increase. Within this free association of businesses, Overseas Chinese will remain indistinguishable in any important regard from other merchant princes of the East. But to focus so exclusively upon what may or may not happen—or upon what should or should not happen—in China, will only re-create the myth of a Golden Cathay or throw up demonic shadows: imaginings which magnify legitimate interests and concerns into blind obsessions and fears made still more vivid by the pretence at 'science' and by a faith in the deterministic force of culture. A jewelled arc of economic powers stretching from India to Vietnam, Thailand, Malaysia, Singapore, Indonesia, the Philippines, South Korea and Japan would constitute the best protection and opportunity, no matter whether China allows the intellectuals' fabricated 'Chineseness' to become a vehicle for aggressive resentments; chooses once again the cold darkness of isolation, the misery of poverty and the hopelessness of chaos; or finally and irrevocably opens itself to the world's trade winds and the new life which they offer.

Notes

1. The term private traders is used in this book to refer to all individual and independent organisations engaged in trade within all economic sectors in urban and rural areas. In the Chinese-language literature private traders commonly fall into three categories: (i) private trade organisations, which include those with eight or more employees (these are often referred to as *private enterprises*); and organisations with less than eight employees and often based upon the family, but which may comprise just one or two individuals; (ii) farmers engaged in trade with the non-farming population; and (iii) individuals and organisations operating inside *free markets*.

Private traders participate in activities which may be broadly termed industrial and commercial. However, there are several other categories which are sometimes regarded as subcategories and sometimes regarded as additional categories. *Free markets* (or *country fairs*) may also encompass other classifications of market-places which may sometimes be regarded as subclassifications and sometimes as additional classifications. In some cases a distinction is also made between urban and rural free markets; in other instances country fair trade is understood to include trade between farmers and non-farmers, whereas statistics on sales by farmers to non-farmers usually exclude country fair trade.

2. The definitional points noted above are of more than purely semantic interest. They have implications for the monitoring and control of pri-

vate traders; and they help emphasise that published statistics—on, for instance, the number of private traders or the value of their sales—should be viewed with a great deal of caution. The figures given in the text for the number of private trade organisations and, more generally, for private traders are probably far too low. It is impossible to know if the different statistical categories of private traders, types of activity, types of goods and places of operation are mutually exclusive or if they overlap to some extent. Nor is it known if the interaction of state, collective and private units influences or is excluded from these figures. For example, state and free market-places are often situated next to each other or even merge; and representatives of state and collective units operate in the free markets, and private traders in the state markets. Moreover, published estimates for the number of private trade organisations tend to be based on the number of licences issued. Yet many private traders operate without licences or register their organisations improperly as collectives. It is also worth noting that many tens of millions of people in China are unemployed and effectively itinerant labourers. Figures that are available for the number of private traders, then, are indicative of trends, and should be interpreted as such.

3. Questions surrounding the existence of a structured geographical pattern and the value of such a concept to an analysis of trade and markets are discussed elsewhere (R.N.W. Hodder, 1993b).

4. See also Kraus (1991).

5. Yet another has been the debate on 'new authoritarianism'. See White (1994).

6. Predictably, some academics in Hong Kong have been quick to follow with their own variation of this particular brand of cultural one-upmanship. Confucianism, it is declared, advocated private enterprise and minimum government long before the appearance of 'Western' economists and the American constitution (see, for instance, Chang Kuo-sin, 1995).

7. See also Pearson (1994).

8. See also Chapter 4.

9. Or perhaps Wang has in mind Freedman's (1979) belief that the Chinese Diaspora is an illusion that will disappear in time?

10. See also Xiang Dayou (1993).

11. See, for example, *ZHGSSB*, 8 September 1993, p. 6.

12. M. Smith (1991) estimates that the value of legal trade between China and Burma was US$1.5 billion in 1988. By 1991, China accounted for 40% of Burma's foreign trade; much of it is modern weaponry.

Bibliography and historical maps

Amer, R. (1992a) 'Vietnam', in Minority Rights Group (ed.) *The Chinese of Southeast Asia*, Manchester Free Press, Manchester, pp. 26–28.

—— (1992b) 'Cambodia', *ibid.*, pp. 28–30.

—— (1992c) 'Laos', *ibid.*, p. 31.

Amoyen, N.D., Angeles, T.N. and Lacuesta, M.C. (1992) 'A profile of the urban poor in Davao City', *Tambara*, XI, pp. 95–124.

Amyot, J. (1973) *The Manila Chinese: familism in the Philippine environment*, Institute of Philippine Culture (IPC) Monograph No. 2, IPC, Manila.

An Zhiwen (1985) 'Chongkai chengmen, chongfen fahui chengshi de gongneng', *Hong Qi (Red Flag)*, 14, pp. 18–21.

Armstrong, D.M. (1973) *Belief, Truth and Knowledge*, Cambridge University Press, Cambridge.

Ash, R.F. and Kueh, Y.Y. (1993) 'Economic integration within Greater China: trade and investment flows between China, Hong Kong and Taiwan', *The China Quarterly*, 136, pp. 711–745.

Aye, E.U. (1967) *Old Calabar through the Centuries*, Hope Waddell Press, Calabar.

Bagehot, W. (1892) *Lombard Street*, Kegan Paul, Truber and Co., London.

Baker, H.D.R. (1966) 'The five great clans of the New Territories', *Journal of the Hong Kong Branch of the Royal Asiatic Society*, 6, pp. 25–47.

—— (1968) *A Chinese Lineage Village: Sheung Shui*, Stanford University Press, Stanford.

Banton, M. (ed.) (1965) *The Relevance of Models for Social Anthropology*, Tavistock, London.

Barnett, M.L. (1960) 'Kinship as a factor affecting Cantonese economic adaptation in the United States', *Human Organisation*, 19, pp. 40–46.

Barton, C.A. (1983) 'Trust and credit: some observations regarding business strategies of Overseas Chinese traders in South Vietnam', in L. Lim and P. Gosling (eds) *The Chinese in Southeast Asia*, Volume I, Maruzen Asia, Singapore, pp. 46–64.

---

* Reference numbers for those maps held by the Hong Kong Government, Lands Department.

Bauer, P.T. (1991) *The Development Frontier*, Harvester Wheatsheaf, London.

Bauer, P.T. and Yamey, B.S. (1957) *The Economics of Under-developed Countries*, Cambridge University Press, Cambridge.

Belcher, E., Cpt. Sir (1841) *Victoria Harbour in 1841* (HK 19)*

Benedict, R. (1946) *The Chrysanthemum and the Sword*, Meridian, New York.

Berger, P. (1988) 'An East Asia development model', in P. Berger and M.H. Hsiao (eds) *In Search of an East Asian Development Model*, Transaction Books, New Brunswick, New Jersey, pp. 3–11.

Berger, P. and Hsiao, M.H. (1988) *In Search of an East Asian Development Model*, Transaction Books, New Brunswick, New Jersey.

Blaker, J.R. (1965) 'The Chinese newspaper in the Philippines: toward the definition of a tool', *Asian Studies*, 3, pp. 243–261.

Blau, P.M. (1964) *Exchange and Power in Social Life*, John Wiley and Sons, New York.

Bond, M.H. (ed.) (1986) *The Psychology of the Chinese People*, Oxford University Press, Hong Kong.

Bond, M.H. and Cheung, T.S. (1983) 'The spontaneous self-concept of college students in Hong Kong, Japan and the United States', *Journal of Cross-cultural Psychology*, 4, pp. 153–171.

Bond, M.H. and Hofstede, G. (1990) 'The cash value of Confucian values', in S.R. Clegg and S.G. Redding (eds) *Capitalism in Contrasting Cultures*, Walter de Gruyter, Berlin, pp. 383–390.

Bond, M.H. and Hwang, K.K. (1986) 'The social psychology of the Chinese people', in M.H. Bond (ed.) *The Psychology of the Chinese People*, Oxford University Press, Hong Kong, pp. 213–266.

Bond, M.H. and Lee, P.W.H. (1984) 'Face-saving in Chinese culture: a discussion and experimental study of Hong Kong students', in A.Y.C. King and R.P.L. Lee (eds) *Social Life and Development in Hong Kong*, the Chinese University Press, Hong Kong, pp. 289–304.

Bondi, L. and Domosh, M. (1994) 'Editorial', *Gender, Place and Culture*, 1, pp. 3–4.

Bottomore, T.B. (1964) *Elites and Society*, Penguin, Harmondsworth.

—— (1972) *Sociology*, Allen and Unwin, London.

Bourdieu, P. (1977) *Outline of a Theory of Practice*, Cambridge University Press, Cambridge (trans. R. Nice).

Bristow, R. (1987) *Land-use Planning in Hong Kong*, Oxford University Press, Hong Kong.

British Insurance Company (1850) *Copy of Plan of Victoria* (HG 20/1)*

Broadbridge, S. (1966) *Industrial Dualism in Japan*, Frank Cass and Co., London.

Bullock, A. (1955) *Men, Chance and History*, The Essex Hall Lecture, 1955, Lindsey Press, London.

Burnett, A. (1993) *The Western Pacific: the challenge of sustainable growth*, Earthscan, London.

Carino, T. (ed.) (1985) *Chinese in the Philippines*, Chinese Studies Programme, De La Salle University, Manila.

Carrithers, M. (1992) *Why Humans Have Cultures: explaining anthropology and social diversity*, Oxford University Press, Oxford.

Carrithers, M., Collins, S. and Lukes, S. (eds) (1985) *The Category of the Person: anthropology, philosophy, history*, Cambridge University Press, Cambridge.

Carstens, S.A. (1975) *Chinese Associations in Singapore Society*, Institute of Southeast Asian Studies, Singapore.

—— (1983) 'Pulai Hakka Chinese Malaysians: a labyrinth of cultural identities', in L. Lim and L. Gosling (eds) *The Chinese in Southeast Asia*, Volume II, Maruzen Asia, Singapore, pp. 79–98.

Chan, Wing-tsit (1946) 'The spirit of Oriental philosophy', in C.A. Moore (ed.) *Philosophy—East and West*, Princeton University Press, New Jersey, pp. 137–167.

—— (1967) 'Syntheses in Chinese metaphysics', in C.A. Moore (ed.) *The Chinese Mind: essentials of Chinese philosophy and culture*, East-West Center Press, Honolulu, pp. 132–147.

Chang Kuo-sin (1995) 'Confucianist teaching is older than government itself', *Hong Kong Standard*, February 6, 1995, p. 15.

Chang, M.H. (1992) 'China's future: regionalism, federation or disintegration', *Studies in Comparative Communism*, 25, pp. 211–227.

Chang Xiuze (1990) 'Zhongguo guojia suoyouzhi qiye zhidu gaige lungao', *Nankai Jingji Yanjiu (Nankai Economic Research)*, 5, pp. 3–13.

Chang Yunshik (1991) 'The personalistic ethic and the market in Korea', *Comparative Studies in Society and History*, 33, pp. 106–129.

Chapman, R.A. (1969) *Decision Making: a case study of the decision to raise the bank rate in September 1957*, Routledge and Kegan Paul, London.

Chao, K. (1977) *The Development of Cotton Textile Production in China*, Harvard University Press, Cambridge.

Cheng Lim-keak (1985) *Social Change and the Chinese in Singapore: a socio-economic geography with special reference to the bang structure*, Singapore University Press, Singapore.

Clegg, S.R. and Redding, S.G. (eds) (1990) *Capitalism in Contrasting Cultures*, Walter de Gruyter, Berlin.

Clifford, M. (1994) 'Heir force', *Far Eastern Economic Review*, 17 November, pp. 78–79.

Cohen, A. (1974) *Two-dimensional Man*, University of California Press, Berkeley.

—— (1981) *The Politics of Elite Culture*, University of California Press, Berkeley.

Collinson, R.E., Ltn. (1845) *The Ordnance Map of Hong Kong* (HE 4).*

Corciño, E. (1945) *Davao, its Land and People*, Souvenir Book of Davao Fair and Exposition, Davao.

Corner, E.J.H. (1981) *The Marquis: a tale of Syonan-to*, Heinemann Asia, Singapore.

Coughlin, R.J. (1960) *Double Identity: the Chinese in modern Thailand*, Hong Kong.

Cragg, C. (1995) *The New Taipans*, Century Business Books, Random House, London.

Crissman, L.W. (1967) 'The segmentary structure of urban overseas Chinese communities', *Man*, 2, pp. 185–204.

—— (1973) 'Town and countryside: central-place theory and Chinese marketing systems with particular reference to southwestern Changhua *xian*, Taiwan', Ph.D. thesis, Cornell University.

Cushman, J.W. (1991) *Family and State: the formation of a Sino-Thai tin-mining dynasty, 1797–1932*, Oxford University Press, Singapore.

Cushman, J. and Wang Gungwu (eds) (1988) *Changing Identities of the Southeast Asian Chinese since World War II*, Hong Kong University Press, Hong Kong.

Dalu Siying Qiye Jingxuan, 1991, Jingji Kexue Chubanshe, Beijing.

Darby, H.C. (1947) *The Theory and Practice of Geography*, An Inaugural Lecture, The University Press of Liverpool, Hodder and Stoughton, London.

Davis, S.G. (1949) *Hong Kong in its Geographical Setting*, Collins, London.

Davis, W.G. (1973) *Social Relations in a Philippine Market*, University of California Press, Berkeley.

Davao City Government (1993) *Development Framework of Davao City, 1993–95*, Davao.

DeGlopper, D.R. (1972) 'Doing business in Lukang', in W.E. Willmott (ed.) *Economic Organisation in Chinese Society*, Stanford University Press, California, pp. 297–326.

Dewey, A. (1962) *Peasant Marketing in Java*, Glencoe, Free Press.

Deyo, F. (1983) 'Chinese management practices and work commitment in comparative perspective', in L. Lim and L. Gosling (eds) *The Chinese in Southeast Asia*, Volume II, Maruzen Asia, Singapore, pp. 215–230.

Dixon, C. (1991) *Southeast Asia in the World Economy*, Cambridge University Press, Cambridge.

Dixon, C. and Drakakis-Smith, D. (eds) (1993) *Economic and Social Development in Pacific Asia*, Routledge, London.

Donnithorne, A. (1983) 'Hong Kong as an economic model for the great cities of China', in A.J. Youngson (ed.) *China and Hong Kong: the economic nexus*, Oxford University Press, Hong Kong, pp. 282–310.

Drummond, L. (1983) 'Jonestown: a study in ethnographic discourse', *Semiotica*, 46, pp. 167–209.

Dumont, L. (1970) *Homo Hierarchicus*, University of Chicago Press, Chicago.

East Asia Analytical Unit (1995) *Overseas Chinese Business Networks in Asia*, Department of Foreign Affairs and Trade, Canberra.

Economist, The (1994a) 'Asian values', 28 May, pp. 9–10.

—— (1994b) 'Democracy and growth', 27 August, pp. 15–17.

—— (1994c) 'The Chinese takeover of Hong Kong Inc.', 7 May, pp. 27–28.

Eitzen, D.S. (1968) 'Two minorities: the Jews of Poland and the Chinese of the Philippines', *Jewish Journal of Sociology*, 10, pp. 221–240.

Elster, J. and Moene, K. (eds) (1989) *Alternatives to Capitalism*, Cambridge University Press, Cambridge.

Elvin, M. (1972) 'The high-level equilibrium trap: the causes of decline of invention in the traditional Chinese textile industries', in W.E. Willmott (ed.) *Economic Organisation in Chinese Society*, Stanford University Press, California, pp. 137–172.

Emerson, R. (1937) *Malaysia: a study in direct and indirect rule*, Macmillan, New York.

Endacott, G.B. (1958) *A History of Hong Kong*, Oxford University Press, London.

Etzioni, A. (ed.) (1969) *A Sociological Reader on Complex Organisation*, Holt, Rinehart and Winston, New York (2nd edn).

Evans, D.M.E. (1975) 'The foundations of Hong Kong: a chapter of accidents', in M. Topley (ed.) *Hong Kong: the interaction of tradition and life in the towns*, Hong Kong Branch of the Royal Asiatic Society, Hong Kong, pp. 11–41.

Evans-Pritchard, E.E. (1956) *Nuer Religion*, Oxford University Press, Oxford.

—— (1963) *The Comparative Method in Social Anthropology*, L.T. Hobhouse Memorial Trust Lecture, No. 33, Athlone Press, University of London.

Faure, D. (1984a) 'The Tangs of Kam Tin—a hypothesis on the rise of a gentry family', in D. Faure, J. Hayes and A. Birch (eds) *From Village to City: studies in the traditional roots of Hong Kong society*, Centre of Asian Studies, University of Hong Kong, Hong Kong, pp. 24–42.

—— (1984b) 'Notes on the history of Tsuen Wan', *Journal of the Hong Kong Branch of the Royal Asiatic Society*, 24, pp. 46–104.

Faure, D., Hayes, J. and Birch, A. (eds) (1984) *From Village to City: studies in the traditional roots of Hong Kong society*, Centre of Asian Studies, University of Hong Kong, Hong Kong.

Fei, X.T. (1948) *Rural China*, Shanghai.

Felix, A. Jnr (ed.) (1969) *The Chinese in the Philippines, 1770–1898* (2 vols), Historical Conservation Society, Solidaridad Publishing House, Manila.

Ferris, P. (1960) *The City*, Penguin, Harmondsworth.

Firth, R. (1956) *Elements of Social Organisation*, Watts and Co., London.

Foronda, M.A. Jnr (1985) 'The Chinese in the Philippines: an oral history approach', in T. Carino (ed.) *Chinese in the Philippines*, Chinese Studies Programme, De La Salle University, Manila, pp. 74–81.

Fortes, M. (1967) *The Web of Kinship among the Tallensi (Second Part)*, Oxford University Press, London.

Foster, B.L. (1982) *Commerce and Ethnic Differences: the case of the Mons in Thailand*, Papers in International Studies, Southeast Asia series No. 59, Athens, Ohio.

Franke, W. (1972) *The Reform and Abolition of the Traditional Chinese Examination System*, Harvard East Asian Monographs, 10, Harvard University Press, Cambridge, Mass.

—— (1989) *Sino-Malaysiana: selected papers on Ming and Qing history and on the Overseas Chinese in Southeast Asia, 1942–1988*, South Seas Society, Singapore.

Freedman, M. (1957) *Chinese Family and Marriage in Singapore*, Colonial Office, Colonial Research Studies No. 20, Her Majesty's Stationery Office, London.

—— (1958) *Lineage Organisation in Southeast China*, London School of Economics, Monograph on Social Anthropology, No. 18, Athlone Press, University of London.

—— (1959) 'The handling of money: a note on the background to the economic sophistication of Overseas Chinese', *Man*, April, pp. 64–65.

—— (1960a) 'Immigrants and associations: Chinese in nineteenth-century Singapore', *Comparative Studies in Society and History*, 3, pp. 25–48.

—— (1960b) 'Overseas Chinese associations: a comment', *Comparative Studies in Society and History*, 3, pp. 478–480.

—— (1966) *Chinese Lineage and Society: Fukien and Kwangtung*, London School of Economics, Monograph on Social Anthropology, No. 33, Athlone Press, University of London.

—— (ed.) (1970a) *Family and Kinship in Chinese Society*, Stanford University Press, Stanford, California.

—— (1970b) 'Ritual aspects of Chinese kinship and marriage', in M. Freedman (ed.) *Family and Kinship in Chinese Society*, Stanford University Press, Stanford, California, pp. 163–187.

—— (1979a) *The Study of Chinese Society: essays by Maurice Freedman*, Stanford University Press, Stanford, California (selected and introduced by G.W. Skinner).

—— (1979b) 'The Chinese in Southeast Asia: an epicycle of Cathay', *ibid.*, pp. 39–57.

—— (1979c) 'The Chinese in Southeast Asia: a longer view', *ibid.*, pp. 3–21.

Fried, M.H. (1953) *Fabric of Chinese Society: a study of the social life of a Chinese county seat*, Octagon Books, New York (reprint 1969).

Friedman, E. (1989) 'Modernisation and democratisation in Leninist states: the case of China', *Studies in Comparative Communism*, 22, pp. 251–264.

Fudan Daxue Jingji Yanjiu Zhongxin (1987) 'Shanghai gongye qiye hengxiang jingji lianhe diaocha baogao', *Gongye Jingji (Industrial Economics)*, 8, pp. 101–110.

Fujian Duiwai Jingmao (Fujian Foreign Economic Relations and Trade), Fujian.

Fujian Jingji Bao (Fujian Economic Daily), Fujian.

Fok, K.C. (1990) *Lectures on Hong Kong History*, The Commercial Press, Hong Kong.

Fong, M.K.L. (1979) 'Directions in urban growth in Hong Kong', in C.K. Leung, J.W. Cushman and G.W. Wang (eds) *Hong Kong: dilemmas of growth*, Centre of Asian Studies Occasional Papers and Monographs, No. 45, Australian University and University of Hong Kong, Canberra, pp. 267–287.

Gao Shangquan (1984) 'Dapo tiaokuai fenge fazhan hengxiang jingji lianxi', *Jingji Yanjiu (Economic Research)*, 11, pp. 3–8.

Gardner, H. (1983) *Frames of Mind: the theory of multiple intelligences*, Basic Books, New York.

Garver, J.W. (1992) 'China's push through the South China Sea: the interaction of bureaucratic and national interests', *The China Quarterly*, 132, pp. 999–1028.

Geertz, C. (1962) 'The rotating credit association: a middle rung in development', *Economic Development and Cultural Change*, 10, pp. 241–264.

—— (1963) *Peddlers and Princes: social change and economic modernisation in two Indonesian towns*, University of Chicago Press.

—— (1973) *The Interpretation of Cultures*, Basic Books, New York.

Gereffi, G. (1989) 'Developmental strategies and the global factory', in P.A. Gourevitch (ed.) 'The Pacific Region: challenges to policy and theory', Special Edition, *Annals of the American Academy of Political and Social Science*, 505 (September), pp. 92–104.

Gerth, H.M. and Wright, M.C. (eds and trans.) (1947) *From Max Weber: essays in socialism*, Kegan Paul, London.

Giddens, A. (1987) *Social Theory and Sociology*, Stanford University Press, Stanford, California.

Goffman, E. (1955) 'On face-work: an analysis of ritual elements in social interaction', *Psychiatry*, 18, pp. 213–231.

—— (1959) *The Presentation of Self in Everyday Life*, Doubleday, New York.

Gold, T.B. (1989) 'Urban private business in China', *Studies in Comparative Communism*, 22, pp. 187–201.

Goldberg, M.A. (1985) *The Chinese Connection: getting plugged in to Pacific Rim real estate, trade and capital markets*, University of British Columbia Press, Vancouver.

Goldstein, C. (1994) 'Activist unbowed', *Far Eastern Economic Review*, September 15, pp. 25–26.

—— 'The morning after', *ibid.*, pp. 26–27.

—— 'Innocents abroad', *ibid.*, pp. 22–24.

Gongshang Guanliju (Tianjin) (1990) *Guanyu Cujin Woshi Geti Siying Jingji Fazhan de Yijian*, Tianjin.

Gosling, L. (1983a) 'Chinese crop dealers in Malaysia and Thailand: the myth of the merciless monopsonistic middleman', in L. Lim and L. Gosling (eds) *The Chinese in Southeast Asia*, Volume I, Maruzen Asia, Singapore, pp. 131–170.

—— (1983b) 'Changing Chinese identities in Southeast Asia: an introductory review', in L. Lim and L. Gosling (eds) *The Chinese in Southeast Asia*, Volume II, Maruzen Asia, Singapore, pp. 1–14.

Gourevitch, P.A. (1989) (ed.) 'The Pacific Region: challenges to policy and theory', Special Edition, *Annals of the American Academy of Political and Social Science*, 505 (September).

Graham, A.C. (1988) 'Confucianism', in R.C. Zaehner (ed.) *The Hutchinson Encyclopaedia of Living Faiths*, Hutchinson, London, pp. 357–373.

Grassby, R. (1970) 'English merchant capitalism in the late seventeenth century', *Past and Present*, 46, pp. 87–107.

Greenblatt, S.L., Wilson, R.W. and Wilson, A.A. (eds) (1979) *Organisational Behaviour in Chinese Society*, Praeger, New York.

—— (1982) *Social Interaction in Chinese Society*, Praeger, New York.

Gregory, C.A. (1982) *Gifts and Commodities*, Academic Press, London.

Grimm, T. (1985) 'State and power in juxtaposition: an assessment of Ming despotism', in S. Schram (ed.) *The Scope of State Power in China*, European Science Foundation, School of Oriental and African Studies, and the Chinese University of Hong Kong, Hong Kong, pp. 27–50.

Groves, R.G. (1964) 'The origins of two market towns in the New Territories', *Aspects of Social Organisation in the New Territories*, Weekend Symposium, 9–10 May, Royal Asiatic Society, Hong Kong Branch, Cathay, Hong Kong.

Guangdong Duiwai Jingmao (Guangdong Foreign Economic Relations and Trade), Guangdong.

Guangdong Qiaobao (Guangdong Overseas Chinese Daily), Guangdong.

Gullick, J.M. (1958) *Indigenous Political Systems of Western Malaya*, London School of Economics, Monographs on Social Anthropology, No. 17, Athlone Press, University of London.

Guojia Gongshang Xingzheng Guanliju Geti Jingjisi (1987) *Geti Gongshang Jiben Qingkuang Tongji*, Beijing.

Hacking, I. (1983) *Representing and Intervening: introductory topics in the philosophy of natural science*, Cambridge University Press, Cambridge.

Hafner, J.A. (1983) 'Market gardening in Thailand: the origins of an ethnic Chinese monopoly', in L. Lim and L. Gosling (eds) *The Chinese in Southeast Asia*, Volume I, Maruzen Asia, Singapore, pp. 30–45.

Hall, E.T. (1959) *The Silent Language*, Doubleday, New York.

—— (1976) *Beyond Culture*, Anchor Press, New York.

Hallowell, A.I. (1955) *Culture and Experience*, University of Pennsylvania Press, Philadelphia.

Harrell, S. (1982) *Ploughshare Village: culture and context in Taiwan*, University of Washington Press.

Hayes, J. (1984) 'Hong Kong Island before 1841', *Journal of the Hong Kong Branch of the Royal Asiatic Society*, 24, pp. 105–142.

Heng, P.K. (1988) *Chinese Politics in Malaysia: a history of the Malaysian Chinese Association*, Oxford University Press, Singapore.

—— (1992) 'The Chinese business elite of Malaysia', in R. McVey (ed.) *Southeast Asian Capitalists*, Southeast Asia Program, Cornell University, Ithaca, New York, pp. 127–144.

Hennessy, P. (1990) *Whitehall*, Fontana, London.

Herschede, F. (1991) 'Trade between China and ASEAN: the impact of the Pacific Rim era', *Pacific Affairs*, 64, pp. 179–193.

Hicks, G. (1989) 'The Four Little Dragons: an enthusiast's reading guide', *Asian-Pacific Economic Literature*, 3, pp. 35–49.

Hicks, G. and Mackie, J.A.C. (1994) 'A question of identity', *Far Eastern Economic Review*, 14 July, pp. 46–48.

Hicks, G.L. and Redding, S.G. (1983a) 'Uncovering the sources of Southeast Asian economic growth', mimeographed paper.

—— (1983b) 'The story of the East Asian economic miracle', *Euro-Asia Business Review*, 2(3), pp. 24–32 (Part I); (4), pp. 18–22 (Part II).

—— (1983c) 'Culture, causation and Chinese management', Mong Kwok Ping Management Data Bank, Working Paper, Department of Management Studies, University of Hong Kong.

Higgott, R. and Robison, R. (eds) (1985) *Southeast Asia: essays in the political economy of structural change*, Routledge and Kegan Paul, London.

Hirschman, C. (1988) 'Chinese identities in Southeast Asia: alternative perspectives', in J. Cushman and G.W. Wang (eds) *Changing Identities of the Southeast Asian Chinese since World War II*, Hong Kong University Press, Hong Kong, pp. 22–31.

Hirschon, R. (ed.) (1984) *Woman and Property, Woman as Property*, St Martin's Press, London.

Ho, P.T. (1954) 'The salt merchants of Yang Chou: a study of commercial capitalism in eighteenth-century China', *Harvard Journal of Asiatic Studies*, 17, pp. 130–168.

—— (1966) *Zhongguo Huiguan Shilun*, Taipei.

Ho, Y.P. and Kueh, Y.Y. (1993) 'Whither Hong Kong in an open-door, reforming Chinese economy', *The Pacific Review*, 6, pp. 333–351.

Hodder, B.W. (1953) 'Racial groupings in Singapore', *The Malayan Journal of Tropical Geography*, 1, pp. 25–36.

—— (1980) *Economic Development in the Tropics*, Methuen, London (3rd edn).

Hodder, B.W. and Ukwu, U.I. (1969) *Markets in West Africa*, Ibadan University Press, Ibadan.

Hodder, I. (1986) *Reading the Past*, Cambridge University Press, Cambridge.

Hodder, R.N.W. (1991) 'Planning for development in Davao City, the Philippines', *Third World Planning Review*, 13, pp. 105–128.

—— (1992) *The West Pacific Rim: an introduction*, Belhaven Press, London.

—— (1993a) 'Exchange and reform in the economy of Shanghai', *Annals of the Association of American Geographers*, 83, pp. 303–319.

—— (1993b) *The Creation of Wealth in China: domestic trade and material progress in a communist state*, Belhaven, London.

Hofstede, G. (1980) *Culture's Consequences: international differences in work-related values*, Sage, Beverly Hills (abridged edition).

Hong Changshi (1984) 'Shanghai jingjiqu gongye shengchan buju tedian, fazhan yinsu yu cunzai de jige wenti', *Jingji Dili (Economic Geography)*, 3, pp. 185–190.

Hong Kong Government (1859) *Victoria Hong Kong, Population in 1859* (HD 33)*

—— (1859) *A Portion of the City of Victoria, Hong Kong, 1859* (HG 35-3)*

—— (1870) *Colony of Hong Kong, Dependency of British Kowloon, 1870* (HL 53)*

—— (1887) *Plan of Kowloon, 1887* (HG 10)*

—— (1889) *City of Victoria, Hong Kong, 1889* (HG 28-H06)*

—— (1889) *Plan of the City of Victoria, Hong Kong* (HG 28)*

—— (1892) *Kowloon* (HG 11)*

—— (1893) *Tai Kok Tsui Village, 1893* (HL 32)*

—— (1893) *Fuk Tsuen Village, 1893* (HL 31)*

—— (1893) *Fuk Tsun Heung, Leased Lots, 1893* (HL 30)*

—— (1893) *Survey of Mati Village, 1983, South Portion* (HL 38)*

—— (1896) *Key Plan of Tai Kok Tsui Squatters, 1896* (HL 45)*

—— (1896) *Yaumati–Kowloon* (HG 42)*

—— (1910) *City of Victoria, Hong Kong, 1910, District 9 and 7* (HG 31-4/5)*

—— (1920) *Kowloon Peninsula* (HG 1-3, HG 1-1)*

—— (1922) *City of Victoria* (HD 1939)*

Hong Kong Guide, The (1893) Oxford University Press, Hong Kong.

Housing and Urban Development Coordinating Council (1989) *A Shelter Strategy for Region XI*, Davao City.

Hsiao, K.C. (1960) *Rural China: Imperial control in the nineteenth century*, Seattle, Washington.

Hsieh, J. (1977) 'Internal structure and socio-cultural change: a Chinese case in the multi-ethnic society of Singapore', Ph.D thesis, University of Pittsburgh, University Microfilms International, Ann Arbor, Michigan.

Hsieh, Yu-wei (1967) 'Filial piety and Chinese society', in C.A. Moore (ed.) *The Chinese Mind: essentials of Chinese philosophy and culture*, East-West Center Press, Honolulu, pp. 167–187.

Hsu, F.L.K. (1963) *Clan, Caste, and Club*, D. Van Nostrand Company Inc., Princeton, New Jersey.

Hu Bing (1987) 'Shanghai gongye xitong qiye jituan fazhan zhong de jige wenti', *Gongye Jingji (Industrial Economics)*, 8, pp. 111–114.

Hu Chuangnuan (1981) 'Jige xingcheng zhong de yingli wenti', Zhongguo Jiage Xuehui, Bejingshi de Wujiaju, *Jiage Zhuanti Jiangzuo*, pp. 233–257.

Hu, H.C. (1944) 'The Chinese concept of "face"', *American Anthropologist*, 46, pp. 45–64.

—— (1948) *The Common Descent Group in China and its Functions*, Viking Fund Publications in Anthropology, 10, New York.

Huang Dejun (1989) 'Siying qiye: xianzhuang, tedian yu weilai', *Nankai Jingji Yanjiu (Nankai Economic Research)*, 3, pp. 67–72.

Huang, L.C. and Harris, M.B. (1973) 'Conformity in Chinese Americans: a field experiment', *Journal of Cross-cultural Psychology*, 4, pp. 427–434.

Huang Xunping (1992) 'Xinjiapo shichang', *Guangdong Duiwai Jingmao (Guangdong Foreign Trade)*, 6, pp. 6–7.

Hunt, C. and Walker, L. (1974) 'Marginal trading peoples: Chinese in the Philippines and Indians in Kenya', *Ethnic Dynamics*, pp. 93–127.

Huo, Y.P. and Steers, R.M. (1993) 'Cultural influences on the design of incentive systems: the case of East Asia, *Asia–Pacific Journal of Management*, 10, pp. 71–85.

Hurley, R.C. (1920) *Hand Book of the British Crown Colony of Hong Kong and Dependencies*, Kelly and Walsh, Hong Kong.

Hutterer, K.L. (ed.) (1977) *Economic Exchange and Social Interaction in Southeast Asia: perspectives from prehistory, history and ethnography*, Michigan Papers on South and Southeast Asia, No. 13, Ann Arbor, Michigan.

Hwang, K.K. (1983) 'Business organisational patterns and employees' working morale in Taiwan', *Bulletin of the Institute of Ethnology, Academica Sinica*, 56, pp. 85–133.

Inquirer, Manila.

Ishihara, H. *et al.* (1980) *Traditional Periodic Markets and Modern Shopping Centres in Nagoya and Okazaki*, 24th International Geographical Congress, August, Nagoya.

Jacobs, J.B. (1979) 'A preliminary model of particularistic ties in Chinese political alliances: *kan-ch'ing* and *kuan-hsi* in a rural Taiwanese township', *The China Quarterly*, 78, pp. 236–273.

—— (1980) *Local Politics in a Rural Chinese Cultural Setting: a field study of Mazu Township, Taiwan*, Contemporary China Centre, Research School of Pacific Studies, Australian National University, Canberra.

—— (1982) 'The concept of *guanxi* and local politics in a rural Chinese cultural setting', in S.L. Greenblatt, R.W. Wilson and A.A. Wilson (eds) *Social Interaction in Chinese Society*, Praeger, New York, pp. 209–236.

Jayasankaran, S. (1994) 'You gotta believe', *Far Eastern Economic Review*, 1 December, pp. 72–74.

Jingji Ribao (Economic Daily), Beijing.

Jones, S. Mann (1972) 'Finance in ningpo: the "ch'ien chuang", 1750–1880', in W.E. Willmott (ed.) *Economic Organisation in Chinese Society*, Stanford

University Press, California, pp. 47–77.
—— (1974) 'The Ningpo *pang* at Shanghai: the changing organisation of financial power', in G.W. Skinner and M. Elvin (eds) *The Chinese City between Two Worlds*, Stanford University Press, California, pp. 73–96.

Kahn, H. (1979) *World Development*, Croom Helm, London.

Karp, J. (1993) 'The new insiders', *Far Eastern Economic Review*, 27 May, pp. 62–66.

Keswick, M. (ed.) (1982) *The Thistle and the Jade*, Octopus Books, London.

King, A.Y.C. (1987) 'The transformation of Confucianism in the post-Confucian era: the emergence of rationalistic Confucianism in Hong Kong', (unpublished paper).

King, A.Y.C. and Lee, R.P.L. (eds) (1984) *Social Life and Development in Hong Kong*, The Chinese University Press, Hong Kong.

King, A.Y.C. and Leung, D.H.K. (1975) 'The Chinese touch in small industrial organisations', *Working Paper*, Social Research Centre, The Chinese University of Hong Kong.

Kotkin, J. (1993) *Tribes: how race, religion and identity determine success in the new global economy*, Random House, New York.

Kraar, L. (1993) 'The importance of Chinese in Asian business', *Journal of Asian Business*, 9, pp. 87–94.

Kraus, W. (1991) *Private Business in China: revival between ideology and pragmatism*, Hurst and Co., London (trans. E. Holz).

Kroebner, A.L. and Parsons, T. (1958) 'The concept of culture and social systems', *American Sociological Review*, 23, pp. 583–594.

Kueh, Y.Y. (1992) 'Foreign investment and economic change in China', *The China Quarterly*, 131, pp. 637–689.

La Piana, G. (1927) 'Foreign groups in Rome during the first centuries of the empire', *Harvard Theological Review*, 29, pp. 425–464.

Lai, D.C.Y. (1963) 'Some geographical aspects of the industrial development of Hong Kong since 1841', MA thesis (unpublished), The Chinese University of Hong Kong.

Lai, D.C.Y. and Sit, V. (1984) 'Factories in domestic premises: an urban problem in Hong Kong', *Asian Geographer*, 3, pp. 1–14.

Lança, P. (1994) 'Down with the feminazis', *The Salisbury Review*, 12 (4), pp. 21–25.

Landa, J.T. (1983) 'The political economy of the ethnically homogeneous Chinese middleman group in Southeast Asia: ethnicity and entrepreneurship in a plural society', in L. Lim and L. Gosling (eds) *The Chinese in Southeast Asia*, Volume I, Maruzen Asia, Singapore, pp. 86–116.

Lau, S.K. (1978) 'From traditional familism to utilitarianistic familism: the metamorphosis of familial ethos among the Hong Kong Chinese', *Occasional Paper No.78*, Social Research Centre, The Chinese University of Hong Kong.

—— (1982) *Society and Politics in Hong Kong*, The Chinese University Press, Hong Kong.

Lee-in Chen Chiu and Chin Chung (1993) 'An assessment of Taiwan's indirect investment towards Mainland China', *Asian Economic Journal*, 7, pp. 41–70.

Leeming, F. (1975) 'The earlier industrialisation of Hong Kong', *Modern Asian Studies*, 9, pp. 337–342.

—— (1977) *Street Studies in Hong Kong: localities in a Chinese city*, Oxford University Press, Hong Kong.

Leung, C.K. (1993) 'Personal contacts, subcontracting linkages, and development in the Hong Kong–Zhjiang delta region', *Annals of the Association of American Geographers*, 83, pp. 272–302.

Leung, C.K., Cushman, J.W. and Wang G.W. (eds) (1979) *Hong Kong: dilemmas of growth*, Centre of Asian Studies Occasional Papers and Monographs, No.45, Australian University and University of Hong Kong, Canberra.

Levy, M.J. (1949) *The Family Revolution in Modern China*, Harvard University Press, Cambridge.

Levy, M.J. and Shih, K.H. (1949) *The Rise of the Modern Chinese Business Class*, Institute of Pacific Relations, New York.

Liang Yingming (1993) 'Zhongguo gaige kaifang yu dongnan ya huaren', *Huaqiao Huaren Lishi Yanjiu (Overseas Chinese Historical Research)*, 4, pp. 16–21.

Lim, D. (ed.) (1975) *Readings on Malaysian Economic Development*, Oxford University Press, Kuala Lumpur.

Lim Eng-chye (1979) 'Malaysia's New Economic Policy: an analysis of the objectives of restructuring Malaysian society', Graduation Essay, Department of Geography, Nanyang University, Singapore.

Lim, L. (1983a) 'Chinese economic activity in Southeast Asia: an introductory review', in L. Lim and L. Gosling (eds) *The Chinese in Southeast Asia*, Volume I, Maruzen Asia, Singapore, pp. 1–29.

—— (1983b) 'Chinese business, multinationals and the state: manufacturing for export in Malaysia and Singapore', in L. Lim and L. Gosling (eds) *The Chinese in Southeast Asia*, Volume I, Maruzen Asia, Singapore, pp. 245–274.

—— (1992) 'The emergence of a Chinese economic zone in Asia?', *Journal of Southeast Asian Business*, 8, pp. 41–46.

Lim, L. and Gosling, L. (eds) (1983) *The Chinese in Southeast Asia* (2 vols), Maruzen Asia, Singapore.

Lim, M.H. (1981) *Ownership and Control of the One Hundred Largest Corporations in Malaysia*, Oxford University Press, Kuala Lumpur.

—— (1983) 'The ownership and control of large corporations in Malaysia: the role of Chinese businessmen', in L. Lim and L. Gosling (eds) *The Chinese in Southeast Asia*, Volume I, Maruzen Asia, Singapore, pp. 275–315.

Limlingan, V.S. (1986) *The Overseas Chinese in ASEAN: business strategies and management practices*, Vita, Manila.

Lin Jinzhi (1993) '1979–1992 nian haiwai huaren zai zhongguo dalu touzi de xianzhuang jiqi jinhou fazhan qushi', *Huaqiao Huaren Lishi Yanjiu (Overseas Chinese Historical Research)*, 1, pp. 1–14.

Lintner, B. (1994) 'Enter the dragon', *Far Eastern Economic Review*, 22 December, pp. 22–24.

Litterer, J.A. (1965) *The Analysis of Organisations*, John Wiley and Sons, New York.

Liu Rong (ed.) (1993) *Zhongguo Jingji Xiezuo Xitonglun*, Sanlian, Xianggang.

Lo, H.L. (1963) *Hong Kong and Its External Communications before 1842*, Institute of Chinese Culture, Hong Kong.

Loewe, M. (1985) 'Attempts at economic coordination during the Western Han dynasty', in S. Schram (ed.) *The Scope of State Power in China*, European Science Foundation, School of Oriental and African Studies, and the Chinese University of Hong Kong, Hong Kong, pp. 237–268.

Loh, K.W. (1983) 'The transformation from class to ethnic politics in an opposition area: a Malaysian case study', in L. Lim and L. Gosling (eds) *The Chinese in Southeast Asia*, Volume II, Maruzen Asia, Singapore, pp. 189–214.

Lonner, W.J. (1980) 'The search for psychological universals', in H.C. Triandis and W.W. Lambert (eds) *Handbook of Cross-cultural Psychology*, Volume 1, Allyn and Bacon, Boston, pp. 143–204.

Lupton, T. and Wilson, C.S. (1959) 'The social background and connections of

"Top Decision Makers'", *Manchester School*, 28, pp. 30–51.

McGee, T.G. (1970) *Hawkers in Selected Asian Cities*, Centre of Asian Studies, University of Hong Kong.

—— (1973) *Hawkers in Hong Kong*, Centre of Asian Studies, University of Hong Kong.

McGee, T.G. and Yeung Y.M. (1977) *Hawkers in Southeast Asian Cities*, International Development Research Centre, Ottawa.

Mackie, J.A.C. (ed.) (1976) *The Chinese in Indonesia*, Heinemann Educational Books (Asia), Hong Kong.

—— (1988) 'Changing economic roles and ethnic identity of the Southeast Asian Chinese: a comparison of Indonesia and Thailand', in J. Cushman and G.W. Wang (eds) *Changing Identities of the Southeast Asian Chinese since World War II*, Hong Kong University Press, Hong Kong, pp. 217–260.

—— (1992a) 'Changing patterns of Chinese big business in Southeast Asia', in R. McVey (ed.) *Southeast Asian Capitalists*, Southeast Asia Program, Cornell University, Ithaca, New York, pp. 161–190.

—— (1992b) 'Overseas Chinese entrepreneurship', *Asian–Pacific Economic Literature*, 6, pp. 41–64.

McVey, R. (ed.) (1992) *Southeast Asian Capitalists*, Southeast Asia Program, Cornell University, Ithaca, New York.

Mah, F.H. (1972) *The Foreign Trade of Mainland China*, Edinburgh University Press, Edinburgh.

Mair, L. (1961) *Studies in Applied Anthropology*, London School of Economics, Monographs on Social Anthropology, No. 16, Athlone Press, University of London.

Maquet, J.J. (1961) *The Premise of Inequality in Ruanda*, Oxford University Press, London.

Mauss, M. (1985) 'A category of the human mind: the notion of person; the notion of self', in M. Carrithers, S. Collins and S. Lukes (eds) *The Category of the Person: anthropology, philosophy, history*, Cambridge University Press, Cambridge, pp. 1–25.

Mei, Y.P. (1967) 'The status of the individual in Chinese social thought and practice', in C.A. Moore (ed.) *The Chinese Mind: essentials of Chinese philosophy and culture*, East-West Center Press, Honolulu, pp. 323–339.

Meskill, J.M. (1970) 'The Chinese genealogy as a research source', in M. Freedman (ed.) *Family and Kinship in Chinese Society*, Stanford University Press, Stanford, California, pp. 139–161.

Metzger, T.A. (1970) 'The state and commerce in Imperial China', *African and Asian Studies*, 6, pp. 23–46.

Minority Rights Group (ed.) (1992) *The Chinese of Southeast Asia*, Manchester Free Press, Manchester.

Mitchell, R.E. and Lo, I. (1968) 'Implications of changes in family authority relations for the development of independence and assertiveness in Hong Kong children', *Asian Survey*, 8, pp. 309–322.

Mok, V. (1973) 'The organisation and management of factories in Kwun Tong', *Occasional Paper*, Social Research Centre, The Chinese University of Hong Kong.

Moore, C.A. (ed.) (1946) *Philosophy—East and West*, Princeton University Press, New Jersey.

—— (ed.) (1962) *Philosophy and Culture: East and West*, University of Hawaii Press, Honolulu.

—— (ed.) (1967) *The Chinese Mind: essentials of Chinese philosophy and culture*, East-West Center Press, Honolulu.

Morse, H.B. (1966) *The Trade and Administration of the Chinese Empire*, Ch'eng-wen Publishing, Taipei.

Murphey, R. (1970) *The Treaty Ports and China's Modernisation: what went wrong?* Centre for Chinese Studies, University of Michigan, Ann Arbor, Michigan.

—— (1977) *The Outsiders: the Western experience in India and China*, University of Michigan Press, Ann Arbor, Michigan.

Myers, R. (1989) 'Confucianism and economic development: Mainland China, Hong Kong and Taiwan', in Chung-hua Institute for Economic Research, Taipei, Conference Series 13.

Nakamura, H. (1964) *Ways of Thinking of Eastern Peoples*, East-West Centre Press, Honolulu.

Nathan, A.J. (1993) 'Is Chinese culture distinctive?', *Journal of Asian Studies*, 52, pp. 923–936.

National Economic and Development Authority (1989) *Southern Mindanao Development Statistics, 1988*, NEDA Region XI, Davao City.

Needham, J. (1956) *Science and Civilization in China*, Volume II, Cambridge University Press, Cambridge, Mass.

—— (1978) *The Shorter Science and Civilization in China*, Cambridge University Press, Cambridge (abridged by C.A. Ronan).

Ng, C.K. (1983) *Trade and Society: the Amoy network on the China coast, 1683–1735*, Singapore University Press, Singapore.

Ng, M. (1977) 'A critical study of Freud's theory of guilt in society', *Working Paper*, Centre of Asian Studies, University of Hong Kong.

Ng, N.Y.T. and Ng, L.N.H. (1992) 'The dynamics of market town growth in the New Territories, with special reference to the Tai Po markets', *Geographic Review* (occasional publication, The Chinese University of Hong Kong), pp. 26–35.

Ng, P.L.Y. (1961) 'The 1819 edition of the Hsin-An Hsien-Chih: a critical examination with translations and notes (Hong Kong, Kowloon and the New Territories, 1644 –1842)', MA thesis (unpublished), University of Hong Kong.

Nonini, D.M. (1983) 'The Chinese truck transport "industry" of a Peninsular Malaysia market town', in L. Lim and L. Gosling (eds) *The Chinese in Southeast Asia*, Volume I, Maruzen Asia, Singapore, pp. 171–206.

Northrop, F.S.C. (1946) 'The complementary emphases of eastern intuitive and western scientific philosophy', in C.A. Moore (ed.) *Philosophy—East and West*, Princeton University Press, New Jersey, pp. 168–234.

Northrup, D. (1978) *Trade without Rulers*, Clarendon Press, Oxford.

Nurge, E. (1965) *Life in a Leyte Village*, University of Washington Press, Seattle.

Office of City Planning and Development Coordination (1989a) *Davao City Development Profile*, Davao City.

—— (1989b) *Comprehensive Development Plan of Davao City, 1979–2000* (2 vols), Davao City.

Okechuku, C. and Yee W.M.V. (1991) 'Comparison of managerial traits in Canada and Hong Kong', *Asia–Pacific Journal of Management*, 8, pp. 223–235.

Olsen, S.M. (1972) 'The inculcation of economic values in Taipei business families', in W.E. Willmott (ed.) *Economic Organisation in Chinese Society*, Stanford University Press, California, pp. 261–295.

Olson, M. (1993) 'Dictatorship, democracy and development', *American Political Science Review*, 87, pp. 567–576.

Omohundro, J.T. (1977) 'Trading patterns of Philippine Chinese: strategies of sojourning middlemen', in K.L. Hutterer (ed.) *Economic Exchange and Social Interaction in Southeast Asia: perspectives from prehistory, history, and ethnography*, Michigan Papers on South and Southeast Asia, No. 13, Ann Arbor, Michigan, pp. 113–136.

—— (1981) *Chinese Merchant Families in Iloilo: commerce and kin in a central Philippine city*, Ateneo de Manila University Press, Manila, and Ohio University Press, Athens, Ohio.

—— (1983) 'Social networks and business success for the Philippine Chinese', in L. Lim and L. Gosling (eds) *The Chinese in Southeast Asia*, Volume I, Maruzen Asia, Singapore, pp. 65–85.

Orme, G.N. (1912) *Report on the New Territories, 1899–1912*, Hong Kong Government Sessional Papers.

Palmier, L.H. (1960) *Social Status and Power*, London School of Economics, Monographs on Social Anthropology, No. 20, Athlone Press, University of London.

Pan, L. (1989) 'Playing the identity card', *Far Eastern Economic Review*, 9 February, pp. 30–32.

Pan, L. (1990) *Sons of the Yellow Emperor*, Secker and Warburg, London.

Panglaykim, J. and Palmer, I. (1970) 'Study of entrepreneurship in developing countries: the development of one Chinese concern in Indonesia', *Journal of Southeast Asian Studies*, 1, pp. 85–95.

Parris, K. (1993) 'Local initiative and national reform: the Wenzhou model of development', *The China Quarterly*, 134, pp. 242–263.

Parry, G. (1969) *Political Elites*, Allen and Unwin, London.

Pearson, M.M. (1994) 'The Janus face of business associations in China: socialist corporatism in foreign enterprises', *The Australian Journal of Chinese Affairs*, 31, pp. 27–46.

Pelzel, J.C. (1970) 'Japanese kinship: a comparison', in M. Freedman (ed.) *Family and Kinship in Chinese Society*, Stanford University Press, Stanford, California, pp. 227–248.

Phillips, A. (ed.) (1953) *Survey of African Marriage and Family Life*, International African Institute, Oxford University Press, London.

Pirenne, H. (1925) *Medieval Cities: their origins and the revival of trade*, Princeton University Press, Princeton.

Plano del Pueblo de Davao (Vergera), Cabeza del 4° Distrito de la Jola de Mindanao (1862), Ayer List No. 290, Ateneo College Library, Davao.

Plattner, S. (ed.) (1985) *Markets and Marketing*, University Press of America, Lanham.

Polanyi, K. (1944) *The Great Transformation*, Beacon Press, Boston.

Polanyi, K., Arensberg, C.M. and Pearson, H.W. (eds) (1957) *Trade and Markets in Early Empires*, The Free Press, New York.

Polanyi, M. (1958) *Personal Knowledge*, Routledge and Kegan Paul, London.

Potter, J.M. (1970) 'Land and lineage in traditional China', in M. Freedman (ed.) *Family and Kinship in Chinese Society*, Stanford University Press, Stanford, California, pp. 121–138.

Price, J. (1989) 'What did merchants do? Reflections on British overseas trade, 1660–1790', *Journal of Economic History*, 49, pp. 267–284.

Public Works Department (Hong Kong) (1901) *Victoria Hong Kong* (HH22/26)*

Purcell, V. (1965) *The Chinese in Southeast Asia*, Royal Institute of International Affairs, Oxford University Press, London (2nd edn).

Qi Luo and Howe, C. (1993) 'Direct investment and economic integration in the Asia Pacific: the case of Taiwanese investment in Xiamen', *The China Quarterly*, 136, pp. 746–769.

Qian Wenyuan (1985) *The Great Inertia: scientific stagnation in traditional China*, Croom Helm, London.

Ralston, D.A., Gustafson, D.J., Mainiero, L. and Umstot, D. (1993a) 'Strategies of upward influence: a cross-national comparison of Hong Kong and American managers', *Asian Pacific Journal of Management*, 10, pp. 157–175.

Ralston, D.A., Gufstafson, D.J., Tenpstra, R.H., Hott, D.H., Cheung, F. and Ribbens, B.A. (1993b) 'The impact of managerial values on decision-making behaviour: a comparison of the United States and Hong Kong', *Asia Pacific Journal of Management*, 10, pp. 21–37.

Rao Meijiao (1993) 'Dongmeng guojia yu zhongguo zhi jingji guanxi', Liu Rong, *Zhongguo Jingji Xiezuo Xitonglun*, pp. 137–163.

Redding, S.G. (1980) 'Cognition as an aspect of culture and its relation to management processes: an exploratory view of the Chinese case', *The Journal of Management Studies*, 17, pp. 127–148.

—— (1988) 'The role of the entrepreneur in the new Asian capitalism', in P. Berger and M.H. Hsiao (eds) *In Search of an East Asian Development Model*, Transaction Books, New Brunswick, New Jersey, pp. 99–111.

—— (1990) *The Spirit of Chinese Capitalism*, Walter de Gruyter, Berlin.

Redding, S.G. and Ng, M. (1982) 'The role of face in the organisational perception of Chinese managers', *Organisational Studies*, 3, pp. 201–219.

Redding, S.G. and Wong, G.Y.Y. (1986) 'The psychology of Chinese organisational behaviour', in M.H. Bond (ed.) *The Psychology of the Chinese People*, Oxford University Press, Hong Kong, pp. 267–295.

Regional Development Council (1989) *Regional Accelerated Programme for Investment and Development: vision, strategy and prospects*, Davao City.

Régnier, P. (1993) 'Spreading Singapore's wings worldwide: a review of traditional and new investment strategies', *The Pacific Review*, 6, pp. 305–312.

Reid, A. (1982) 'East Point', in M. Keswick (ed.) *The Thistle and the Jade*, Octopus Books, London, pp. 196–197.

Reischauer, E. (1977) *The Japanese*, Harvard University Press.

Revel, J.F. (1994) 'Minimal government', *Far Eastern Economic Review*, 24 November, pp. 48–49.

Reynolds, H.R. (1966) 'Continuity and change as shown by attitudes of two generations of Chinese in the Ilocos Province, Philippines', *Silliman Journal*, 13, pp. 12–21.

Richards, P. (1972) 'A quantitative analysis of the relationship between language tone and melody in a hausa song', *African Language Studies*, XIII, pp. 137–161.

Rigg, J. (1991) *Southeast Asia: region in transition*, Unwin Hyman, London.

Robison, R. (1986) *Indonesia: the rise of capital*, Allen and Unwin, Sydney.

Rosenbaum, A.L. (ed.) (1992) *State and Society in China: the consequences of reform*, Westview Press, Boulder, Colorado.

Ruel, M. (1969) *Leopards and Leaders: constitutional politics among a Cross River people*, Tavistock, London.

Ryan, E.J. (1961) 'The value system of a Chinese community in Java', Ph.D thesis (unpublished), Harvard University.

Sahlins, M.D. (1961) 'The segmentary lineage: an organisation of predatory expansion', *American Anthropologist*, 63, pp. 322–345.

Said, E. (1991) *Orientalism*, Penguin, London.

Salmon, C. (1981) 'The contribution of the Chinese to the development of Southeast Asia: a new appraisal', *Journal of Southeast Asian Studies*, 12, pp. 260–275.

Sampson, A. (1962) *Anatomy of Britain*, Harper and Row, New York and Evanston.

Sangren, P. (1984) 'Traditional Chinese corporations: beyond kinship', *Journal of Asian Studies*, 43, pp. 391–415.

—— (1987) *History and Magical Power in a Chinese Community*, Stanford University Press, Stanford, California.

Santangelo, P. (1985) 'The Imperial factories of Suzhou: limits and characteristics of state intervention during the Ming and Qing dynasties', in S. Schram (ed.) *The Scope of State Power in China*, European Science Foundation, School of Oriental and African Studies, and the Chinese University of Hong Kong, Hong Kong, pp. 269–294.

Sayer, G.R. (1937) *Hong Kong: birth, adolescence and coming of age*, Oxford University Press, London.

Schermerhorn, J.R. Jnr and Bond, M.H. (1991) 'Upward and downward influence tactics in managerial networks: a comparative study of Hong Kong Chinese and Americans', *Asia Pacific Journal of Management*, 8, pp. 147–158.

Schram, S. (ed.) (1985) *The Scope of State Power in China*, European Science Foundation, School of Oriental and African Studies, and The Chinese University of Hong Kong, Hong Kong.

Seagrave, S. (1995) *Lords of the Rim: the invisible empire of the Overseas Chinese*, Bantam, London.

See, Chinbin (1981) 'Chinese clanship in the Philippine setting', *Journal of Southeast Asian Studies*, 12, pp. 224–247.

—— (1988) 'Chinese organisation and ethnic identity in the Philippines', in J. Cushman and G.W. Wang (eds) *Changing Identities of the Southeast Asian Chinese since World War II*, Hong Kong University Press, Hong Kong, pp. 319–334.

Segal, G. (1994) *China Changes Shape: regionalisation and foreign policy*, Adelphi Paper 287, March, International Institute of Strategic Studies, London.

Sender, H. (1994) 'Chinese chequered', *Far Eastern Economic Review*, 10 February, pp. 50–53.

Shanghai Jingji Xinxi (Shanghai Economic News), Shanghai.

Siaw, L.K.L. (1981) 'The legacy of Malaysian Chinese social structures', *Journal of Southeast Asian Studies*, 12, pp. 395–402.

Silin, R.H. (1965) 'Trust and confidence in a Hong Kong wholesale vegetable market', MA thesis, University of Hawaii, Honolulu.

—— (1972) 'Marketing and credit in a Hong Kong wholesale market', in W.E. Willmott (ed.) *Economic Organisation in Chinese Society*, Stanford University Press, California, pp. 327–352.

—— (1976) *Leadership and Values*, Harvard University Press, Cambridge, Mass.

Silverman, G. (1994) 'Almost family', *Far Eastern Economic Review*, 17 November, pp. 85–86.

Siow, M. (1983) 'The problems of ethnic cohesion among Chinese in Peninsular Malaysia: intra-ethnic divisions and inter-ethnic accommodation', in L. Lim and L. Gosling (eds) *The Chinese in Southeast Asia*, Volume II, Maruzen Asia, Singapore, pp. 170–188.

Sit, V.F.S., Wong, S.L. and Kiang, T.S. (1979) *Small-scale Industry in a Laissez-faire Economy*, Centre of Asian Studies, University of Hong Kong, Hong Kong.

Siu, H.F. (1990) 'Recycling tradition: culture, history and political economy in the chrysanthemum festivals of South China', *Comparative Studies in Society and History*, 32, pp. 765–794.

Siu, K.K. (1984) 'The Hong Kong region before and after coastal evacuation in the early Ch'ing dynasty', in D. Faure, J. Hayes and A. Birch (eds) *From Village to City: studies in the traditional roots of Hong Kong society*, Centre of Asian Studies, University of Hong Kong, Hong Kong, pp. 1–9.

Skeldon, R. (1990) 'Emigration and the future of Hong Kong', *Pacific Affairs*, 63, pp. 500–523.

Skinner, G.W. (1957) *Chinese Soiety in Thailand: an analytical history*, Cornell University Press, Ithaca, New York.

—— (1958) *Leadership and Power in the Chinese Community of Thailand*, Cornell University Press, Ithaca, New York.

—— (1960) 'Change and persistence in Chinese culture overseas: a comparison of Thailand and Java', *Journal of the South Seas Society*, 16, pp. 86–100.

—— (1964) 'Markets and social structure in rural China, Part I', *Journal of Asian Studies*, 24, pp. 3–43.

—— (1965a) 'Markets and social structure in rural China, Part II', *Journal of Asian Studies*, 24, pp. 195–228.

—— (1965b) 'Markets and social structure in rural China, Part III', *Journal of Asian Studies*, 24, pp. 363–399.

—— (1971) 'Chinese peasants and the closed community: an open and shut case', *Comparative Studies in Society and History*, 13, pp. 270–281.

—— (1985a) 'Rural marketing in China: repression and revival', *The China Quarterly*, 91, pp. 393–413.

—— (1985b) 'Rural marketing in China: revival and reappraisal', in S. Plattner (ed.) *Markets and Marketing*, University Press of America, Lanham, pp. 7–47.

Skinner, G.W. and Elvin, M. (eds) (1974) *The Chinese City between Two Worlds*, Stanford University Press, California.

Skinner, G.W. and Winckler, E.A. (1969) 'Compliance succession in rural Communist China: a cyclical theory', in A. Etzioni (ed.) *A Sociological Reader on Complex Organisation*, Holt, Rinehart and Winston, New York (2nd edn), pp. 410–438.

Smart, A. (1993) 'Gifts, bribes and *guanxi*: a reconsideration of Bourdieu's social capital', *Cultural Anthropology*, 8, pp. 388–408.

Smart, J. and Smart, A. (1991) 'Personal relations and divergent economies: a case study of Hong Kong investment in south China', *International Journal of Urban and Regional Research*, 15, pp. 216–233.

Smelsner, N.J. (1963) *The Sociology of Economic Life*, Prentice-Hall, New Jersey.

Smith, C.T. (1984) 'Shamshuipo: from proprietary village to industrial–urban complex', in D. Faure, J. Hayes and A. Birch (eds) *From Village to City: studies in the traditional roots of Hong Kong society*, Centre of Asian Studies, University of Hong Kong, Hong Kong, pp. 73–105.

Smith, M. (1991) *Burma: insurgency and the politics of ethnicity*, Zed Books, London.

—— (1992) 'Burma', Minority Rights Group (ed.) *The Chinese in Southeast Asia*, Manchester Free Press, Manchester, pp. 32–33.

Solinger, D (1984) *Chinese Business Under Socialism*, University of California Press, California.

—— (1992) 'Urban entrepreneurs and the state: the merger of state and society', in A.L. Rosenbaum (ed.) *State and Society in China: the consequences of reform*, Westview Press, Boulder, Colorado, pp. 121–142.

Somers Heidhues, M. (1974) *Southeast Asia's Chinese Minorities*, Longman, Melbourne.
—— (1992a) 'Malaysia', in Minority Rights Group (ed.) *The Chinese in Southeast Asia*, Manchester Free Press, Manchester, pp. 12–14.
—— (1992b) 'Indonesia', *ibid.*, pp. 14–17.
—— (1992c) 'Singapore', *ibid.*, p. 19.
—— (1992d) 'Thailand', *ibid.*, pp. 20–21.
—— (1992e) 'Philippines', *ibid.*, pp. 22–23.
—— (1992f) 'Brunei', *ibid.*, p. 23.
Sommers, C.H. (1993) 'Sister soldiers', *The Salisbury Review*, 11 (3), pp. 8–13.
Steel, R.W. and Prothero, R.M. (eds) (1964) *Geographers and the Tropics: Liverpool essays*, Longman, London.
Sterba, R.L.A. (1978) 'Clandestine management in the imperial Chinese bureaucracy', *Academy of Management Review*, 3, pp. 69–78.
Stough, J. (1983) 'Chinese and Malay factory workers: desire for harmony and experience of discord', in L. Lim and L. Gosling (eds) *The Chinese in Southeast Asia*, Volume II, Maruzen Asia, Singapore, pp. 231–265.
Stover, L.E. (1974) *The Cultural Ecology of Chinese Civilization*, New American Library, New York.
Strathern, M. (1983) 'Subject or object? Women and the circulation of valuables in Highland New Guinea', in R. Hirschon (ed.) *Women and Property, Women as Property*, St Martin's Press, London, pp. 158–175.
Sun Qian (1993) 'Zhongguo chuantong wenhua yu huaqiao huaren jingji', *Huaqio Huaren Lishi Yanjiu (Overseas Chinese Historical Research)*, 4, pp. 45–52.
Sung, Yun-wing (1992) 'Non-institutional economic integration via cultural affinity: the case of mainland China, Taiwan and Hong Kong', *Occasional Paper, No. 13*, Hong Kong Institute of Asia–Pacific Studies, The Chinese University of Hong Kong.
Surveyor General Office (Hong Kong) (1863a) *Victoria Hong Kong* (HD 22-244)*
—— (1863b) *Kowloon as Planned in 1863* (HD 23-2)*
—— (1873) *Plan of Victoria Hong Kong* (HG 24)*
Suryadinata, L. (1985) *China and the ASEAN States*, Singapore University Press, Singapore.
—— (1988) 'Chinese economic elites in Indonesia: a preliminary study', in J. Cushman and G.W. Wang (eds) *Changing Identities of Southeast Asian Chinese since World War II*, Hong Kong University Press, Hong Kong, pp. 261–288.
—— (1989) *The Ethnic Chinese in the ASEAN States*, Institute of Southeast Asian Studies, Singapore.
Swift, M.G. (1965) *Malay Peasant Society in Jelebu*, London School of Economics, Monographs on Social Anthropology, No. 29, Athlone Press, University of London.
Szczepanik, E. (1958) *The Economic Growth of Hong Kong*, Oxford University Press, London.
Tan, A.S. (1972) *The Chinese in the Philippines, 1838–1935*, Quezon City.
Tan, Chee-beng (1983) 'Acculturation and the Chinese in Melaka: the expression of *baba* identity today', in L. Lim and L. Gosling (eds) *The Chinese in Southeast Asia*, Volume II, Maruzen Asia, Singapore, pp. 56–78.
—— (1985) 'The development and distribution of *dejiao* associations in Malaysia and Singapore: a study of a Chinese religious organisation', *Occasional Paper No. 79*, Institute of Southeast Asian Studies, Singapore.
T'ang, Chün-I (1962) 'The development of ideas of spiritual value in Chinese

philosophy', in C.A. Moore (ed.) *Philosophy and Culture: East and West*, Princeton University Press, New Jersey, pp. 225–244.

Thee, Kien Wie (1991) 'The surge of Asian NIC investment into Indonesia', *Bulletin of Indonesian Economic Studies*, 27, pp. 55–88.

Thomas, N. (1991) *Entangled Objects: exchange, material culture, and colonialism in the Pacific*, Harvard University Press, Cambridge, Mass.

Thompson, L.A. (1989) *Romans and Blacks*, Routledge and Oklahoma University Press, London.

T'ien, Ju-k'ang (1956) *The Chinese of Sarawak: a study of social structure*, Monograph on Social Anthropology, No.12, London School of Economics, London.

Tiglao, R. (1994a) 'Lean machine', *Far Eastern Economic Review*, 27 October, p. 60.

—— (1994b) 'Focus. Philippines: trade and investment', *Far Eastern Economic Review*, 16 June, pp. 47–55.

—— (1994c) 'Paralysed by politics', *Far Eastern Economic Review*, 12 May, pp. 22–30.

—— (1994d) 'Not strictly flamenco', *Far Eastern Economic Review*, 17 November, pp. 82–84.

Topley, M. (ed.) (1975) *Hong Kong: the interaction of tradition and life in the towns*, Hong Kong Branch of the Royal Asiatic Society, Hong Kong.

Tregear, T.R. and Berry, L. (1959) *The Development of Hong Kong and Kowloon as Told in Maps*, Hing Kam Chiu Press, Hong Kong.

Triandis, H.C. and Lambert, W.W. (eds) (1980) *Handbook of Cross-cultural Psychology*, Allyn and Bacon, Boston.

Tse, F.Y. (1974) *Market and Street Trading: a conceptual framework*, Social Research Centre, The Chinese University of Hong Kong.

Twang Peck-yang (1979) 'Political attitudes and allegiances in the *totok* business community, 1950–54', *Indonesia*, 28, pp. 65–83.

Usher, A.B. (1943) *The Early History of Deposit Banks in Mediterranean Europe*, Volume I, Harvard University Press, Cambridge.

Wade, R. (1990) *Governing the Market: economic theory and the role of government in East Asian industrialisation*, Princeton University Press.

Wang Gungwu (1958) 'The nanhai trade: a study of the early history of Chinese trade in the South China Sea', *Journal of the Malayan Branch of the Royal Asiatic Society*, 31 (2), No. 182, Kuala Lumpur.

—— (1981) *Community and Nation: essays on Southeast Asia and the Chinese*, Heinemann for Asian Studies Association of Australia, Singapore.

—— (1988) 'The study of Chinese identities in Southeast Asia', in J. Cushman and G.W. Wang (eds) *Changing Identities of the Southeast Asian Chinese since World War II*, Hong Kong University Press, Hong Kong, pp. 1–21.

—— (1991) *China and the Chinese Overseas*, Times Academic Press, Singapore.

—— (1993) 'Greater China and the Overseas Chinese', *The China Quarterly*, 136, pp. 926–948.

Wang Zhihua (1990) 'Tizhi he jiegou zhuanbianqi de caizheng zhengce', *Nankai Jingji Yanjiu (Nankai Economic Research)*, 1, pp. 10–15.

War Office (Hong Kong) (1890) *Kowloon Frontier*, (HM 7a)*

Ward, B. (1965) 'Varieties of the conscious model: the fishermen of South China', in M. Banton (ed.) *The Relevance of Models for Social Anthropology*, Tavistock, London, pp. 113–137.

—— (1972) 'A small factory in Hong Kong: some aspects of its internal organisation', in W.E. Willmott (ed.) *Economic Organisation in Chinese Society*, Stanford University Press, California, pp. 353–385.

Weber, M. (1968) *Economy and Society: an outline of interpretive sociology*, Bedminster Press, New York.
Wei Jie (1991) 'Touzi zhong de jihua yu shichang xiang jiehe', *Caijing Wenti Yanjiu (Financial and Economic Problems Research)*, 6, pp. 1–5.
Weidenbaum, M. (1992) 'The changing pattern of the world economy—the shifting role of business and government', *Wei Lun Lecture Series IV*, Chinese University Bulletin, Supplement 28, pp. 20–29.
Weightman, G.H. (1985) 'The Philippine Chinese: from aliens to cultural minority', *Journal of Comparative Family Studies*, 16, pp. 161–179.
Wertheim, W.F (1964) *East–West Parallels: sociological approaches to modern Asia*, N.V. Uyitgeverij W. van Hoeve, The Hague.
Wheatley, P. (1964) 'The land of Zanj: exegetical notes on Chinese knowledge of East Africa prior to A.D. 1500', in R.W. Steel and R.M. Prothero (eds) *Geographers and the Tropics: Liverpool essays*, Longman, London, pp. 39–87.
White, G. (1993) 'Prospects for civil society in China: a case study of Xiaoshan City', *Australian Journal of Chinese Affairs*, 29, pp. 63–87.
—— (1994) 'Democratization and economic reform in China', *Australian Journal of Chinese Affairs*, 31, pp. 73–92.
White, G. and Gray, J. (eds) (1988) *Developmental States in East Asia*, Institute of Development Studies, Macmillan, Basingstoke.
White, G. and Wade, R. (1988) 'Developmental states and markets in East Asia: an introduction', in G. White and J. Gray (eds) *Developmental States in East Asia*, Institue of Development Studies, Macmillan, Basingstoke, pp. 1–29.
Whyte, W.H. (1956) *The Organisation Man*, Doubleday Anchor, New York.
Wickberg, E. (1965) *The Chinese in Philippine Life, 1850–1898*, Yale University Press, New Haven.
—— (1988) 'Chinese organisations and ethnicity in Southeast Asia and North America since 1945: a comparative analysis', in J. Cushman and G.W. Wang (eds) *Changing Identities of the Southeast Asian Chinese Since World War II*, Hong Kong University Press, Hong Kong, pp. 303–334.
Wilfred, B. (1969) *The Impact of Chinese Secret Societies in Malaya*, Oxford University Press, London.
Wilks, S. (1990) 'The embodiment of industrial culture in bureaucracy and management', in S.R. Clegg and S.G. Redding (eds) *Capitalism in Contrasting Cultures*, Walter de Gruyter, Berlin, pp. 131–152.
Williams, L.E. (1952) 'Chinese entrepreneurs in Indonesia', *Explorations in Entrepreneurial History*, 1, pp. 34–60.
Williamson, J. (ed.) (1994) *The Political Economy of Policy Reform*, Institute for International Economics, Washington, DC.
Willmott, D.E. (1960) *The Chinese of Semarang*, Cornell University Press, New York.
Willmott, W.E. (ed.) (1972) *Economic Organisation in Chinese Society*, Stanford University Press, California.
Wilson, G. and Wilson, M. (1965) *The Analysis of Social Change: based on observations in Central Africa*, Cambridge University Press, Cambridge.
Wilson, R.W. and Pusey, A.W. (1982) 'Achievement motivation and small business relationship patterns in Chinese society', in S.L. Greenblatt, R.W. Wilson and A.A. Wilson (eds) *Social Interaction in Chinese Society*, Praeger, New York, pp. 195–208.
Winchester, S. (1991) *The Pacific*, Hutchinson, London.
Winzeler, R. (1983) 'The ethnic status of the rural Chinese of the Kelantan Plain',

in L. Lim and L. Gosling (eds) *The Chinese in Southeast Asia*, Volume II, Maruzen Asia, Singapore, pp. 34–55.

Wong, K. (1994) 'Family fortunes', *Sunday Morning Post*, 6 March, p. 4.

Wong, Siu-lun (1983) 'Business ideology of Chinese industrialists in Hong Kong', *Journal of the Hong Kong Branch of the Royal Asiatic Society*, 23, pp. 137–171.

—— (1985) 'The Chinese family firm: a model', *British Journal of Sociology*, 36, pp. 58–72.

—— (1988a) *Emigrant Entrepreneurs: Shanghai industrialists in Hong Kong*, Oxford University Press, Hong Kong.

—— (1988b) 'The applicability of Asian family values to other socio-cultural settings', in P. Berger and M.H.H. Hsiao (eds) *In Search of an East Asian Development Model*, Transaction Books, New Brunswick, New Jersey, pp. 134–152.

Wood, A.W. (1940) *A Brief History of Hong Kong*, South China Morning Post, Hong Kong.

World Bank (1991) *World Development Report, 1991: the challenge of development*, Washington.

—— (1993) *The East Asian Miracle: economic growth and public policy*, Oxford University Press, New York.

Wu, D.Y.H. (1982) *The Chinese in Papua New Guinea: 1880–1980*, The Chinese University Press, Hong Kong.

Wu Yuanli and Wu Chun-hsi (1980) *Economic Development in Southeast Asia: the Chinese dimension*, Hoover Institution Press, Stanford, California.

Xiang Dayou (1993) 'Lun zhongguo gaige kaifang yu huazitou xiang qushi', *Huaqiao Huaren Lishi Yanjiu (Overseas Chinese Historical Research)*, 4, pp. 9–15.

Yahuda, M. (1990) 'China', in World of Information, *Asia Pacific Review*, pp. 53–61.

—— (1993) 'The foreign relations of Greater China', *The China Quarterly*, 136, pp. 687–710.

Yang Chengxun (1984) 'Chengshi jingji tizhi gaige yingyi fada de shehuizhuyi shangpin jingji wei jidian', *Jingji Yanjiu (Economic Research)*, 11, pp. 17–21.

Yang, K.S. (1986) 'Chinese personality and its change', in M.H. Bond (ed.) *The Psychology of the Chinese People*, Oxford University Press, Hong Kong, pp. 106–170.

Yang, M.M.H. (1989) 'The gift economy and state power in China', *Comparative Studies in Society and History*, 31, pp. 25–54.

Yang, M.M.H. (1994) *Gifts, Favours and Banquets: the art of social relationships in China*, Cornell University Press, Ithaca, New York.

Yao Souchou (1984) 'Why Chinese voluntary associations: structure or function?', *Journal of the South Seas Society*, 39, pp. 75–99.

Ye Zhen, Zheng Jingping, Zhang Yingxiang (1993) 'Guanyu cujin geti siying jingji fazhan de duice jianyi', *Jingji Guanli (Economic Management)*, 3, pp. 8–11.

Ye Zhenpeng (1991) 'Jihua yu shichang de kuangniu shi shehui caili de fenpei', *Gaige (Reform)*, 1, pp. 72–77.

Yen Ching-hwang (1981) 'Early Chinese clan organisations in Singapore and Malaya, 1819–1911', *Journal of Southeast Asian Studies*, 12, pp. 62–92.

Yengoyan, A.A. (1983) 'The buying of futures: Chinese merchants and the fishing industry in Capiz, Philippines', in L. Lim and L. Gosling (eds) *The Chinese in Southeast Asia*, Volume I, Maruzen Asia, Singapore, pp. 117–130.

Yeung Yue-man (1971) 'Commercial patterns in Singapore's public housing estates', *Town Planning Review*, 43, pp. 56–70.

—— (1989) 'Fifty years of public housing in Hong Kong: retrospect and prospect', *Occasional Paper No. 97*, Department of Geography, The Chinese University of Hong Kong.

Yoshino, M.Y. (1971) *The Japanese Marketing System: adaptation and innovation*, MIT, Massachusetts.

Young, J. A. (1971) 'Interpersonal networks and economic behaviour in a Chinese market town', Ph.D Thesis, Stanford, University Microfilms, Ann Arbor, Michigan.

Young, M. and Willmott, P. (1957) *Family and Kinship in East London*, Routledge and Kegan Paul, London.

Youngson, A.J. (ed.) (1983) *China and Hong Kong: the economic nexus*, Oxford University Press, Hong Kong.

Yu Ying-shih (1967) *Trade and Expansion in Han China*, University of California Press, Berkeley.

Zaehner, R.C. (ed.) (1990) *Hutchinson Encyclopedia of Living Faiths*, Hutchinson, London.

Zang Xiaowei (1991) 'Provincial elites in post-Mao China', *Asian Survey*, 31, pp. 512–525.

Zhang Fubao (1984) 'Guanyu Shanghai jingjiqu shiguanxian de yixie wenti', *Jingji Dili (Economic Geography)*, 4, pp. 299–303.

Zhao Qingyi (1994) 'Shanghai liutong tizhi shuaixian gaige de ruogan sikao', *Chengshi Jingji Yanjiu (Urban Economic Research)*, 3, pp. 7–9.

Zhonggong Tianjinshi (1988) *Tianjin Siying Jingji Fazhan de Xianzhuang yu Duice*, Tianjin.

Zhongguo Jiage Xuehui, Bejingshi de Wujiaju (1981) *Jiage Zhuanti Jiangzuo*, Beijing.

Zhonghua Gongshang Shibao (China Business Times), Beijing.

Zhou Diankun (1986) 'Shilun chengshi jingjiqu de liutong wangluo tixi', *Jingji Dili (Economic Geography)*, 4, pp. 260–265.

Index